Liapunov Theory for Integral Equations with Singular Kernels and Fractional Differential Equations

T. A. Burton

Northwest Research Institute
732 Caroline St.
Port Angeles, WA 98362
taburton@olypen.com

732 Caroline St.
Port Angeles, WA 98362
August 23, 2012

Preface

This book is concerned with the construction of Liapunov functionals for integral equations, fractional differential equations, and integrodifferential equations. There is a major focus on the difficulties generated by singular kernels. The Liapunov functionals are constructed in such a way that qualitative properties of solutions are readily inferred. The first section of Chapter 0 gives a brief history of Liapunov theory and how this material fits into the long development of the subject. We will repeat none of that here.

We begin with theory which was developed for equations with continuous kernels from 1992 to 2008. That work was published in a preliminary version with much elementary material for use in a course which I taught at the University of Memphis in the Spring of 2009 as a visiting professor. That elementary version is still available. Work continued and in 2010 the first papers were published containing Liapunov theory for integral equations and fractional differential equations with singular kernels. From then on there has been much progress and that material occupies more than half of the book. Selected material from the aforementioned book is used here in the way of introduction and it serves another useful purpose. Unlike Liapunov theory for differential equations, different kernels can be added and the Liapunov functionals associated with each of those kernels can be added to produce a new Liapunov functional. Thus, we need to supply Liapunov functionals for equations with continuous kernels so that we can handle problems in which there is a sum of kernels, one continuous and one singular.

While we do develop some basic material on existence and resolvent theory, this is definitely not to be viewed as a text on integral equations. We hope it will not be judged in that way. It is written for exactly one purpose and that is to expound the application of Liapunov's direct method to integral and fractional differential equations. In large measure the development stems from work of Krasovskii in the 1950s. We hope that its application will be as successful as that of Krasovskii.

I am greatly indebted to Leigh Becker, David Dwiggins, Tetsuo Furumochi, John Haddock, John Purnaras, and Bo Zhang in so many ways in the development of this material. Every one of them collaborated in at least one research paper on which the book is based. Moreover, Leigh Becker, John Purnaras, and Bo Zhang have spent countless hours reading the manuscript. It is impossible to adequately express my gratitude to them.

This manuscript is being translated into the Russian language by Prof. Sergey I. Kumkov of the N. N. Krasovskii Institute of Mathematics and Mechanics, Ekaterinburg, Russia. The translation editing will be performed by Prof. Nikolay Yurievich Lukoyanov. The translation is to be published by the Autonomous Nonprofit Organization, Izhevsk Institute of Computer Science, Universitetskaya, I, Izhevsk, 426034, Russia.

T. A. Burton
Northwest Research Institute
Port Angeles, Washington
taburton @olypen.com

Using the subject index

The subject index gives several guided tours through the book. The main tour is a list of the Liapunov functionals found mainly in Chapter 2. They are long entries and they are numbered as: "Twenty-seventh Liapunov functional, singular, periodic." It is the twenty-seventh main Liapunov functional and it concerns a singular kernel, which gives rise to a singular resolvent. That resolvent must be decomposed by a process called iterated kernels, and that resolvent then becomes the kernel of a Liapunov functional for a nonlinear singular equation. If one reads through that sequence of index entries while consulting the statements of the theorems with the Liapunov functional, then the plan of the book unfolds in a simple way. Most of the entries are not as complex as the one discussed here.

There are numerous other tours. Throughout the index we see entries of the form "separating $a(t)g(t,x)$." This indicates a major problem with much room for research. Most of the nonlinear Liapunov functionals have a derivative which contains a product $a(t)g(t,x)$ which must be replaced by $a(t)g(t,x) \leq f(t) + H(t,x)$ so that we can integrate the derivative and obtain an integral bound on the solution, $x(t)$. We offer several methods, including a sophisticated form of Young's inequality and a particular strategy in Chapter 4. But without a good method of separation we are forced to severely restrict the nonlinearity of the problem. That is to be avoided since this is an elementary type of algebraic problem and it limits the utility of the Liapunov functional. A clever idea here can greatly enhance almost half of the Liapunov functionals we consider.

We also see many entries of the type $x \to a$ or $C(t,s) \to R(t,s)$. These can be followed through the book and one finds a recurring theme that unknown functions can be well-approximated by functions occurring in the integral equation. Section 2.9 gives the most transparent discussion of these properties, but they occur throughout the book and are prime targets for research. It is known from work of Ritt that the resolvent, $R(t,s)$, is arbitrarily complicated; yet, under many conditions it is well-approximated by the kernel of the integral equation. On the other hand, for fractional differential equations it is also known that those two functions are worlds apart. It is a real victory to establish either case.

In the classical theory of integral equations, and certainly in the theory of integrodifferential equations using a Razumikhin technique, the kernel, $C(t, s)$, of the equation is virtually always integrated with respect to s and properties of solutions and resolvents are given in terms of that integral. It is introduced in Theorem 2.1.3 (see Theorem 2.6.3.1 for a parallel proof) and the entire Section 3.9 contrasts the Razumikhin conditions with the Liapunov conditions. The case is made throughout the book that s is the L^∞ coordinate and t is the L^p coordinate, although an L^p conclusion is frequently parlayed into an L^∞ conclusion. In the subject index we frequently see a condition, say M, followed by the abbreviation "wrt" either s or t. This indicates that an integration was performed "with respect to" either s or t. Working through these properties and enlarging the set is a major area of investigation and will have significant impact on the classical theory.

There is a long list of results containing the assumption that the kernel satisfies a relation $C(t, t) > K$ and another long list containing the assumption $C(t, t) < k$. While the proofs are quite different, the conclusions can be very similar. The intriguing question is that perhaps at least some of these results can be combined resulting in a theorem which says nothing about the size of the kernel.

This is a selection of examples of tours found in the index.

Contents

Chapter 0

Introduction and Overview

0.1 History

In 1892 Liapunov (1992) published a major work in which he advanced an idea for studying qualitative properties of solutions of an ordinary differential equation,

$$x' = F(t, x)$$

where $' = \frac{d}{dt}$ and $F : [0, \infty) \times \Re^n \to \Re^n$ is usually at least continuous. Thus, for each $(t_0, x_0) \in [0, \infty) \times \Re^n$ there is at least one solution $x(t) = x(t, t_0, x_0)$ through (t_0, x_0) defined on some interval $[t_0, t_0 + \alpha)$, $\alpha > 0$. His theory ranged from utter simplicity to mathematically deep and intricate properties, typified by the span from simple boundedness to periodicity of solutions.

While it started as a technique to study ordinary differential equations, it progressed smoothly to functional differential equations, control theory, partial differential equations, and difference equations. In many ways it has provided a unifying link between these seemingly different subjects.

Here is its simplest form. We mainly follow Yoshizawa (1966; p. 2) and say that a *Liapunov function* is a differentiable scalar function. Suppose we construct a differentiable scalar function $V : [0, \infty) \times \Re^n \to [0, \infty)$ with the property that $V(t, x) \to \infty$ as $|x| \to \infty$ uniformly for $0 \le t < \infty$. Now $V(t, x)$ and $x(t) = x(t, t_0, x_0)$ are totally unrelated, but if we write $V(t, x(t))$ then they are related and by the chain rule, for example, we can compute

$$\frac{dV(t, x(t))}{dt} = \frac{\partial V}{\partial t} + \frac{\partial V}{\partial x_1}\frac{dx_1}{dt} + \cdots + \frac{\partial V}{\partial x_n}\frac{dx_n}{dt},$$

where the dx_i/dt are just the components of the KNOWN function F.

Here is the first of two critical points. An existence theorem tells us that the solution $x(t)$ exists; but virtually never can we find that solution. We do not need to know the solution because we can compute the derivative of V DIRECTLY from the differential equation itself. Hence, the name "Liapunov's direct method."

The second critical point is this. That derivative which we just obtained may yield no information at all unless it is of a special form. For example, if we have shrewdly constructed $V(t, x)$ we may find that

$$\frac{dV(t, x(t))}{dt} \le 0.$$

This will mean that for $t \ge t_0$ then

$$V(t, x(t)) \le V(t_0, x_0)$$

and so $x(t)$ is bounded since $V(t, x) \to \infty$ as $|x| \to \infty$. This is the crudest qualitative property and much more can be obtained with further study. For example, much later we learned to obtain the boundedness from V' alone, without the need for the unboundedness of V. See Burton, Eloe, and Islam (1990).

We have achieved a great victory and it hinged completely on being able to relate $V(t, x)$ to $x(t)$ by the chain rule.

The task, then, is to learn to construct suitable functions, $V(t, x)$, called Liapunov functions. Early on Barbashin (1968) constructed an immense set of Liapunov functions for a great variety of problems. It turned out that investigators were able to modify those functions to make them applicable to more problems and that process continues to this day. From thousands of pages of literature on such constructions and intricate theorems we select a few more examples. Perhaps the oldest is the Liapunov function consisting of the sum of the kinetic and potential energy of a Liénard equation. This was advanced to a delay equation by Krasovskii (1963, p. 173) using a simple device which converted a functional to a function. Krasovskii's work showed us how to generally advance our Liapunov functions for differential equations to Liapunov functionals for many functional differential equations.

That sum of kinetic and potential energy was also advanced to the control problem of Lurie and then on to the damped wave equation. That single Liapunov functional spanned and brought some unification to three distinct areas of mathematics. It unified areas in the sense that an investigator moved from one area to the other with the same tools and applicable experience.

Much of the work in constructing Liapunov functions was *ad hoc*, but from earliest investigations there a was general theory of linear equations of the form $x' = Ax$, A being a constant $n \times n$ matrix, using a Liapunov function of the form $V = x^T Bx$ and a scheme is given for constructing B. In recent years Bo Zhang (2001, 2005) presented formal theory for constructing Liapunov functions for systems of linear ordinary differential equations, delay equations, and partial differential equations. He was joined by Graef and Qian (2004) in advancing the theory to difference equations.

Two of the most prominent classical works with transition from ordinary differential equations to functional differential equations are Krasovskii (1963) and Yoshizawa (1966). The transition of the aforementioned sum of kinetic and potential energy from a Liénard equation to a delay equation, on to the damped wave equation, and then to a Lurie control problem can be seen in Barbashin (1968; p. 1102), Krasovskii (1963; p. 173), Burton (1991), and Burton and Somolinos (2007). An introductory treatment showing a smooth transition of Liapunov functions from ordinary to partial differential equations is found in Henry (1981).

0.2 The Present Problem

We reiterate that the success of Liapunov's direct method is tied to our ability to unite the differential equation $x' = F(t, x)$ to a totally independent function, $V(t, x)$.

In the same general category of equations under discussion above is the integral equation

$$x(t) = f(t) + \int_0^t g(t, s, x(s))ds.$$

Investigators have long desired to advance Liapunov's direct method to such, but the difficulty is that we do not know how to unite a Liapunov function, $V(t, x)$, to this equation.

The book by R. K. Miller (1971a) presents an excellent introduction to integral equations from a classical point of view. On p. 337 he begins Chapter VI as follows: "If a system of integral equations of the form

$$x(t) = f(t) + \int_0^t g(t, s, x(s))ds$$

can be written in differentiated form

$$x'(t) = f'(t) + g(t, t, x(t)) + \int_0^t g_t(t, s, x(s))ds, x(0) = f(0),$$

then it may be possible to analyze the behavior of the solution $x(t)$ by means of Lyapunov's second method." Miller has plainly stated the prevailing view.

In large measure the present book begins at this point. Frequently we do find it profitable to differentiate an integral equation, as we will do many times in this book. However, there is a long list of reasons why we do not want to differentiate the integral equation. That list includes:

1. The functions may not be differentiable, so it is impossible.

2. Integration smooths, but differentiation produces roughness; thus, differentiation can introduce complications and great disorder.

3. If the functions are differentiable, then we can prove existence and uniqueness of solutions. But the derived equation may have functions which are, at most, continuous. Thus, the derived equation may have nonunique solutions and we may have introduced extraneous solutions. This means that the properties we prove for the derived equation are actually absent in the original equation.

4. The problem arises as an integral equation and it would seem to be a matter of integrity to deal with it in that category.

5. There is a vast category of problems in the form

$$x(t) = f(t) + \int_0^t [C(t,s)x(s) + g(t,s,x(s))]ds$$

in which there is no hope for differentiation. But it is possible to extract the resolvent problem

$$R(t,s) = C(t,s) - \int_s^t C(t,u)R(u,s)du$$

which we may treat with Liapunov functionals and reformulate the original equation with $R(t,s)$ about which we have very precise information.

Thus, our problem here has three parts. First, we must learn to construct suitable Liapunov functionals for a basic set of integral equations. We must learn how to unite them with the integral equation in the absence of a chain rule. We must learn how to account for the singularity in the Liapunov functional, yet still be able to differentiate that functional. The aforementioned functional which Krasovskii added to a Liapunov function in order to study delay equations turns out to be the key to Liapunov functionals for integral equations.

0.3 Real-world Problems: uncertainties, singularities

This book is concerned with qualitative behavior of solutions of integral equations which are often of the form

$$x(t) = a(t) - \int_0^t C(t,s)g(s,x(s))ds$$

where a may be continuous and g is almost always continuous, but $C(t, s)$ may be weakly singular in the sense of Definition 1.2.1 of Chapter 1, Section 2. If we derive an equation from first principles, such as Newton's laws of motion, we have an idealized problem. There are inherent perturbations and uncertainties in real-world analogs. Reality demands that we provide for these uncertainties. Liapunov functionals for integral equations provide a very simple way of doing exactly that. This interesting property of Liapunov theory for integral equations is seldom seen in differential equations. If we have a Liapunov functional for each of two integral equations with different kernels, then we can add the kernels and add the Liapunov functionals, obtaining a Liapunov functional for the equation with the sum of kernels. (See Theorem 2.5.6 and Theorem 2.6.4.1.) Thus, it often happens that one kernel is singular while the other is continuous. It is then imperative that we deal separately with integral equations with continuous kernels in preparation for that addition. Moreover, following Miller's program of differentiating the equation, we may start with a continuous kernel and pass to a singular kernel of an integrodifferential equation. The reversal can often be achieved using the Tonelli-Hobson test.

All of this means that a faithful representation of the title of this book will involve integral equations with either continuous or singular kernels and integrodifferential equations with singular kernels.

We construct many Liapunov functionals for a variety of equations including infinite delay, constant delay, and neutral equations. But the following discussion will give a good look at the kinds of problems treated in this book.

0.4 Fractional Differential Equations

A broad example of weakly singular kernels in modern literature is found in fractional differential equations. While there has been a recent surge of interest in the subject, it is very old and histories are found in Miller and Ross (1993) and in Oldham and Spanier (1974, 2002). Difficulties with the classical theory in suiting the initial conditions (see Mainardi (2010, p. 73)) to natural problems led to Caputo theory which is presented in Diethelm (2004), Lakshmikantham-Leela-Vasundhara Devi (2009), Mainardi (2010), and Podlubny (1999). We will present none of the underlying theory here because in one step we transform it into a very standard integral equation for which there is more than a century of well-developed theory on which the investigator can constantly draw. Indeed, this is consistent with the statement of Oldham and Spanier (2002, p. xii).

For more complete details and definition, see Section 2.5.1. With our opening equation in mind we consider the fractional differential equation of Caputo type and order q

$$^{c}D^{q}x(t) = f(t) - g(t, x(t)),\ x(0) \in \Re,\ 0 < q < 1.$$

If f and g are continuous then it is inverted as

$$x(t) = x(0) - \frac{1}{\Gamma(q)} \int_0^t (t-s)^{q-1}[g(s, x(s)) - f(s)]ds \tag{2.5.1.7}$$

where Γ is the Euler gamma function. We refer the reader to Lakshmikantham, Leela, and Vasundhara Devi (2009; p. 54) or to Chapter 6 of Diethelm (2004; pp. 78, 86, 103) for proofs of the inversion. For the important case of $xg(t, x) \geq 0$, in Theorem 2.5.1.2 we offer a first Liapunov functional for this equation for a general $q \in (0, 1)$.

The kernel $C(t, s) = \frac{1}{\Gamma(q)}(t-s)^{q-1}$ for $0 < q < 1$ is the epitome of weakly singular kernels. Even the value $q = 1/2$ gives rise to an incredible array of extremely important real-world problems. A sample of these will be given later. Moreover, that kernel is a simple example of a more general class of convex kernels with singularities at $t = s$. This establishes the generality of convex kernel problems, but there is much more to come. Problems in viscoelasticity, circulating fuel nuclear reactors, neural networks, and many other applications are modeled by convex kernels either with or without singularities and in both integral and integrodifferential equations. Indeed, Volterra (1928) noted that many such problems were being modeled using convex kernels and that continues to this day. That single Liapunov functional of Theorem 2.5.1.2 covers a wide array of real-world problems. Even when the function f is large, the Liapunov functional provides a measure of the growth of the solution.

This kernel is much more than just convex with singularities. It is completely monotone and integral equation theory says much about the resolvent generated by such kernels. For the case of $g(t, x) = x + G(t, x)$ it gives rise to an important parallel theory. The linear part,

$$y(t) = y(0) - \int_0^t C(t-s)y(s)ds,$$

has the resolvent equation

$$R(t) = C(t) - \int_0^t C(t-s)R(s)ds$$

with R also being completely monotone and

$$0 < R(t) \leq C(t),\ \int_0^\infty R(s)ds = 1.$$

Our equation is then transformed into a new integral equation having R as its kernel and giving rise to an entirely new class of Liapunov functionals.

0.5 Examples with singular kernels

To emphasize how broad the example is we offer the following list of applications for the single case of $q = 1/2$. In addition to these, the reader may wish to consult Chapter Five of the book by Oldham and Spanier which is much concerned with applications when $q = 1/2$.

Consiglio (1940) studied turbulence using

$$y(t) = (1/K)\left(1 - \int_0^t (t-s)^{-1/2} y^2(s)ds\right).$$

In 1951 Mann and Wolf (1951) and Roberts and Mann (1951) studied the temperature $u(x,t)$ in a semi-infinite rod by means of the integral equation

$$u(0,t) = \int_0^t \frac{G(u(0,s))}{\pi^{1/2}(t-s)^{1/2}} ds$$

where $G(1) = 0$, G is continuous, and $G(u)$ is strictly increasing. By a translation we would have a forcing function and a new G with $xG(x) \geq 0$, a condition we will see later. The problem was generalized and studied by Padmavally (1958) using a variable heating source described by

$$y(t) = \int_0^t \frac{G(s, y(s))}{\pi(t-s)^{1/2}} ds$$

under considerably more complicated conditions on G.

Miller (1971a, pp. 68 ff and 208 ff) continues that study dealing with the equation

$$y(t) = -(\pi K)^{-1/2} \int_0^t (t-s)^{-1/2} g(s, y(s))ds$$

and a more general equation with forcing function under the assumption that $yg(t,y) \geq 0$. An elementary derivation of such equations is given by Weinberger (1965).

Nicholson and Shain (1964; pp. 72-3) study stationary electrode polography and obtain

$$x(bt) = e^{bt-u(t)}\left(1 - \int_0^{bt} (bt-s)^{-1/2} x(s)ds\right)$$

where $u(t)$ is given and b is a positive constant.

More recently, Kirk and Olmstead (2002) studied blow-up in a reactive diffusive medium with a moving heat source. Their equation is

$$F(t, x_0(t)) = h(t) + \int_0^t \frac{e^{-\frac{[x_0(t)-x_0(s)]^2}{4(t-s)}}}{2(\pi(t-s))^{1/2}} ds.$$

Their study continues in (2005).

0.6 Nonsingular Examples

When we take into account uncertainties, then we add a kernel which is small in some sense and which does not have a designated sign. Often, we will choose a continuous kernel because we have no idea where a singularity might occur in the uncertainty. Thus, we will certainly need to construct Liapunov functionals for a variety of continuous, but small, kernels. In addition, there is a multitude of problems, some of which were noted as far back as 1928 with Volterra, with continuous, convex, or even completely monotone, kernels. Today we see these kernels in models of problems found in biology, neural networks, viscoelasticity (often one-dimensional with a truncated integral), nuclear reactors of more than one type, to mention just a few. See, for example Burton (1993b) and (2008c), Burton and Dwiggins (2010), Levin (1963), (1965), 1968), Levin and Nohel (1963) and (1964), Londen (1972), Volterra (1928), Zhang (2009ab) for work on integral and integrodifferential equations with continuous convex kernels modeling problems of the type just mentioned. Finally, these models are frequently found in feedback control systems. The problem of Lurie is given as an example near the end of Chapter 1.

0.7 Some of the Main Goals of Liapunov Theory

We discuss linear problems here, but the theory extends to nonlinear and singular problems as well. Suppose we are given

$$x(t) = a(t) - \int_0^t C(t, s)x(s)ds, \tag{0.7.1}$$

its resolvent equation

$$R(t, s) = C(t, s) - \int_s^t C(t, u)R(u, s)du, \tag{0.7.2}$$

the variation of parameters formula

$$x(t) = a(t) - \int_0^t R(t, s)a(s)ds, \tag{0.7.3}$$

and suppose that we successfully construct two Liapunov functionals $V_1(t,x)$ for (0.7.1) and $V_2(t,R)$ for (0.7.2).

Liapunov theory has always been used to find limit sets of solutions and growth rates of the solution. A quick reading of the statements of the theorems in Section 2.9 will supply more detail for the following sketch.

First, in a natural way our Liapunov functional $V_1(t,x)$ has a derivative which will locate a limit function, say $b(t)$, of the solution of (0.7.1) so that if $a \in L^p[0,\infty)$ then we will conclude that $x-b \in L^q[0,\infty)$ for some positive integers p and q. With added effort we will show that $x(t)-b(t) \to 0$ as $t \to \infty$. Should a fail to be in L^p then the derivative of that Liapunov functional will yield what we will call a **Liapunov growth** condition for the solution. That growth is clearly given for a general fractional differential equation of Caputo type.

Next, in a natural way the derivative of our Liapunov functional $V_2(t,R)$ will show that

$$\int_s^t |R(u,s)-C(u,s)|^p du \le \int_s^t |C(u,s)|^q du$$

and with additional effort we show that $R(t,s)-C(t,s) \to 0$ as $t \to \infty$ for fixed s. Note that in the convolution case it will say that $C \in L^q$ implies $R(t)-C(t) \in L^p$ and that $R(t)-C(t) \to 0$ as $t \to \infty$.

The reader's first guess is then correct. In (0.7.3) we can replace the unknown $R(t,s)$ with the clearly given $C(t,s)$ with measurable errors.

Finally, the bounds on the solutions are used repeatedly to prove the existence of periodic and asymptotically periodic solutions using fixed point theory.

Chapter 1

Linear Equations, Existence, and Resolvents

1.1 A Qualitative Approach

We are concerned with a linear integral equation written

$$x(t) = a(t) - \int_0^t C(t,s)x(s)ds \tag{1.1.1}$$

where x and a are n-vectors, $n \geq 1$, while C is an $n \times n$ matrix. Usually, we ask that a be at least continuous on $[0, \infty)$ but a remark in the next section shows how to avoid that condition. It is assumed that C is either continuous, weakly singular on $[0, \infty) \times [0, \infty)$, or differentiable. There is different existence theory, resolvent theory, and Liapunov theory for each type and we present it in that order. The first surprise awaiting the student of differential equations is that the scalar case, $n = 1$, is fundamental since a linear n^{th} order normalized ordinary differential equation can be expressed in that way. By contrast, an integral equation of the form of (1.1.1) can be expressed as a differential equation only under very special conditions.

We strive to make this book accessible to a very wide audience. With one exception, throughout the book we deal with Riemann integrals, including convergent improper integrals. When we write $f \in L^1[0, \infty)$, we mean that the improper Riemann integral of $|f|$ on $[0, \infty)$ converges to a finite number. The single exception is described in the final section of this chapter. It often happens that we encounter the convolution of two functions, each of which is in $L^1[0, \infty)$. In this one case (which occurs many times in his book), we will mean the Lebesgue integral.

Along with (1.1.1) we study the resolvent equations which are written in two ways (usually equivalent)

$$R(t,s) = C(t,s) - \int_s^t C(t,u)R(u,s)du$$

$$R(t,s) = C(t,s) - \int_s^t R(t,u)C(u,s)du \tag{1.1.2}$$

and the variation of parameters formula

$$x(t) = a(t) - \int_0^t R(t,s)a(s)ds. \tag{1.1.3}$$

Under conditions in this book, it is shown in Miller (1971a, p. 200) as well as in Burton and Dwiggins (2011), that the same unique function $R(t,s)$ satisfies both forms of (1.1.2). We prefer the first because it is in the classical form of an integral equation and it is the form best suited to Liapunov's direct method. The second form is best suited to the Razumikhin technique and we do use it at least one time in this book, namely in the proof of Theorem 2.1.3. The resolvent, $R(t,s)$, depends only on $C(t,s)$ and not on $a(t)$. Once we have determined certain basic properties of R, the equation (1.1.3) will speak volumes about (1.1.1) for a wide variety of functions a.

In addition to the aforementioned surprise, two more await us. While $R(t,s)$ depends only on $C(t,s)$, under a variety of conditions on C we can identify vector spaces of functions ϕ such that $\int_0^t R(t,s)\phi(s)ds$ is a good copy of ϕ. We will, for example, give conditions on C so that if $\phi' \in L^p[0,\infty)$ for some $p > 0$, then in (1.1.3) we have

$$x(t) = \phi(t) - \int_0^t R(t,s)\phi(s)ds \in L^p[0,\infty);$$

the function $\phi(t) = (t+1)^\beta$ for $0 < \beta < 1$ is included.

This gives a glimpse into the types of conditions and conclusions to be studied. No integrations are involved, no graphs are plotted, and no initial conditions are provided. Instead, a vector space of functions ϕ with $\phi' \in L^p$ is specified and the conclusion is that the solution is in L^p. Thus, we know that the solution resides near 0 much of the time although it may have spikes. Also, there are large functions $a = \phi$ which seem to affect the solution very little; they are harmless perturbations.

In each of the problems we discuss we present three packages: P_1, P_2, P_3. P_1 states conditions on C. P_2 describes the vector space in which a resides. P_3 describes the behavior of all solutions generated by elements of the vector space. These properties are in the way of stability results. Because of uncertainties or even stochastic elements, a may not be known. Yet we find vast vector spaces containing functions a which would yield acceptable solutions. Along the same lines, the vector spaces can be greatly expanded to include functions b which may not satisfy the exact technical conditions such as $b' \in L^p$, but if $a' \in L^p$ and $|a(t) - b(t)|$ is bounded, then b may be included.

There is an example which is ever before us that can be stated as the following question. If $C(t, s)$ is a "nice" function, what is the difference between the solution of

$$x(t) = t + \sin t + (t+1)^{1/2} \sin(t+1)^{1/3} - \int_0^t C(t, s)x(s)ds \tag{1.1.4}$$

and

$$y(t) = \sin t - \int_0^t C(t, s)y(s)ds? \tag{1.1.5}$$

There would be no surprise and no story to tell unless the answer were that, in spite of the large difference in size of the forcing functions, the difference in solutions is an L^p-function. This is the third surprise. Obviously, much is hidden in the package P_1 which we call "nice" and that, itself, is a long problem to unravel.

Here, we find two curious properties. The first is that, on average,

$$\int_0^t R(t, s)[(s+1)^{1/2} \sin(s+1)^{1/3}]ds$$

is a good copy of $(t+1)^{1/2} \sin(t+1)^{1/3}$ for large t; a surprising fact since the function oscillates unboundedly. By contrast, $\int_0^t R(t, s) \sin s\, ds$ is utterly unable to copy the small and nicely behaved function $\sin t$ and (1.1.5) has an asymptotically periodic solution. Substituting these functions into (1.1.3), together with a bit more unstated work, leaves us with both (1.1.4) and (1.1.5) having an asymptotically periodic solution.

The example leads us into a main project of this book and it tells us many important things. First, take courage when encountering a large function $a(t)$ in a delicate physical problem; such a function may offer little cause for worry. On the other hand, never dismiss small terms as they can have absolutely disastrous effects and must be studied with respect. But far more than that is the idea that we are looking for broad qualitative properties. We are more than satisfied with the information that a solution is in $L^p[0, \infty)$ or that it is asymptotically periodic.

Before we become immersed in technical details let us look at two examples which set the tone for the book. We will consider (1.1.1) for $n = 1$, the scalar case, and give two simple results. The first uses contraction mappings and is very old. While the proof given here is newer, the result itself is probably one of the first ever obtained for integral equations and it is named accordingly. The second result is obtained by means of a Liapunov functional and it is very new. Taken together these results lead us to understand the richness of the nonconvolution case.

In preparation for the first result and, indeed the entire book, we define a complete metric space and state the contraction mapping principle.

Definition 1.1.1. *A pair* $(\mathcal{S}, \rho)$ *is a* metric space *if* $\mathcal{S}$ *is a set and* $\rho : \mathcal{S} \times \mathcal{S} \to [0, \infty)$ *such that when* y, z, *and* u *are in* $\mathcal{S}$ *then*

(a) $\rho(y, z) \geq 0$, $\rho(y, y) = 0$, *and* $\rho(y, z) = 0$ *implies* $y = z$,

(b) $\rho(y, z) = \rho(z, y)$, *and*

(c) $\rho(y, z) \leq \rho(y, u) + \rho(u, z)$.

The metric space is complete *if every Cauchy sequence in* $(\mathcal{S}, \rho)$ *has a limit in that space.*

Definition 1.1.2. *Let* $(\mathcal{S}, \rho)$ *be a metric space and* $A : \mathcal{S} \to \mathcal{S}$. *The operator* A *is a* contraction operator *if there is an* $\alpha \in (0, 1)$ *such that* $x \in \mathcal{S}$ *and* $y \in \mathcal{S}$ *imply*

$$\rho[A(x), A(y)] \leq \alpha\rho(x, y).$$

Theorem 1.1.1. Contraction Mapping Principle *Let* $(\mathcal{S}, \rho)$ *be a complete metric space and* $A : \mathcal{S} \to \mathcal{S}$ *a contraction operator. Then there is a unique* $\phi \in \mathcal{S}$ *with* $A(\phi) = \phi$. *Furthermore, if* $\psi \in \mathcal{S}$ *and if* $\{\psi_n\}$ *is defined inductively by* $\psi_1 = A(\psi)$ *and* $\psi_{n+1} = A(\psi_n)$, *then* $\psi_n \to \phi$, *the unique fixed point. Finally, if* ϕ *is the fixed point and if* x *is any other point then* $\rho(x, \phi) \leq \rho(x, Ax)/(1 - \alpha)$.

Proof. Let $x_0 \in \mathcal{S}$ and define a sequence $\{x_n\}$ in $\mathcal{S}$ by $x_1 = A(x_0)$, $x_2 = A(x_1) \stackrel{\text{def}}{=} A^2x_0, \ldots, x_n = Ax_{n-1} = A^n x_0$. To see that $\{x_n\}$ is a Cauchy sequence, note that if $m > n$, then

$$\begin{aligned}
\rho(x_n, x_m) &= \rho(A^n x_0, A^m x_0) \\
&\le \alpha\rho(A^{n-1}x_0, A^{m-1}x_0) \\
&\vdots \\
&\le \alpha^n \rho(x_0, x_{m-n}) \\
&\le \alpha^n \big\{\rho(x_0, x_1) + \rho(x_1, x_2) + \cdots + \rho(x_{m-n-1}, x_{m-n})\big\} \\
&\le \alpha^n \big\{\rho(x_0, x_1) + \alpha\rho(x_0, x_1) + \cdots + \alpha^{m-n-1}\rho(x_0, x_1)\big\} \\
&= \alpha^n \rho(x_0, x_1)\{1 + \alpha + \cdots + \alpha^{m-n-1}\} \\
&\le \alpha^n \rho(x_0, x_1)\{1/(1-\alpha)\}\,.
\end{aligned}$$

Because $\alpha < 1$, the right side tends to zero as $n \to \infty$. Thus, $\{x_n\}$ is a Cauchy sequence, and because $(\mathcal{S}, \rho)$ is complete, it has a limit $x \in \mathcal{S}$. Now A is certainly continuous, so

$$A(x) = A(\lim_{n\to\infty} x_n) = \lim_{n\to\infty} A(x_n) = \lim_{n\to\infty} x_{n+1} = x\,,$$

and x is a fixed point. To see that x is unique, consider $A(x) = x$ and $A(y) = y$. Then

$$\rho(x, y) = \rho\big(A(x), A(y)\big) \le \alpha\rho(x, y)\,,$$

and because $\alpha < 1$, we conclude that $\rho(x, y) = 0$, so that $x = y$. For the last part,

$$\rho(x, \phi) \le \rho(x, Ax) + \rho(Ax, A\phi) \le \rho(x, Ax) + \alpha\rho(x, \phi),$$

from which the conclusion follows. This completes the proof.

Frequently the complete metric space will be a Banach space and the metric, ρ, will be the norm. Much of the power of the contraction mapping theorem lies in a careful choice of metric. We begin with perhaps the simplest, the supremum norm. The following result is sometimes called the Adam and Eve theorem since it seems to appear in the very earliest treatments of integral equations.

Theorem 1.1.2. *Suppose that C is continuous and that there are positive numbers K and α, $\alpha < 1$, such that*

$$|a(t)| \leq K \text{ and } \sup_{t\geq 0} \int_0^t |C(t,s)|ds \leq \alpha.$$

Then there is a unique solution of (1.1.1) and it is bounded and continuous on $[0,\infty)$.

Proof. Let $(X, \|\cdot\|)$ be the Banach space of bounded continuous functions $\phi : [0,\infty) \to \Re^n$ with the supremum norm. Define a mapping $P : X \to X$ by $\phi \in X$ implies

$$(P\phi)(t) = a(t) - \int_0^t C(t,s)\phi(s)ds.$$

Notice that if $\phi, \eta \in X$ then

$$|(P\phi)(t) - (P\eta)(t)| \leq \int_0^t |C(t,s)||\phi(s) - \eta(s)|ds \leq \alpha\|\phi - \eta\|.$$

Hence, P is a contraction and there is a unique $\phi \in X$ with

$$(P\phi)(t) = \phi(t) = a(t) - \int_0^t C(t,s)\phi(s)ds.$$

Notice that we obtain existence, uniqueness, and boundedness all at the same time. Our next example is a contrast in several ways. First, we need an existence theorem to even begin. In the next section we will prove such a result and we do so by means of a contraction mapping with a weighted metric. It is a very interesting construction. The next contrast is that we integrate the first coordinate of C. Finally, the conclusion is that $x \in L^1[0,\infty)$, rather than being bounded. Such a result is always just the first step. Given that the solution is L^1 the enterprising investigator may go back to (1.1.1) and parlay that into boundedness and, often, decay to zero. The reader is welcome to try that on the example below.

The idea of a Liapunov functional is marvelous. It converts a functional equation into an algebraic equation. Suppose, for example, that we have a linear algebraic equation

$$x = a(t) + b(t)x$$

and we want to show that x is bounded using the facts that $a(t)$ is bounded and $|b(t)| \leq \alpha < 1$. Then we might write

$$|x| \leq |a(t)| + |b(t)||x| \leq |a(t)| + \alpha|x|$$

or

$$|x|(1-\alpha) \leq |a(t)|$$

so that

$$|x| \leq |a(t)|/(1-\alpha).$$

Obviously, we can not do that for (1.1.1) since $x(s)$ is in the integral. But a carefully chosen Liapunov functional allows us to do precisely that. Here, we use a rather pedestrian Liapunov functional, but as we proceed through the book we will see more intricate and useful choices.

Notice! In the proof below we will see for the first time how we can unite the Liapunov functional with the integral equation without using a chain rule.

Theorem 1.1.3. *Suppose that C is continuous and that there is an $\alpha < 1$ with*

$$\int_0^\infty |C(u+t,t)|du \leq \alpha.$$

If $a \in L^1[0,\infty)$, so is any solution of (1.1.1).

Proof. We begin exactly as we did with our algebraic argument and write

$$|x(t)| \leq |a(t)| + \int_0^t |C(t,s)||x(s)|ds.$$

Assume that a solution of (1.1.1) exists on $[0,\infty)$ and contrive to construct a functional having a derivative containing that last term. One simple choice is

$$V(t) = \int_0^t \int_{t-s}^\infty |C(u+s,s)|du|x(s)|ds$$

so that

$$V'(t) = \int_0^\infty |C(u+t,t)|du|x(t)| - \int_0^t |C(t,s)||x(s)|ds.$$

Substituting our first inequality into this expression yields

$$V'(t) \leq \int_0^\infty |C(u+t,t)|du|x(t)| - |x(t)| + |a(t)| \leq (\alpha-1)|x(t)| + |a(t)|.$$

Integrate both sides and obtain

$$0 \le V(t) \le V(0) + \int_0^t |a(s)|ds - (1-\alpha)\int_0^t |x(s)|ds$$

from which we obtain

$$\int_0^t |x(s)|ds \le (1/(1-\alpha))\int_0^t |a(s)|ds,$$

as required.

Remark Once we have defined weak singularities (see Definition 1.2.1) the reader can readily prove Theorem 1.1.2 for kernels of this type. But the corresponding extension of Theorem 1.1.3 must wait.

1.2 Existence and Resolvents

The first few pages of this section will contain the basic relations which will be needed throughout the book. This is the material which the reader will frequently consult. We then give full proofs of existence, uniqueness, resolvent properties, and variation of parameters formulae for integral equations with kernels which are either continuous, weakly singular (see Definition 1.2.1), or differentiable.

We begin with a study of the linear integral equation

$$x(t) = a(t) - \int_0^t C(t,s)x(s)ds \tag{1.2.1}$$

where $a : [0,\infty) \to \Re^n$ is at least continuous, while C is an $n \times n$ matrix of functions which are either continuous or weakly singular for $0 \le s \le t < \infty$.

Along with (1.2.1) is the resolvent equation

$$R(t,s) = C(t,s) - \int_s^t C(t,u)R(u,s)du. \tag{1.2.2}$$

There is then the variation of parameters formula

$$x(t) = a(t) - \int_0^t R(t,s)a(s)ds \tag{1.2.3}$$

so that for x defined by (1.2.3), then x is the unique solution of (1.2.1) on $[0,\infty)$.

If C, and sometimes a, is differentiable then there are other forms of (1.2.1) which can be very useful. The most obvious is obtained by differentiating (1.2.1) to obtain the integrodifferential equation

$$x'(t) = a'(t) - C(t,t)x(t) - \int_0^t C_t(t,s)x(s)ds \tag{1.2.4}$$

where C_t or C_1 denotes $\frac{\partial C(t,s)}{\partial t}$.

There is a less-known form obtained as follows. It requires differentiability of C, but not of a. In (1.2.1) we integrate by parts and have

$$x(t) = a(t) - C(t,t)\int_0^t x(u)du + \int_0^t C_s(t,s)\int_0^s x(u)duds. \tag{1.2.5}$$

If we let $y(t) = \int_0^t x(u)du$, $y(0) = 0$, then (1.2.5) becomes

$$y'(t) = a(t) - C(t,t)y(t) + \int_0^t C_s(t,s)y(s)ds. \tag{1.2.6}$$

Thus, either (1.2.4) or (1.2.6) leads us to

$$x'(t) = A(t)x(t) + \int_0^t B(t,u)x(u)du + f(t). \tag{1.2.7}$$

Along with (1.2.7) is Becker's resolvent equation

$$\frac{\partial}{\partial t}Z(t,s) = A(t)Z(t,s) + \int_s^t B(t,u)Z(u,s)du,\ Z(s,s) = I, \tag{1.2.8}$$

written less formidably as

$$z'(t) = A(t)z(t) + \int_s^t B(t,u)z(u)du \tag{1.2.9}$$

where $Z(t,s)$ is the $n \times n$ matrix whose columns are solutions of (1.2.8) on $[s,\infty)$ with $Z(s,s) = I$. We will show that $Z(t,s)$ exists for $0 \leq s \leq t < \infty$ and that for a given constant vector $x(0)$, there is a unique solution of (1.2.7) on $[0,\infty)$ given by the variation of parameters formula

$$x(t) = Z(t,0)x(0) + \int_0^t Z(t,s)f(s)ds. \tag{1.2.10}$$

A more detailed form is also true. The solution of

$$x'(t) = A(t)x(t) + \int_\tau^t B(t,u)x(u)du + f(t),\ x(\tau) = x_0, \tag{1.2.11}$$

is given by

$$x(t) = Z(t,\tau)x_0 + \int_\tau^t Z(t,s)f(s)ds. \tag{1.2.12}$$

Now (1.2.9) is the resolvent equation of Becker (1979) (2006) and we much prefer it to the classical resolvent equation of Grossman and Miller (1970) which is

$$\frac{\partial H(t,s)}{\partial s} = -H(t,s)A(s) - \int_s^t H(t,u)B(u,s)du, \quad H(t,t) = I. \tag{1.2.13}$$

We will show that $Z(t,s) = H(t,s)$ and this will be important for the following reason. As the solution of (1.2.1) is also a solution of (1.2.4) and the solution of (1.2.1) is expressed as (1.2.3), while solutions of (1.2.4) are expressed as (1.2.10), it follows that (1.2.10) and (1.2.3) are related. Now if $H = Z$ then it follows that $Z_s(t,s)$ is continuous and that will allow us to integrate that integral in (1.2.10) by parts. We take $f(t) = a'(t)$ and $x(0) = a(0)$ in (1.2.10) and have

$$\begin{aligned} x(t) &= Z(t,0)a(0) + Z(t,s)a(s)\big|_0^t - \int_0^t Z_s(t,s)a(s)ds \\ &= Z(t,0)a(0) + Z(t,t)a(t) - Z(t,0)a(0) - \int_0^t Z_s(t,s)a(s)ds \end{aligned}$$

so that

$$x(t) = a(t) - \int_0^t Z_s(t,s)a(s)ds, \tag{1.2.14}$$

a variation of parameters formula for (1.2.1) when C_t is continuous.

APPLICATION See Theorems 2.1.3, 3.2.10, and 2.6.2.1 for use of this form. In Burton (2005b, p. 321) we offer a theorem which can be used with many of the Liapunov functionals which we will present in Chapter 2 and exploit (1.2.14) in just this same way.

This is the collection of results which we will need to study (1.2.1). The remainder of this section will be devoted to their properties.

For an $n \times n$ matrix A, we define the operator norm of A as follows. Let $|\cdot|$ denote any vector norm on $\Re^n$ and for any $n \times n$ matrix A let

$$|A| = \sup\{|Ax| : |x| \leq 1, x \in \Re^n\}. \tag{1.2.15}$$

This is equivalent to

$$|A| = \inf_{0<M<\infty} \{M : |Ax| \leq M|x|, \text{ for all } x \in \Re^n\}. \tag{1.2.16}$$

By (1.2.16),

$$|Ax| \leq |A||x|. \tag{1.2.17}$$

We now show that if A and B are $n \times n$ matrices, then $|AB| \leq |A||B|$. To this end, we note that by (1.2.17) we have

$$|(AB)x| = |A(Bx)| \leq |A||Bx| \leq |A||B||x|.$$

By (1.2.17) again we have

$$|(AB)x| \leq |AB||x|.$$

Finally, by (1.2.16) we then have

$$|AB| \leq |A||B|.$$

For existence results under very weak conditions see, for example, Corduneanu (1991), Miller (1971a), and Tricomi (1985). Our choice of existence theorem is dictated by the types of real-world problems for which we can construct Liapunov functionals. These are typified by kernels of the form $(t-s)^{q-1}$ with $0 < q < 1$, so that there is a single singularity at $t = s$, although the work here is much more general. Before we get into those problems we offer a very simple result based on a continuous kernel. From the conditions there and the steps in the proof the reader will see the motivation of several conditions in the general result.

Theorem 1.2.1. *Let $a : [0, \infty) \to \Re^n$ be continuous and let $C(t, s)$ be an $n \times n$ matrix of functions continuous for $0 \leq s \leq t < \infty$. Then there is one and only one continuous function $x : [0, \infty) \to \Re^n$ satisfying (1.2.1).*

Proof. Let $b > 0$ and denote by X the vector space of continuous functions $\phi : [0, b] \to \Re^n$. If r is a fixed positive number then we define the norm on X by

$$|\phi|_r := \sup\{|\phi(t)|e^{-rt} : 0 \leq t \leq b\}.$$

Then $(X, |\cdot|_r)$ is a Banach space.

Let b be any fixed positive number. We now show that the unique solution of (1.2.1) exists on $[0, b]$ and, as b is arbitrary, the theorem will follow.

Define $r := \sup_{0\leq s\leq t\leq b}|C(t,s)|+1$ and let $P : X \to X$ be defined by $\phi \in X$ implies that

$$(P\phi)(t) = a(t) - \int_0^t C(t,s)\phi(s)ds.$$

Then for $\phi, \eta \in X$ we have

$$\begin{aligned}
|(P\phi)(t) - (P\eta)(t)|e^{-rt} &\leq e^{-rt}\int_0^t |C(t,s)||\phi(s)-\eta(s)|ds \\
&= \int_0^t e^{-r(t-s)}|C(t,s)|e^{-rs}|\phi(s)-\eta(s)|ds \\
&\leq \int_0^t e^{-r(t-s)}|C(t,s)|ds|\phi-\eta|_r \\
&\leq \int_0^t e^{-r(t-s)}(r-1)ds|\phi-\eta|_r \\
&= \frac{r-1}{r}|\phi-\eta|_r.
\end{aligned}$$

It follows that P is a contraction and there is a unique fixed point satisfying (1.2.1). This completes the proof.

Our next objective is a solution of (1.2.2) and it proceeds exactly as in Theorem 1.2.1, taking one column of $R(t,s)$ at a time. Notice that if Q^i denotes the ith column of an $n \times n$ matrix Q, then (1.2.2) represents n integral equations

$$R^i(t,s) = C^i(t,s) - \int_s^t C(t,u)R^i(u,s)du.$$

Notice also that the C in the integral has no superscript. The next result is found in Burton and Dwiggins (2011).

Theorem 1.2.2. *If $C(t,s)$ is an $n \times n$ matrix of scalar functions $c_{ij}(t,s)$ which are jointly continuous in (t,s) for $0 \leq s \leq t < \infty$, then there is a unique matrix $R(t,s)$ of scalar functions jointly continuous in (t,s) on that same interval which satisfies (1.2.2). If $c = \sup_{0\leq s\leq t}|C(t,s)|$ then $|R(t,s)| \leq nce^{c(t-s)}$.*

Proof. Let $b > 0$ be arbitrary, let

$$r = \sup_{0 \le s \le t \le b} |C(t,s)| + 1, \;\; \alpha = \frac{r-1}{r},$$

and let $(X, |\cdot|_r)$ denote the Banach space of vector valued functions $\phi(t,s)$ jointly continuous in (t,s) for $0 \le s \le t \le b$ where $\phi \in X$ implies

$$|\phi|_r = \sup_{0 \le s \le t \le b} e^{-r(t-s)} |\phi(t,s)|.$$

Now, for each $i = 1, \cdots, n$ define a mapping $P : X \to X$ by $\phi \in X$ implies

$$(P\phi)(t,s) = C^i(t,s) - \int_s^t C(t,u)\phi(u,s)du.$$

If $\phi, \eta \in X$ then

$$\begin{aligned} &|(P\phi)(t,s) - (P\eta)(t,s)|e^{-r(t-s)} \\ &\qquad \le \int_s^t e^{-r(t-s)+r(u-s)-r(u-s)}(r-1)|\phi(u,s) - \eta(u,s)|du \\ &\qquad \le |\phi - \eta|_r \frac{r-1}{r} e^{-r(t-u)}\Big|_s^t \\ &\qquad \le \alpha|\phi - \eta|_r, \end{aligned}$$

and so P is a contraction with unique fixed point which gives $R^i(t,s)$ as the unique solution, jointly continuous in (t,s). Then $R(t,s) = (R^1, \ldots, R^n)$ is the unique $n \times n$ matrix which satisfies (1.2.2).

We now find a bound for R. Let $t > 0$ be fixed and $c = \sup_{0 \le s \le t} |C(t,s)|$. We will find a bound on that fixed point $\phi(s)$ for $0 \le s \le t$. Designate the columns of R more fully as

$$\phi^i(t,s) = C^i(t,s) - \int_s^t C(t,u)\phi^i(t,u)du$$

and then let $\phi(t,s)$ be any of the ϕ^i. We have

$$|\phi(t,s)| \le c + \int_s^t |\phi(t,u)|cdu.$$

Separate variables and integrate from s to t obtaining

$$\int_s^t \frac{-|\phi(t,v)|cdv}{c + \int_v^t |\phi(t,u)|cdu} \ge -c(t-s)$$

and after integration and rearranging we have

$$|\phi(t,s)| \leq c + \int_s^t c|\phi(t,u)|du \leq e^{c(t-s)}.$$

That is a bound on $\phi(t,s)$ for $0 \leq s \leq t$. There is one such bound for each i, which represents the i-th column of $R(t,s)$ and so a bound for $R(t,s)$ is $nce^{c(t-s)}$. If the need arises, one can improve that bound.

The work just done in this proof will also prove an old and useful result.

Theorem 1.2.3 (Gronwall's Inequality). *Let $f, g : [0,\alpha] \to [0,\infty)$ be continuous and let c be a nonnegative number. If*

$$f(t) \leq c + \int_0^t g(s)f(s)\,ds\,, \quad 0 \leq t < \alpha\,,$$

then

$$f(t) \leq c \exp \int_0^t g(s)\,ds\,, \quad 0 \leq t < \alpha\,.$$

Proof. Suppose first that $c > 0$. Divide by $c + \int_0^t g(s)f(s)\,ds$ and multiply by $g(t)$ to obtain

$$f(t)g(t) \Big/ \left[c + \int_0^t g(s)f(s)\,ds\right] \leq g(t)\,.$$

An integration from 0 to t yields

$$\ln\left\{\left[c + \int_0^t g(s)f(s)\,ds\right] \Big/ c\right\} \leq \int_0^t g(s)\,ds$$

or

$$f(t) \leq c + \int_0^t g(s)f(s)\,ds \leq c \exp \int_0^t g(s)\,ds\,.$$

If $c = 0$, take the limit as $c \to 0$ through positive values. This completes the proof.

Discontinuous $a(t)$

Discontinuities in $a(t)$ occur frequently and usually pose no difficulty. They are so common in control theory. Tricomi (1957; pp. 10-13) offers a proof of the existence of L^2 solutions. There is an iteration procedure which bring such equations into the scope of Theorem 1.2.1. Let $a(t)$ be discontinuous and write (1.2.1) as

$$\begin{aligned} z(t) := x(t) - a(t) &= -\int_0^t C(t,s)[x(s) - a(s) + a(s)]ds \\ &= -\int_0^t C(t,s)a(s)ds - \int_0^t C(t,s)z(s)ds \end{aligned}$$

or

$$z(t) = a^*(t) - \int_0^t C(t,s)z(s)ds.$$

If a^* is continuous, then Theorem 1.2.1 yields a unique continuous solution, $z(t)$, and we have

$$x(t) = a(t) + z(t)$$

which is, of course, discontinuous. If a^* is not continuous one may repeat the process and refine the result. We will see a similar procedure when C is not continuous and we seek to obtain a resolvent. In that case we are assured that repeating the process enough times will produce continuity.

The next result was formulated in Burton and Dwiggins (2011), verifying (1.2.3). It differs from so many similar results which begin with a different form of the resolvent equation, namely

$$R(t,s) = C(t,s) - \int_s^t R(t,u)C(u,s)du,$$

which turns out to have the same solution, R, under general conditions. It is the preferred form for obtaining the variation of parameters formula, but (1.2.2) is used more widely in later applications. Usually, then, investigators have to prove their equivalence.

Theorem 1.2.4. *If C is continuous and if $R(t,s)$ is the unique continuous solution of (1.2.2) then the unique solution of (1.2.1) can be expressed as (1.2.3).*

Proof. First observe that (1.2.1) can be rewritten as

$$\int_0^t C(t,s)x(s)ds = a(t) - x(t),$$

which illustrates that verifying a given expression for $x(t)$ solves (1.2.1) means one should first calculate the integral of $x(s)$ against $C(t,s)$. Assume that $R(t,s)$ solves (1.2.2) and let $y(t) = a(t) - \int_0^t R(t,s)a(s)ds$. Then

$$\begin{aligned}
\int_0^t C(t,s)y(s)ds &= \int_0^t C(t,s)\left[a(s) - \int_0^s R(s,u)a(u)du\right]ds \\
&= \int_0^t C(t,s)a(s)ds - \int_0^t \int_0^s C(t,s)R(s,u)a(u)duds \\
&= \int_0^t C(t,s)a(s)ds - \int_0^t \int_u^t C(t,s)R(s,u)ds\, a(u)du \\
&= \int_0^t C(t,s)a(s)ds - \int_0^t \big[C(t,u) - R(t,u)\big]a(u)du \\
&= \int_0^t R(t,u)a(u)du = a(t) - y(t).
\end{aligned}$$

Note in the above string of equalities we interchanged the order of integration in the third line, and we used the assumption that R solves (1.2.2) in the fourth line. Thus we have $y(t)$ as a solution to (1.2.1), and so by uniqueness of solutions $x(t) = y(t)$, as required.

The next definition and theorem were adapted from Becker (2011).

Definition 1.2.1. *Let $\Omega_T := \{(t,s) : 0 \leq s \leq t \leq T\}$. The kernel C of (1.2.1) is weakly singular on the set Ω_T if it is discontinuous in Ω_T; but for each $t \in [0,T]$, $C(t,s)$ has at most finitely many discrete discontinuities in the interval $\{s : 0 \leq s \leq t\}$ and for every continuous function $\phi : [0,T] \to \Re^n$*

$$\int_0^t C(t,s)\phi(s)ds \tag{i}$$

and

$$\int_0^t |C(t,s)|ds \tag{ii}$$

both exist and are continuous on $[0,T]$. If $C(t,s)$ is weakly singular on Ω_T for every $T > 0$, then it is weakly singular on the set $\Omega := \{(t,s)\colon 0 \leq s \leq t < \infty\}$.

Theorem 1.2.5 is the main existence result for this book. It is stated for the linear equation (1.2.1) but it is not difficult to sift through the proof and see that if (1.2.1) is changed to

$$x(t) = a(t) - \int_0^t C(t,s)g(s,x(s))ds \tag{1.2.1N}$$

where $g : [0,\infty) \times \Re \to \Re$ is continuous in both variables and satisfies a global Lipschitz condition in x, then the proof is readily changed to yield a unique solution of that equation. Thus, later in this chapter we will be considering equations of that type and will assume that there is a unique solution without further proof.

The next lemma, found in Becker (2011), is essential and is certainly to be expected; however, it is not at all easy to prove. Theorem 1.2.5 is also from the same paper.

Lemma 1.2.1. *Let $C(t,s)$ be weakly singular on Ω_T. If $\phi : [0,T] \to \Re^n$ is continuous, then*

$$\left|\int_{p_1}^{p_2} C(t,s)\phi(s)ds\right| \le \int_{p_1}^{p_2} |C(t,s)\phi(s)|ds \le \int_{p_1}^{p_2} |C(t,s)||\phi(s)|ds$$

for $0 \le p_1 \le p_2 \le T$.

Theorem 1.2.5. *Let $a : [0,T] \to \Re^n$ be continuous. Let $C(t,s)$ be an $n \times n$ matrix of scalar functions. If $C(t,s)$ is weakly singular on Ω_T and if constants $\gamma > 0$ and $k \in (0,1)$ exist such that*

$$\int_0^t e^{-\gamma(t-s)}|C(t,s)|\, ds \le k$$

for $t \in [0,T]$, then (1.2.1) has a unique continuous solution on $[0,T]$.

Proof. Let $(X_T, |\cdot|_\gamma)$ denote the Banach space of all continuous functions $\phi : [0,T] \to \Re^n$, where γ is the constant of the theorem and $|\cdot|_\gamma$ denotes the norm defined by

$$|\phi|_\gamma := \sup\{\, e^{-\gamma t}|\phi(t)| : 0 \le t \le T\,\}.$$

Define a mapping P on X_T as follows: for $\phi \in X_T$

$$(P\phi)(t) := a(t) + \int_0^t C(t,s)\phi(s)\, ds.$$

By the continuity of a and $\int_0^t C(t,s)\phi(s)ds$, $P\colon X_T \to X_T$. Furthermore, P is a contraction mapping on $(X_T, |\cdot|_\gamma)$. To see this, let $\phi, \eta \in X_T$. By Lemma 1.2.1 we have

$$|(P\phi)(t) - (P\eta)(t)| \le \int_0^t |C(t,s)||\phi(s) - \eta(s)|\, ds.$$

Hence,

$$\begin{aligned}|(P\phi)(t) - (P\eta)(t)|e^{-\gamma t} &\le \int_0^t e^{-\gamma(t-s)}|C(t,s)|e^{-\gamma s}|\phi(s) - \eta(s)|\, ds \\ &\le |\phi - \eta|_\gamma \int_0^t e^{-\gamma(t-s)}|C(t,s)|\, ds.\end{aligned}$$

By the integral assumption of the theorem,

$$|(P\phi)(t) - (P\eta)(t)|e^{-\gamma t} \le k|\phi - \eta|_\gamma$$

for $t \in [0,T]$, from which it follows that

$$|P\phi - P\eta)|_\gamma \le k|\phi - \eta|_\gamma.$$

Thus, P is a contraction mapping as $k < 1$. By Banach's contraction mapping principle, P has a unique fixed point $x \in X_T$. In other words, x is the unique continuous solution of (1.2.1) on $[0,T]$.

Existence is one thing, but as we move on to two forms of the resolvent and differentiation of Liapunov functionals we are guided in large measure by this classical result (See Natanson (1960; Vol. II, p. 93).)

Tonelli-Hobson test If $f(x,y)$ is measurable on $R(a \le x \le b, c \le y \le d)$ and if

$$\int_a^b dx \int_c^d |f(x,y)|dy < \infty,$$

then the double integral exists and equals the common value

$$\int_a^b dx \int_c^d f(x,y)dy = \int_c^d dy \int_a^b f(x,y)dx.$$

In words, if $f(x,y)$ is measurable and if either of the iterated integrals of $|f(x,y)|$ exists, then the double integral and both iterated integrals of $f(x,y)$ exist and they are all equal.

Preparing the Resolvent In the proof of Theorem 1.2.5 we saw how the solution (1.2.1) is shielded from the discontinuities in $C(t,s)$ by the integral. But in (1.2.2) there is no way to shield $R(t,s)$ from discontinuities of $C(t,s)$. It seems clear that $R(t,s)$ will have the same kind of discontinuities and in the same place as those of $C(t,s)$. We can proceed exactly as we did for a discontinuous $a(t)$.

We follow Becker (2011). Let $C(t,s)$ be weakly singular and from (1.2.2) define

$$R(t,s) =: C(t,s) - R_1(t,s). \tag{1.2.18}$$

Then rewrite (1.2.2) as

$$C(t,s) - R_1(t,s) = C(t,s) - \int_s^t C(t,u)[C(u,s) - R_1(u,s)]du$$

or

$$R_1(t,s) = \int_s^t C(t,u)C(u,s)du - \int_s^t C(t,u)R_1(u,s)du.$$

We define

$$C^*(t,s) = \int_s^t C(t,u)C(u,s)du \tag{1.2.19}$$

and our long-term focus will be on

$$R_1(t,s) = C^*(t,s) - \int_s^t C(t,u)R_1(u,s)du. \tag{1.2.20}$$

If $C^*(t,s)$ is continuous, then very slight modifications of the proof of Theorem 1.2.5 will yield a unique continuous solution, $R_1(t,s)$, and we will have $R(t,s) = C(t,s) - R_1(t,s)$. Then R will, indeed, have precisely the same singularities as C. Moreover, when we apply the proof of Theorem 1.2.4, the Tonelli-Hobson test will hold and we will get the variation of parameters formula (1.2.3). On the other hand, if $C^*(t,s)$ is not continuous then we can make the same change of variable in the C^* equation and obtain a new C^{**} and a new resolvent equation for an R_2. If C^{**} is continuous then the process can stop when we have found the continuous function R_2. In this process we are obtaining a series solution for $R(t,s)$ and the process will work, although we do not prove that part.

We now present a result of Becker (2011). It is predicated on the assumption that C^* is continuous. If it is not continuous, the idea is to keep on iterating until we arrive at a function which is continuous. Following this theorem we will offer a long and important example illustrating the process.

Note In this chapter we offer existence theorems for integrodifferential equations by writing them as integral equations. But in the case of a singular kernel various questions arise which we do not want to treat here. Instead, we offer a separate existence result for integrodifferential equations with singular kernel in Section 3.9, together with explanation of some of the difficulties.

Theorem 1.2.6. *Let $C(t,s)$ be an $n \times n$ matrix function. For a given $T > 0$, let $\Omega_T = \{(t,s) \colon 0 \le s \le t \le T\}$. Let $C(t,s)$ be weakly singular on Ω_T. Suppose that C^* is defined by*

$$C^*(t,s) := \int_s^t C(t,u)C(u,s)\,du, \tag{i}$$

unless (t,t) is a discontinuity of C in which case let

$$C^*(t,t) := \lim_{s \to t^-} \int_s^t C(t,u)C(u,s)\,du, \tag{ia}$$

and it exists and is continuous on Ω_T and that for every continuous function $\phi \colon \Omega_T \to \Re^n$,

$$\int_s^t C(t,u)\phi(u,s)\,du \tag{ii}$$

exists and is continuous on Ω_T. If constants $\gamma > 0$ and $k \in (0,1)$ exist such that

$$\int_s^t e^{-\gamma(t-u)}|C(t,u)|\,du \le k \tag{iii}$$

for $(t,s) \in \Omega_T$, then there is a unique $n \times n$ matrix function $R_1(t,s)$ that is continuous and which solves the matrix equation

$$R_1(t,s) = C^*(t,s) - \int_s^t C(t,u)R_1(u,s)\,du \tag{iv}$$

on Ω_T.

Proof. Equation $R_1(t,s) = C^*(t,s) - \int_s^t C(t,u)R_1(u,s)du$ comprises n vector equations, namely

$$r^i(t,s) = (C^*)^i(t,s) - \int_s^t C(t,u)r^i(u,s)\,du \quad (i = 1,\dots,n), \tag{vi}$$

where r^i and $(C^*)^i$ are the ith column vectors of the matrices R_1 and C^* respectively. Let $(S, |\cdot|_\gamma)$ denote the Banach space of continuous functions $\phi\colon \Omega_T \to \Re^n$, where γ is defined in (iii) and the norm of $\phi \in S$ is

$$|\phi|_\gamma := \sup\{\, e^{-\gamma(t-s)}|\phi(t,s)| \colon (t,s) \in \Omega_T\}. \tag{vii}$$

For each $i \in \{1,\dots,n\}$, define the mapping P^i on S by

$$(P^i\phi)(t,s) := (C^*)^i(t,s) - \int_s^t C(t,u)\phi(u,s)\,du.$$

Since (i) and (ii) are continuous, $P^i\colon S \to S$. Moreover, by Lemma 1.2.1 for each continuous $\phi\colon \Omega_T \to \Re^n$, the existence of (ii) as well as $\int_0^t |C(t,s)|ds$ implies that

$$\left|\int_s^t C(t,u)\phi(u,s)\,du\right| \le \int_s^t |C(t,u)\phi(u,s)|\,du \le \int_s^t |C(t,u)||\phi(u,s)|\,du$$

for $(t,s) \in \Omega_T$. Thus, for any given pair of functions $\phi, \eta \in S$, we have

$$|(P^i\phi)(t,s) - (P^i\eta)(t,s)| \le \int_s^t |C(t,u)||\phi(u,s) - \eta(u,s)|\,du.$$

Hence,

$$\begin{aligned} |(P^i\phi)(t,s) - &(P^i\eta)(t,s)|e^{-\gamma(t-s)} \\ &\le \int_s^t e^{-\gamma(t-u)}|C(t,u)|e^{-\gamma(u-s)}|\phi(u,s) - \eta(u,s)|\,du \\ &\le |\phi - \eta|_\gamma \int_s^t e^{-\gamma(t-u)}|C(t,u)|\,du. \end{aligned}$$

Then by (iii),

$$|(P^i\phi)(t,s) - (P^i\eta)(t,s)|e^{-\gamma(t-s)} \le k|\phi - \eta|_\gamma$$

which implies

$$|P^i\phi - P^i\eta|_\gamma \le k|\phi - \eta|_\gamma.$$

As $k < 1$, P^i is a contraction mapping on S. Therefore, by Banach's contraction mapping principle, there is a unique $r^i \in S$ such that

$$P^i r^i = r^i.$$

That is, there is a unique continuous function $r^i : \Omega_T \to \Re^n$ satisfying that column of (iv). Consequently, the unique solution of (iv) on Ω_T is the $n \times n$ matrix

$$\left[r^1(t,s), \cdots, r^n(t,s)\right] =: R_1(t,s).$$

Under the conditions of Theorem 1.2.6 we obtain a unique solution of (1.2.2) in the form of $R(t,s) = C(t,s) - R_1(t,s)$ where $R_1(t,s)$ is continuous. We turn then to Theorem 1.2.4 aiming to use R to obtain the unique solution $x(t)$ of (1.2.1) in the form of (1.2.3). Going through the proof we see that the interchange of order of integration is the only real problem when a is continuous. We could ask for the conditions of the Tonelli-Hobson test, which is what will usually happen in a particular problem. But there are other tests so we formulate the following corollary and simply ask that the interchange of the order of integration will hold for $C(t,s)$ and the unknown continuous function $R_1(t,s)$.

Corollary 1. *Let the conditions of Theorem 1.2.6 hold. Then the unique solution of (1.2.1) is given by (1.2.3) with $R(t,s) = C(t,s) - R_1(t,s)$ provided that $a(t)$ is continuous, that*

$$\int_0^t \int_0^s C(t,s)C(s,u)a(u)duds = \int_0^t \int_u^t C(t,s)C(s,u)a(u)dsdu,$$

and that for any continuous function $\phi : \Omega_T \to \Re^n$ we have

$$\int_0^t \int_0^s C(t,s)\phi(s,u)a(u)duds = \int_0^t \int_u^t C(t,s)\phi(s,u)a(u)dsdu.$$

A generalization of Theorem 1.2.6 can be found in Becker (2012, Theorem 3.1).

A Fractional Differential Equation

A fractional differential equation can be an integral equation with a parameter and it can represent a vast collection of partial differential equations describing real-world problems. As an example of the resolvent work we have presented here we will offer a fractional differential equation of Caputo type written as

$$ {}^cD^q x(t) = -u(t, x(t)), \quad 0 < q < 1, \quad x(0) \in \Re \tag{1.2.21}$$

where the Caputo fractional derivative of order q of a function x is

$$ {}^cD^q x(t) = \frac{1}{\Gamma(1-q)} \int_0^t (t-s)^{-q} x'(s) ds$$

and where Γ is the Euler gamma function. When $u(t,x)$ is continuous it can be inverted as

$$x(t) = x(0) - \frac{1}{\Gamma(q)} \int_0^t (t-s)^{q-1} u(s, x(s)) ds. \tag{1.2.22}$$

There is a myriad of real-world problems represented by this equation even for the single value $q = 1/2$ which is the value most commonly associated with the heat equation. At the end of Chapter 0 we listed six of these problems which come from very different areas of applied mathematics. An elementary derivation of (1.2.22) from the heat equation is given by Weinberger (1965; p. 357).

To return to the resolvent, we take $u(t,x) = b(t) + x$ so that (1.2.22) becomes

$$x(t) = x(0) - \frac{1}{\Gamma(q)} \int_0^t (t-s)^{q-1} b(s) ds - \frac{1}{\Gamma(q)} \int_0^t (t-s)^{q-1} x(s) ds.$$

Write this as

$$x(t) = a(t) - \frac{1}{\Gamma(q)} \int_0^t (t-s)^{q-1} x(s) ds \tag{1.2.23}$$

which is a linear integral equation with kernel

$$C(t-s) := \frac{1}{\Gamma(q)} (t-s)^{q-1}. \tag{1.2.24}$$

It is readily verified that it is weakly singular. The resolvent equation is

$$R(t-s) = \frac{(t-s)^{q-1}}{\Gamma(q)} - \frac{1}{\Gamma(q)} \int_s^t (t-u)^{q-1} R(u-s) du. \tag{1.2.25}$$

A change of variable leads us to

$$R(t) = \frac{t^{q-1}}{\Gamma(q)} - \frac{1}{\Gamma(q)} \int_0^t (t-v)^{q-1} R(v) dv. \tag{1.2.26}$$

The kernel is not continuous for any value of $q \in (0,1)$. Thus, we are ready to illustrate the construction of C^* and we hope to offer the reader a pleasant surprise. Referring to (1.2.19) and (1.2.24) we see that we must compute

$$C^*(t,s) = \int_s^t \frac{1}{\Gamma^2(q)}(t-u)^{q-1}(u-s)^{q-1}du$$

and we readily understand that such computations look formidable. In fact, they can be simple beyond belief and in the most important case the value of this integral is 1. This proposition is found in Becker, Burton, and Purnaras (2012).

Proposition 1.2.1. *If* $0 < q < 1$ *then*

$$C^*(t,s) = \frac{1}{\Gamma(2q)}(t-s)^{2q-1}.$$

In particular, if $q = 1/2$ *then* $C^*(t,s) = 1$. *Finally, if* $q \geq 1/2$ *then* $C^*(t,s)$ *is continuous and*

$$R_1(t,s) = C^*(t,s) - \int_s^t C(t,u)R_1(u,s)du$$

is continuous. In this case, $R(t,s) = C(t,s) - R_1(t,s)$ *satisfies (1.2.2).*

Proof. We begin with

$$C^*(t,s) = \frac{1}{\Gamma^2(q)}\int_s^t (t-u)^{q-1}(u-s)^{q-1}du$$

and make the change of variable

$$v := (t-s)^{-1}(u-s)$$

so that

$$u = s + (t-s)v, \quad du = (t-s)dv,$$

while

$$t - u = t - s - (t-s)v = (t-s)(1-v).$$

Thus,

$$\begin{aligned} C^*(t,s) &= \frac{1}{\Gamma^2(q)}\int_0^1 (t-s)^{q-1}(1-v)^{q-1}(t-s)^{q-1}v^{q-1}(t-s)dv \\ &= \frac{1}{\Gamma^2(q)}(t-s)^{2q-1}\int_0^1 v^{q-1}(1-v)^{q-1}dv. \end{aligned}$$

Now the beta function is

$$B(p,q) = \int_0^1 v^{p-1}(1-v)^{q-1}dv = \frac{\Gamma(p)\Gamma(q)}{\Gamma(p+q)}, \quad p > 0, q > 0.$$

Thus,

$$B(q,q) = \frac{\Gamma^2(q)}{\Gamma(2q)}.$$

Using this in the display we arrive at the desired conclusion.

Now, for $0 < q < 1/2$, C^* is singular at $t = s$. We then have the new equation

$$R_1(t,s) = C^*(t,s) - \int_s^t C(t,u)R_1(u,s)du$$

and we have $R(t,s) = C(t,s) - R_1(t,s)$ and

$$R_1(t,s) = C^*(t,s) - R_2(t,s)$$

where R_2 is the continuous function introduced earlier in the discussion following (1.2.20). Thus, R and C are put aside and we have

$$C^*(t,s) - R_2(t,s) = C^*(t,s) - \int_s^t C(t,u)[C^*(u,s) - R_2(u,s)]du$$

or

$$-R_2(t,s) = -\int_s^t C(t,u)C^*(u,s)du + \int_s^t C(t,u)R_2(u,s)du$$

or

$$R_2(t,s) = C^{**}(t,s) - \int_s^t C(t,u)R_2(u,s)du$$

with

$$C^{**}(t,s) = \int_s^t C(t,u)C^*(u,s)du = \frac{1}{\Gamma(q)\,\Gamma(2q)} \int_s^t (t-u)^{q-1}(u-s)^{2q-1}du.$$

Proposition 1.2.2. *If $0 < q < 1$ then*

$$C^{**}(t,s) = \frac{1}{\Gamma(3q)}(t-s)^{3q-1}.$$

In particular, if $q = 1/3$, then $C^{**}(t,s) = 1/\Gamma(3q)$. Finally, if $q \geq 1/3$ then C^{**} is continuous and

$$R_2(t,s) = C^{**}(t,s) - \int_s^t C(t,u)R_2(u,s)du$$

is continuous; in this case

$$\begin{aligned} R(t,s) &= C(t,s) - R_1(t,s) \\ &= C(t,s) - [C^*(t,s) - R_2(t,s)] \\ &= \frac{1}{\Gamma(q)}(t-s)^{q-1} - \frac{1}{\Gamma(2q)}(t-s)^{2q-1} + R_2(t,s) \end{aligned}$$

satisfies (1.2.2).

Proof. Making the same change of variable as in the proof of Proposition 1.2.1 we have

$$\begin{aligned} C^{**}(t,s) &= \int_s^t C(t,u)C^*(u,s)du \\ &= \frac{1}{\Gamma(q)\Gamma(2q)}\int_s^t (t-u)^{q-1}(u-s)^{2q-1}du \\ &= \frac{1}{\Gamma(q)\Gamma(2q)}\int_0^1 (t-s)^{q-1}(1-v)^{q-1}(t-s)^{2q-1}v^{2q-1}(t-s)dv \\ &= \frac{(t-s)^{3q-1}}{\Gamma(q)\Gamma(2q)}\int_0^1 (1-v)^{q-1}v^{2q-1}dv \\ &= \frac{(t-s)^{3q-1}}{\Gamma(q)\Gamma(2q)}\frac{\Gamma(q)\Gamma(2q)}{\Gamma(3q)} = \frac{(t-s)^{3q-1}}{\Gamma(3q)}. \end{aligned}$$

The above process can be continued to a continuous function R_3 obtaining

$$\begin{aligned} R(t,s) = {} & \frac{1}{\Gamma(q)}(t-s)^{q-1} - \frac{1}{\Gamma(2q)}(t-s)^{2q-1} \\ & + \frac{1}{\Gamma(3q)}(t-s)^{3q-1} - R_3(t,s). \end{aligned}$$

We are obtaining an alternating series. This example will be seen several times as we proceed through the book. Each time we iterate the kernel we extend the interval for q over which the iterated kernel is continuous. A general solution of a linear fractional differential equation of Caputo type is derived in Becker (2012, Theorem 7.3) using this alternating series for $R(t,s)$.

The nice properties which we see in all of these fractional differential equations follow from the fact that $C(t)$ is completely monotone with an infinite integral. This is discussed around Equation (2.5.1.19). Under these conditions it turns out that R is of convolution type with

$$0 < R(t) \leq C(t) \quad \text{and} \quad \int_0^\infty R(t)dt = 1$$

on which much of the later theory rests.

1.2.1 Integrodifferential equations

Historical Note

The remainder of this section will display work of Leigh C. Becker (1979) concerning the resolvent for integrodifferential equations. Most of that work was part of his doctoral dissertation written in 1979 and available for many years only in the dissertation form. Nevertheless it became widely known and was recognized as one of the most fundamental pieces of work on the subject, forming the basis of a large number of papers by other investigators. In 2005 it was scanned and a pdf file became available on his website. In 2006 it was updated and rewritten for publication in the free Electronic Journal of Qualitative Theory of Differential Equations journal as Becker (2006). We present it here mainly in the form of that paper and is included with permission of both the journal and Becker. That paper contains much more detail and several references not included here. The results are fundamental for the rest of the work in this book.

We first prove that (1.2.11) has a unique solution. That will also yield existence and uniqueness of solutions of (1.2.8) and (1.2.9). Our focus for the remainder of the section will then be on (1.2.9).

Theorem 1.2.1.1. *Let A and f be continuous on $[\tau,\infty)$ and let B be continuous for $\tau \leq u \leq t < \infty$. Then there is a unique differentiable function $x : [\tau,\infty) \to \Re^n$ with $x(\tau) = x_0$ and satisfying (1.2.11) on $[\tau,\infty)$.*

Proof. If there is such a differentiable function then we can integrate (1.2.11) from τ to $t > \tau$ and obtain

$$x(t) = x_0 + \int_\tau^t A(v)x(v)dv + \int_\tau^t \int_\tau^v B(v,u)x(u)dudv + \int_\tau^t f(v)dv.$$

By interchanging the order of integration we have

$$x(t) = x_0 + \int_\tau^t \left[A(u) + \int_u^t B(v,u)dv \right] x(u)du + \int_\tau^t f(u)du. \quad (1.2.19)$$

Let $|\cdot|$ be any norm on $\Re^n$ and choose the operator norm for matrices. Let $b > \tau$ be arbitrary and pick $r > 1$ by

$$\sup_{\tau \le u \le t \le b} \left[|A(u)| + \int_u^t |B(v,u)|dv \right] = r - 1.$$

Let $(X, |\cdot|_r)$ be the Banach space of continuous functions $\phi : [\tau, b] \to \Re^n$ with $|\phi(t)|_r = \sup_{\tau \le t \le b} |\phi(t)| e^{-r(t-\tau)}$. Define $P : X \to X$ by $\phi \in X$ implies that

$$(P\phi)(t) = x_0 + \int_\tau^t \left[A(u) + \int_u^t B(v,u)dv \right] \phi(u)du + \int_\tau^t f(u)du.$$

If $\phi, \eta \in X$ then

$$\begin{aligned}
&|(P\phi)(t) - (P\eta)(t)| e^{-r(t-\tau)} \\
&\le e^{-r(t-\tau)} \left| \int_\tau^t \left[A(u) + \int_u^t B(v,u)dv \right] (\phi(u) - \eta(u))du \right| \\
&\le \int_\tau^t (r-1) e^{-r(t-\tau)+r(u-\tau)} |\phi(u) - \eta(u)| e^{-r(u-\tau)} du \\
&\le |\phi - \eta|_r \int_\tau^t (r-1) e^{-r(t-u)} du \\
&\le |\phi - \eta|_r \frac{r-1}{r}.
\end{aligned}$$

Thus, P is a contraction with unique fixed point ϕ which satisfies $\phi(\tau) = x_0$. Moreover, as it satisfies the integral equation it inherits differentiability and so solves (1.2.11). As b is arbitrary, this completes the proof.

This also gives existence and uniqueness of solutions of (1.2.8) and of (1.2.9), but there will be interpretation to do on that equation.

Having obtained existence, we now turn to the resolvent. The variation of parameters formula

$$x(t) = H(t,0)x_0 + \int_0^t H(t,s)f(s)\,ds \quad (1.2.1.1)$$

gives the unique solution of the linear nonhomogeneous Volterra vector integro-differential equation

$$x'(t) = A(t)x(t) + \int_0^t B(t,u)x(u)\,du + f(t) \tag{1.2.1.2}$$

satisfying the initial condition $x(0) = x_0$. Grossman and Miller (1970) defined the matrix function $H(t,s)$, called the *resolvent*, and used it to derive (1.2.1.1). They formally defined $H(t,s)$ by

$$H(t,s) = I + \int_s^t H(t,u)\Psi(u,s)\,du \quad (0 \le s \le t < \infty) \tag{1.2.1.3}$$

where I is the identity matrix and

$$\Psi(t,s) = A(t) + \int_s^t B(t,v)\,dv. \tag{1.2.1.4}$$

They proved that $H(t,s)$ exists and is continuous for $0 \le s \le t$ and that it satisfies

$$\frac{\partial}{\partial s}H(t,s) = -H(t,s)A(s) - \int_s^t H(t,u)B(u,s)\,du, \quad H(t,t) = I \tag{1.2.1.5}$$

on the interval $[0,t]$, for each $t > 0$. With this they were able to derive the variation of parameters formula (1.2.1.1) (cf. Grossman and Miller (1970; p. 459)).

Despite the prominence of the resolvent $H(t,s)$ in the literature and its indispensability, its definition (1.2.1.3) is not as conceptually simple as one would like. A "linear system of ODEs" point of view was presented in Becker (1979; Ch. II). There the *principal matrix solution* $Z(t,s)$ of the homogeneous Volterra equation

$$x'(t) = A(t)x(t) + \int_s^t B(t,u)x(u)\,du \tag{1.2.1.6}$$

was first introduced. Its definition looks exactly like the classical definition of the principal matrix solution of the homogeneous vector differential equation

$$x'(t) = A(t)x(t).$$

Now $Z(t,s)$ is a matrix solution of (1.2.1.6) with columns that are linearly independent such that $Z(s,s) = I$. Using $Z(t,s)$ instead of $H(t,s)$, the variation of parameters formula

$$x(t) = Z(t,0)x_0 + \int_0^t Z(t,s)f(s)\,ds \tag{1.2.1.7}$$

for (1.2.1.2) is a natural extension of the variation of parameters formula for the nonhomogeneous vector differential equation

$$x'(t) = A(t)x(t) + f(t).$$

The principal matrix version of the resolvent equation (1.2.1.5), namely,

$$\frac{\partial}{\partial t}Z(t,s) = A(t)Z(t,s) + \int_s^t B(t,u)Z(u,s)\,du, \quad Z(s,s) = I \quad (1.2.1.8)$$

has been instrumental in a number of papers for obtaining results that might not have otherwise been obtained with (1.2.1.5) alone.

Not found in Becker (1979) is an alternative to Grossman and Miller's definition of $H(t,s)$. It is this: $H(t,s)$ is the transpose of the principal matrix solution of the adjoint equation

$$y'(s) = -A^T(s)y(s) - \int_s^t B^T(u,s)y(u)\,du \quad (1.2.1.9)$$

for $0 \le s \le t$. The section culminates with the proof that, notwithstanding the difference in their definitions, $Z(t,s)$ and $H(t,s)$ are identical.

1.2.2 Joint Continuity

If we refer back to Theorem 1.2.1.1, replace τ by s, and delete the forcing function f, then we see that for a given $x_0 \in \Re^n$, the homogeneous equation

$$x'(t) = A(t)x(t) + \int_s^t B(t,u)x(u)\,du \quad (1.2.2.1)$$

has a unique solution x_s satisfying the initial condition $x_s(s) = x_0$. Equivalently, by (1.2.19), x_s is the unique continuous solution of

$$x(t) = x_0 + \int_s^t \Phi(t,u)x(u)\,du \quad (1.2.2.2)$$

where

$$\Phi(t,u) := A(u) + \int_u^t B(v,u)\,dv. \quad (1.2.2.3)$$

Up to now the value of s has been fixed. But with that restriction removed, the totality of values $x_s(t)$ defines a function, x say, on the set

$$\Omega := \{\,(t,s) : 0 \le s \le t < \infty\,\}$$

whose value at $(t_1, s_1) \in \Omega$ is the value of the solution x_{s_1} at $t = t_1$.

Definition 1.2.2.1. *For a given $x_0 \in \Re^n$, let x denote the function with domain Ω whose value at (t,s) is*

$$x(t,s) := x_s(t) \tag{1.2.2.4}$$

where x_s is the unique solution of (1.2.2.1) on $[s,\infty)$ satisfying the initial condition $x_s(s) = x_0$.

Since $x(t,s)$ is continuous in t for a fixed s, it is natural to ask if it is also continuous in s for a fixed t and if so, is it jointly continuous in t and s? The next theorem answers both of these in the affirmative. This will play an essential role in the proof of the variation of parameters formula for (1.2.11) that is given in Section 1.2.4.

Theorem 1.2.2.1. *The function $x(t,s)$ defined by (1.2.2.4) is continuous for $0 \le s \le t < \infty$.*

Proof. First extend the domain Ω of the function x to the entire first quadrant by defining $x(t,s) = x_0$ for $s > t$. For any $T > 0$, consider $x(t,s)$ on $[0,T] \times [0,T]$. We will prove $x(t,s)$ is a uniformly continuous function of s for all $t \in [0,T]$, which means that for every $\epsilon > 0$, there exists a $\delta > 0$ such that $|s_1 - s_2| < \delta$ implies that

$$|x(t,s_1) - x(t,s_2)| < \epsilon \tag{1.2.2.5}$$

for all $s_1, s_2 \in [0,T]$ and all $t \in [0,T]$. This and the continuity of $x(t,s)$ in t for each fixed s would establish that $x(t,s)$ is jointly continuous in both variables on the set $[0,T] \times [0,T]$ by the Moore-Osgood theorem (cf. Graves (1946; Thm. 5, p. 102), Hurewicz (1958; p. 13), or Olmsted (1959; Ex. 31, p. 310)).

Proving (1.2.2.5) will require bounds for $x(t,s)$. For a fixed $s \in [0,T]$ and for $t \in [s,T]$, we see from (1.2.2.2) that

$$\begin{aligned} |x(t,s)| &\le |x_0| + \int_s^t |\Phi(t,u)|\,|x(u,s)|\,du \\ &\le |x_0| + \int_s^t \left[|A(u)| + \int_u^t |B(v,u)|\,dv\right] |x(u,s)|\,du \\ &\le |x_0| + \int_s^t k\,|x(u,s)|\,du, \end{aligned} \tag{1.2.2.6}$$

where k is a constant chosen so that

$$|\Phi(t,u)| \le |A(u)| + \int_u^t |B(v,u)|\,dv \le k \tag{1.2.2.7}$$

for all $(t, u) \in [0, T] \times [0, T]$. By Gronwall's inequality (Theorem 1.2.3),

$$|x(t, s)| \leq |x_0| e^{\int_s^t k\, du} = |x_0| e^{k(t-s)}$$

for $0 \leq s \leq t \leq T$. Since $|x(t, s)| = |x_0|$ for $s > t$, we have

$$|x(t, s)| \leq |x_0| e^{kT} \tag{1.2.2.8}$$

for all $(t, s) \in [0, T] \times [0, T]$.

With the aid of (1.2.2.8) we now prove (1.2.2.5). For definiteness, suppose $s_2 > s_1$. For $t \in [0, s_1]$,

$$|x(t, s_1) - x(t, s_2)| = 0 \tag{1.2.2.9}$$

as $x(t, s) = x_0$ for $t \leq s$.

For $t \in (s_1, s_2]$, it follows from (1.2.2.2) and (1.2.2.7) that

$$\begin{aligned} |x(t, s_1) - x(t, s_2)| = |x(t, s_1) - x_0| &\leq \int_{s_1}^{t} |\, \Phi(t, u) \,| \, |x(u, s_1)| \, du \\ &\leq \int_{s_1}^{s_2} |\, \Phi(t, u) \,| \, |x(u, s_1)| \, du \leq \int_{s_1}^{s_2} k |x(u, s_1)| \, du. \end{aligned}$$

Then by (1.2.2.8)

$$|x(t, s_1) - x(t, s_2)| \leq \int_{s_1}^{s_2} k |x_0| e^{kT} du = k |x_0| e^{kT} (s_2 - s_1). \tag{1.2.2.10}$$

For $t \in (s_2, T]$, we have

$$\begin{aligned} |x(t, s_1) - x(t, s_2)| &= \left| \int_{s_1}^{t} \Phi(t, u) x(u, s_1) \, du - \int_{s_2}^{t} \Phi(t, u) x(u, s_2) \, du \right| \\ &= \left| \int_{s_1}^{t} \Phi(t, u) x(u, s_1) \, du - \int_{s_2}^{t} \Phi(t, u) x(u, s_1) \, du \right. \\ &\qquad \left. + \int_{s_2}^{t} \Phi(t, u) x(u, s_1) \, du - \int_{s_2}^{t} \Phi(t, u) x(u, s_2) \, du \right| \\ &\leq \int_{s_1}^{s_2} |\, \Phi(t, u) \,| \, |x(u, s_1)| \, du + \int_{s_2}^{t} |\, \Phi(t, u) \,| \, |x(u, s_1) - x(u, s_2)| \, du \\ &\leq \int_{s_1}^{s_2} k \, |x(u, s_1)| \, du + \int_{s_2}^{t} k \, |x(u, s_1) - x(u, s_2)| \, du. \end{aligned}$$

Applying (1.2.2.8) again,

$$|x(t, s_1) - x(t, s_2)| \leq k |x_0| e^{kT} (s_2 - s_1) + \int_{s_2}^{t} k \, |x(u, s_1) - x(u, s_2)| \, du.$$

By (1.2.2.10), this holds at $t = s_2$ as well. Therefore, for $t \in [s_2, T]$,

$$|x(t, s_1) - x(t, s_2)| \leq k|x_0|e^{kT}(s_2 - s_1)e^{k(t-s_2)} \tag{1.2.2.11}$$

by Gronwall's inequality.

It follows from (1.2.2.9) - (1.2.2.11) that

$$|x(t, s_1) - x(t, s_2)| \leq k|x_0|e^{2kT}(s_2 - s_1) \tag{1.2.2.12}$$

for all $t \in [0, T]$ and $s_2 > s_1$. Of course, it is also true for $s_2 = s_1$.

We conclude

$$|x(t, s_1) - x(t, s_2)| \leq k|x_0|e^{2kT}|s_1 - s_2| \tag{1.2.2.13}$$

for all $s_1, s_2 \in [0, T]$ and $t \in [0, T]$, which implies (1.2.2.5). Therefore, $x(t, s)$ is continuous on $[0, T] \times [0, T]$. Since T is arbitrary, $x(t, s)$ is continuous on $[0, \infty) \times [0, \infty)$, a fortiori, for $0 \leq s \leq t < \infty$.

1.2.3 Principal Matrix Solution

For a fixed $s \geq 0$, let S denote the set of all solutions of (1.2.2.1) on the interval $[s, \infty)$ that correspond to initial vectors. Let $x(t, s)$ and $\tilde{x}(t, s)$ be two such solutions satisfying the initial conditions $x(s, s) = x_0$ and $\tilde{x}(s, s) = x_1$, respectively. Linearity of (1.2.2.1) implies the *principle of superposition*, namely, that the linear combination $c_1 x(t, s) + c_2 \tilde{x}(t, s)$ is a solution of (1.2.2.1) on $[s, \infty)$ for any $c_1, c_2 \in \Re$. Consequently, the set S is a vector space. Note that S comprises all solutions that have their initial values specified at $t = s$, but not those for which an initial function is specified on an initial interval $[s, t_0]$ for some $t_0 > s$.

Theorem 1.2.3.1. *For a fixed $s \in [0, \infty)$, let S be the set of all solutions of (1.2.2.1) on the interval $[s, \infty)$ corresponding to initial vectors. Then S is an n-dimensional vector space.*

Proof. We have already established that S is a vector space. To complete the proof, we must find n linearly independent solutions spanning S. To this end, let $e^1, \ldots, e^n$ be the standard basis for $\Re^n$, where e^i is the vector whose ith component is 1 and whose other components are 0. By Theorem 1.2.1.1, there are n unique solutions $x^i(t, s)$ of (1.2.2.1) on $[s, \infty)$ with $x^i(s, s) = e^i$ $(i = 1, \ldots, n)$. By the usual argument, these solutions are linearly independent.

To show they span S, choose any $x(t,s) \in S$. Suppose its value at $t = s$ is the vector x_0. Let $\xi_1, \ldots, \xi_n$ be the unique scalars such that $x_0 = \xi_1 e^1 + \cdots + \xi_n e^n$. By the principle of superposition, the linear combination

$$\xi_1 x^1(t,s) + \cdots + \xi_n x^n(t,s) = \sum_{i=1}^{n} \xi_i x^i(t,s) \tag{1.2.3.1}$$

is a solution of (1.2.2.1). Since its value at $t = s$ is x_0, the uniqueness part of Theorem 1.2.1.1 implies

$$x(t,s) = \sum_{i=1}^{n} \xi_i x^i(t,s). \tag{1.2.3.2}$$

Hence, the n solutions $x^1(t,s), \ldots, x^n(t,s)$ span S. This and their linear independence make them a basis for S.

If we define an $n \times n$ matrix function $Z(t,s)$ by

$$Z(t,s) := \begin{bmatrix} x^1(t,s) & x^2(t,s) & \cdots & x^n(t,s) \end{bmatrix}, \tag{1.2.3.3}$$

where the columns $x^1(t,s), \ldots, x^n(t,s)$ are the basis for S defined in the proof of Theorem 1.2.3.1, then (1.2.3.2) can be written as

$$x(t,s) = Z(t,s)x_0. \tag{1.2.3.4}$$

Since $x^i(s,s) = e^i$,

$$Z(s,s) = I, \tag{1.2.3.5}$$

the $n \times n$ identity matrix.

If B is the zero matrix, then the columns of $Z(t,s)$ become linearly independent solutions of the ordinary vector differential equation

$$x'(t) = A(t)x(t). \tag{1.2.3.6}$$

This makes $Z(t,s)$ a *fundamental matrix solution* of (1.2.3.6). In fact, because $Z(s,s) = I$, it is the so-called *principal matrix solution.* These terms are also used for the integrodifferential equation (1.2.2.1).

Definition 1.2.3.1. *The* principal matrix solution *of (1.2.2.1) is the $n \times n$ matrix function $Z(t,s)$ defined by (1.2.3.3). In other words, $Z(t,s)$ is a matrix with n columns that are linearly independent solutions of (1.2.2.1) and whose value at $t = s$ is the identity matrix I.*

An alternative term from integral equations is *resolvent*, an apt term in view of (1.2.3.2), which states that every solution of (1.2.2.1) can be resolved into the n columns constituting $Z(t, s)$.

Theorem 1.2.2.1 implies that each of the columns $x^i(t, s)$ of $Z(t, s)$ are continuous for $0 \le s \le t < \infty$. Consequently, we have the following.

Theorem 1.2.3.2. *$Z(t, s)$, the principal matrix solution of equation (1.2.2.1), is continuous for $0 \le s \le t < \infty$.*

Since the ith column of $Z(t, s)$ is the unique solution of (1.2.2.1) whose value at $t = s$ is e^i, $Z(t, s)$ is the unique matrix solution of the initial value problem

$$\frac{\partial}{\partial t} Z(t, s) = A(t)Z(t, s) + \int_s^t B(t, u)Z(u, s)\, du, \quad Z(s, s) = I \quad (1.2.3.7)$$

for $0 \le s \le t < \infty$. Equivalently, it is the unique matrix solution of

$$Z(t, s) = I + \int_s^t \left[A(u) + \int_u^t B(v, u)\, dv \right] Z(u, s)\, du \quad (1.2.3.8)$$

by (1.2.2.2) and (1.2.2.3). Note that this is the principal matrix counterpart of Grossman and Miller's resolvent equation (1.2.1.3).

1.2.4 Variation of Parameters Formula

Let $X(t)$ be any fundamental matrix solution of the homogeneous differential equation

$$x'(t) = A(t)x(t). \quad (1.2.4.1)$$

By definition, the columns of a fundamental matrix solution $X(t)$ are linearly independent solutions of (1.2.4.1). So for $c \in \Re^n$, $x(t) = X(t)c$ is a solution of (1.2.4.1) by the principle of superposition. If $x(\tau) = x_0$, then $X(\tau)c = x_0$. Since $X(\tau)$ is nonsingular (cf. Corduneanu (1977; p. 62)), the unique solution $x(t)$ of (1.2.4.1) satisfying $x(\tau) = x_0$ is

$$x(t) = X(t)X^{-1}(\tau)x_0. \quad (1.2.4.2)$$

Now compare (1.2.4.2) to the unique solution of the nonhomogeneous equation

$$x'(t) = A(t)x(t) + f(t) \quad (1.2.4.3)$$

satisfying $x(\tau) = x_0$. The method of variation of parameters applied to (1.2.4.3) (cf. Corduneanu (1977; p. 65)) yields the following well-known formula for the solution

$$x(t) = X(t)X^{-1}(\tau)x_0 + \int_\tau^t X(t)X^{-1}(s)f(s)\,ds. \tag{1.2.4.4}$$

Of course, (1.2.4.4) reduces to (1.2.4.2) if $f \equiv 0$.

As for the integro-differential equation (1.2.2.1), the counterpart of (1.2.4.2) is (1.2.3.4), which is stated next as a lemma.

Lemma 1.2.4.1. *The solution of*

$$x'(t) = A(t)x(t) + \int_\tau^t B(t,u)x(u)\,du \quad (\tau \geq 0) \tag{1.2.4.5}$$

on $[\tau, \infty)$ *satisfying the initial condition* $x(\tau) = x_0$ *is*

$$x(t) = Z(t,\tau)x_0, \tag{1.2.4.6}$$

where $Z(t,\tau)$ *is the principal matrix solution of (1.2.4.5).*

Suppose $B \equiv 0$ (zero matrix). Then (1.2.4.2), (1.2.4.6), and uniqueness of solutions imply that

$$Z(t,\tau) = X(t)X^{-1}(\tau).$$

Thus, the variation of parameters formula (1.2.4.4) can also be written as

$$x(t) = Z(t,\tau)x_0 + \int_\tau^t Z(t,s)f(s)\,ds. \tag{1.2.4.7}$$

Lemma 1.2.4.1 extends a classical result for the homogeneous differential equation (1.2.4.1) to the homogeneous integro-differential equation (1.2.2.1). This suggests that a variation of parameters formula similar to (1.2.4.7) may also hold for the nonhomogeneous integro-differential equation (1.2.11).

The essential element in the derivation of the variation of parameters formula (1.2.4.4) is the nonsingularity of $X(t)$ for each t. If the same were true of the principal matrix solution $Z(t,s)$ of (1.2.2.1), then (1.2.4.7) would hold for (1.2.2.1) as well. In fact, as Theorem 1.2.4.1 shows, there are examples of (1.2.2.1) other than (1.2.4.1) for which $\det Z(t,s)$ is never zero.

Theorem 1.2.4.1. *(Becker (2007; Cor. 3.4)) Assume $a, b\colon [0,\infty) \to \Re$ are continuous functions and $b(t) \geq 0$ on $[0,\infty)$. Let $x(t)$ be the unique solution of the scalar equation*

$$x'(t) = -a(t)x(t) + \int_s^t b(t-u)x(u)\,du \quad (s \geq 0) \tag{1.2.4.8}$$

on $[s,\infty)$ satisfying the initial condition $x(s) = x_0$. If $x_0 \geq 0$, then

$$x_0 e^{-\int_s^t a(u)\,du} \leq x(t) \leq x_0 e^{-\int_s^t p(u)\,du} \tag{1.2.4.9}$$

for all $t \geq s$, where

$$p(u) := a(u) - \int_0^{u-s} e^{\int_{u-v}^u a(r)\,dr} b(v)\,dv. \tag{1.2.4.10}$$

It follows that the *principal solution* $x(t,s)$ of (1.2.4.8) (i.e., the solution whose value at $t = s$ is 1) is always positive. In our notation, $Z(t,s)$ is the 1×1 matrix $[x(t,s)]$ and so

$$\det Z(t,s) = x(t,s) > 0$$

for all $t \geq s \geq 0$.

However, unlike differential equations, the principal matrix solution of an integrodifferential equation (1.2.2.1) may be singular at points as the next theorem found in Burton (2005b; p. 86) shows.

Theorem 1.2.4.2. *Assume $a \geq 0$ and $b\colon [0,\infty) \to \Re$ is continuous, where $b(t) \leq 0$ on $[0,\infty)$. Let $x(t)$ be the unique solution of*

$$x'(t) = -ax(t) + \int_0^t b(t-u)x(u)\,du \tag{1.2.4.11}$$

satisfying the initial condition $x(0) = 1$. If there exists a $t_1 > 0$ such that

$$\int_{t_1}^t \int_0^{t_1} b(v-u)\,du\,dv \to -\infty \tag{1.2.4.12}$$

as $t \to \infty$, then there exists a $t_2 > 0$ such that $x(t_2) = 0$.

Theorem 1.2.4.2 clearly establishes that the determinant of the principal matrix solution $Z(t,s)$ may vanish. Consequently, we cannot derive (1.2.4.7) in general for the integro-differential equation (1.2.11) by applying the method of variation of parameters to it. However, as Theorem 1.2.4.3 shows, (1.2.4.7) satisfies (1.2.11) irrespective of the values of $\det Z(t,s)$.

Theorem 1.2.4.3. (Variation of Parameters) *The solution of*

$$x'(t) = A(t)x(t) + \int_\tau^t B(t,u)x(u)\,du + f(t) \quad (\tau \geq 0) \tag{1.2.4.13}$$

on $[\tau, \infty)$ *satisfying the initial condition* $x(\tau) = x_0$ *is*

$$x(t) = Z(t,\tau)x_0 + \int_\tau^t Z(t,s)f(s)\,ds, \tag{1.2.4.14}$$

where $Z(t,s)$ *is the principal matrix solution of*

$$x'(t) = A(t)x(t) + \int_s^t B(t,u)x(u)\,du.$$

Proof. By Theorem 1.2.1.1, there is a unique solution $x(t)$ of (1.2.4.13) on $[\tau, \infty)$ such that $x(\tau) = x_0$. Let us show that

$$\varphi(t) := Z(t,\tau)x_0 + \int_\tau^t Z(t,s)f(s)\,ds \tag{1.2.4.15}$$

is also a solution of (1.2.4.13) by differentiating it. To this end, define $Z(t,s) = I$ for $s > t$. Then $Z(t,s)$ is continuous on $[0,\infty) \times [0,\infty)$ by Theorem 1.2.3.2. This and (1.2.3.7) imply the same is true of its partial derivative $Z_t(t,s)$. Consequently, the integral function in (1.2.4.15) is differentiable by Leibniz's rule. Differentiating $\varphi(t)$, we obtain

$$\varphi'(t) = \left[A(t)Z(t,\tau) + \int_\tau^t B(t,u)Z(u,\tau)\,du \right] x_0 \\ + Z(t,t)f(t) + \int_\tau^t \frac{\partial}{\partial t} Z(t,s)f(s)\,ds$$

by (1.2.3.7) and Leibniz's rule. Applying (1.2.3.7) again, we have

$$\begin{aligned} \varphi'(t) &= A(t)Z(t,\tau)x_0 + \int_\tau^t B(t,u)Z(u,\tau)x_0\,du + If(t) \\ &\quad + \int_\tau^t \left[A(t)Z(t,s) + \int_s^t B(t,u)Z(u,s)\,du \right] f(s)\,ds \\ &= f(t) + A(t)\left[Z(t,\tau)x_0 + \int_\tau^t Z(t,s)f(s)\,ds \right] \\ &\quad + \int_\tau^t B(t,u)Z(u,\tau)x_0\,du + \int_\tau^t \int_s^t B(t,u)Z(u,s)f(s)\,du\,ds. \end{aligned}$$

An interchange in the order of integration yields

$$\varphi'(t) = f(t) + A(t)\varphi(t) + \int_\tau^t B(t,u)Z(u,\tau)x_0\,du + \int_\tau^t \int_\tau^u B(t,u)Z(u,s)f(s)\,ds\,du,$$

which simplifies to

$$\begin{aligned}\varphi'(t) &= f(t) + A(t)\varphi(t) \\ &\quad + \int_\tau^t B(t,u)\left[Z(u,\tau)x_0 + \int_\tau^u Z(u,s)f(s)\,ds\right]du \\ &= f(t) + A(t)\varphi(t) + \int_\tau^t B(t,u)\varphi(u)\,du.\end{aligned}$$

Thus, $\varphi(t)$ is a solution on $[\tau, \infty)$. By (1.2.4.15), $\varphi(\tau) = x_0$. Therefore, $x(t) \equiv \varphi(t)$ on $[\tau, \infty)$ by uniqueness of solutions.

Note that (1.2.4.14) reduces to (1.2.4.6) when $f \equiv 0$.

Corollary. *Let $\varphi \in C[0,\tau]$ for any $\tau > 0$. The solution of*

$$x'(t) = A(t)x(t) + \int_0^t B(t,u)x(u)\,du + f(t) \tag{1.2.4.16}$$

on $[\tau, \infty)$ satisfying the condition $x(t) = \varphi(t)$ for $0 \le t \le \tau$ is

$$x(t) = Z(t,\tau)\varphi(\tau) + \int_\tau^t Z(t,s)f(s)\,ds + \int_\tau^t Z(t,s)\left\{\int_0^\tau B(s,u)\varphi(u)\,du\right\}ds. \tag{1.2.4.17}$$

Proof. Since $x(t) \equiv \varphi(t)$ on $[0, \tau]$, we can rewrite (1.2.4.16) as follows:

$$x'(t) = A(t)x(t) + \int_\tau^t B(t,u)x(u)\,du + g(t) \tag{1.2.4.18}$$

where

$$g(t) := f(t) + \int_0^\tau B(t,u)\varphi(u)\,du. \tag{1.2.4.19}$$

By Theorem 1.2.1.1, equation (1.2.4.18) has a unique solution on $[\tau, \infty)$ such that $x(\tau) = \varphi(\tau)$. By the variation of parameters formula (1.2.4.14), the solution is

$$x(t) = Z(t, \tau)\varphi(\tau) + \int_\tau^t Z(t, s)g(s)\, ds,$$

which is (1.2.4.17).

1.2.5 The Adjoint Equation

The differential equation

$$y'(t) = -A^T(t)y(t), \tag{1.2.5.1}$$

where A^T is the transpose of A, is the so-called adjoint to (1.2.4.1). The associated nonhomogeneous adjoint equation (cf. Hartman (1964; p. 62)) is

$$y'(t) = -A^T(t)y(t) - g(t). \tag{1.2.5.2}$$

Let us extend this definition to the integro-differential equation (1.2.4.13).

Definition 1.2.5.1. *The* adjoint *to (1.2.4.13) is*

$$y'(s) = -A^T(s)y(s) - \int_s^t B^T(u, s)y(u)\, du - g(s) \tag{1.2.5.3}$$

where $s \in [0, t]$.

The next theorem establishes that solutions of (1.2.5.3) do exist and are unique.

Theorem 1.2.5.1. *For a fixed* $t > 0$ *and a given* $y_0 \in \Re^n$, *there is a unique solution* $y(s)$ *of*

$$y'(s) = -A^T(s)y(s) - \int_s^t B^T(u, s)y(u)\, du - g(s)$$

on the interval $[0, t]$ *satisfying the condition* $y(t) = y_0$.

Proof. The objective, as it was in proving Theorem 1.2.1.1, is to find a suitable contraction mapping. To this end, integrate (1.2.5.3) from s to t:

$$y(t) - y(s) = -\int_s^t A^T(v)y(v)\,dv - \int_s^t \int_v^t B^T(u,v)y(u)\,du\,dv - \int_s^t g(v)\,dv.$$

Replacing $y(t)$ with y_0 and interchanging the order of integration yields

$$y(s) = y_0 + \int_s^t A^T(v)y(v)\,dv + \int_s^t \int_s^u B^T(u,v)y(u)\,dv\,du + \int_s^t g(v)\,dv$$

or

$$y(s) = y_0 + \int_s^t \left[A^T(u) + \int_s^u B^T(u,v)\,dv\right] y(u)\,du + \int_s^t g(u)\,du. \quad (1.2.5.4)$$

Clearly, the appropriate set of functions on which to define a mapping is

$$C_{y_0}[0,t] := \{\, \phi \in C[0,t] : \phi(t) = y_0 \,\}.$$

Now define the mapping $\tilde{P}$ by

$$(\tilde{P}\phi)(s) := y_0 + \int_s^t \left[A^T(u) + \int_s^u B^T(u,v)\,dv\right] \phi(u)\,du + \int_s^t g(u)\,du$$

for all $\phi \in C_{y_0}[0,t]$. For a given $\phi \in C_{y_0}[0,t]$, it is apparent that $\tilde{P}\phi$ is continuous on $[0,t]$ and that $(\tilde{P}\phi)(t) = y_0$. Thus, $\tilde{P}\colon C_{y_0}[0,t] \to C_{y_0}[0,t]$.

For an arbitrary pair of functions $\phi, \eta \in C_{y_0}[0,t]$,

$$\begin{aligned}
&|(\tilde{P}\phi)(s) - (\tilde{P}\eta)(s)| \\
&\qquad = \left| \int_s^t \left[A^T(u) + \int_s^u B^T(u,v)\,dv\right] (\phi(u) - \eta(u))\,du \right| \\
&\qquad \le \int_s^t \left[|A^T(u)| + \int_s^u |B^T(u,v)|\,dv\right] |\phi(u) - \eta(u)|\,du.
\end{aligned}$$

So if $r > 1$ is chosen so that

$$|A^T(u)| + \int_s^u |B^T(u,v)|\,dv \le r - 1$$

for $0 \le s \le u \le t$, then

$$|(\tilde{P}\phi)(s) - (\tilde{P}\eta)(s)| \le \int_s^t (r-1)\,|\phi(u) - \eta(u)|\,du. \quad (1.2.5.5)$$

The proof of Theorem 1.2.1.1 for τ replaced by s takes place in a Banach space with s fixed and t varying. But now that s varies and t is fixed, let us alter the norm slightly in order to show that $\tilde{P}$ is a contraction mapping: replacing $-r$ with r yields the norm

$$|\phi|_r = \sup_{0 \le s \le t} |\phi(s)| e^{rs}.$$

What remains then is to show that $\tilde{P}$ is a contraction. Returning to (1.2.5.5), we have

$$\begin{aligned} |(\tilde{P}\phi)(s) - (\tilde{P}\eta)(s)| e^{rs} &\le \int_s^t (r-1) e^{rs-ru} \, |\phi(u) - \eta(u)| e^{ru} \, du \\ &\le |\phi - \eta|_r \int_s^t (r-1) e^{r(s-u)} du \\ &\le |\phi - \eta|_r \frac{r-1}{r}. \end{aligned}$$

Therefore, $\tilde{P}$ has a unique fixed point, which translates to the existence of a unique solution of (1.2.5.4) on the interval $[0, t]$.

Definition 1.2.5.2. *The* principal matrix solution *of*

$$y'(s) = -A^T(s) y(s) - \int_s^t B^T(u, s) y(u) \, du \tag{1.2.5.6}$$

is the $n \times n$ matrix function

$$Q(t, s) := \left[\, y^1(t, s) \;\; y^2(t, s) \;\; \cdots \;\; y^n(t, s) \,\right], \tag{1.2.5.7}$$

where $y^i(t, s)$ (t fixed) is the unique solution of (1.2.5.6) on $[0, t]$ that satisfies the condition $y^i(t, t) = e^i$.

By virtue of this definition, $Q(t, s)$ is the unique matrix solution of

$$\frac{\partial}{\partial s} Q(t, s) = -A^T(s) Q(t, s) - \int_s^t B^T(u, s) Q(t, u) \, du, \; Q(t, t) = I \tag{1.2.5.8}$$

on the interval $[0, t]$. Reasoning as in the proof of Theorem 1.2.3.1, we conclude that for a given $y_0 \in \Re^n$, the unique solution of (1.2.5.6) satisfying the condition $y(t) = y_0$ is

$$y(s) = Q(t, s) y_0 \tag{1.2.5.9}$$

for $0 \le s \le t$.

Taking the transpose of (1.2.5.6) and letting $r(s)$ be the row vector $y^T(s)$, we obtain

$$r'(s) = -r(s)A(s) - \int_s^t r(u)B(u,s)\,du.$$

The solution satisfying the condition $r(t) = y_0^T =: r_0$ is the transpose of (1.2.5.9), namely,

$$y^T(s) = y_0^T Q^T(t,s) \tag{1.2.5.10}$$

or

$$r(s) = r_0 H(t,s)$$

where

$$H(t,s) := Q^T(t,s). \tag{1.2.5.11}$$

Consequently, $H(t,s)$ is the principal matrix solution of the transposed equation. As a result, Lemma 1.2.4.1 has the following adjoint counterpart.

Lemma 1.2.5.1. *The solution of*

$$r'(s) = -r(s)A(s) - \int_s^t r(u)B(u,s)\,du \tag{1.2.5.12}$$

on $[0,t]$ *satisfying the condition* $r(t) = r_0$ *is*

$$r(s) = r_0 H(t,s), \tag{1.2.5.13}$$

where $H(t,s)$ *is the principal matrix solution of (1.2.5.12).*

It follows from (1.2.5.8) that $H(t,s)$ is the unique matrix solution of

$$\frac{\partial}{\partial s}H(t,s) = -H(t,s)A(s) - \int_s^t H(t,u)B(u,s)\,du, \quad H(t,t) = I \tag{1.2.5.14}$$

on the interval $[0,t]$. Moreover, it is the unique matrix solution of

$$H(t,s) = I + \int_s^t H(t,u)\left[A(u) + \int_s^u B(u,v)\,dv\right]du \tag{1.2.5.15}$$

for $0 \le s \le t < \infty$, which is derived by integrating (1.2.5.14) from s to t and then interchanging the order of integration.

Now it becomes apparent from comparing (1.2.5.14) and (1.2.5.15) to (1.2.1.5) and (1.2.1.3), respectively, that the principal matrix solution of the adjoint equation (1.2.5.12) is identical to Grossman and Miller's resolvent.

1.2.6 Equivalence of $H(t,s)$ and $Z(t,s)$

The solutions of (1.2.2.1) and its adjoint

$$r'(s) = -r(s)A(s) - \int_s^t r(u)B(u,s)\,du$$

are related via the equation

$$\frac{\partial}{\partial u}[r(u)Z(u,s)] = r(u)\frac{\partial}{\partial u}Z(u,s) + r'(u)Z(u,s) \tag{1.2.6.1}$$

for $0 \leq s \leq u \leq t$. We exploit this to prove that the principal matrix solution and Grossman and Miller's resolvent are one and the same.

Theorem 1.2.6.1. *$H(t,s) \equiv Z(t,s)$.*

Proof. Select any $t > 0$. For a given row n-vector r_0, let $r(s)$ be the unique solution of (1.2.5.12) on $[0,t]$ such that $r(t) = r_0$. Now integrate both sides of (1.2.6.1) from s to t:

$$r(t)Z(t,s) - r(s)Z(s,s) = \int_s^t [r(u)Z_u(u,s) + r'(u)Z(u,s)]\,du.$$

By (1.2.5.12), we have

$$\begin{aligned} r_0Z(t,s) - r(s) = \int_s^t \Bigg[& r(u)Z_u(u,s) - r(u)A(u)Z(u,s) \\ & - \left(\int_u^t r(v)B(v,u)\,dv\right) Z(u,s)\Bigg]\,du. \end{aligned} \tag{1.2.6.2}$$

With an interchange in the order of integration, the iterated integral becomes

$$\begin{aligned} & \int_s^t \left(\int_u^t r(v)B(v,u)\,dv\right) Z(u,s)\,du \\ & \quad = \int_s^t r(v)\left(\int_s^v B(v,u)Z(u,s)\,du\right) dv \\ & \quad = \int_s^t r(u)\left(\int_s^u B(u,v)Z(v,s)\,dv\right) du. \end{aligned} \tag{1.2.6.3}$$

Making this change in (1.2.6.2), we obtain

$$\begin{aligned} r_0Z(t,s) - r(s) = \int_s^t r(u)\Bigg[& Z_u(u,s) - A(u)Z(u,s) \\ - \int_s^u B(u,v)Z(v,s)\,dv\Bigg]\,du. \end{aligned} \tag{1.2.6.4}$$

By (1.2.3.7), the integrand is zero. Hence,

$$r(s) = r_0 Z(t, s). \tag{1.2.6.5}$$

On the other hand,

$$r(s) = r_0 H(t, s)$$

from (1.2.5.13). Therefore, by uniqueness of the solution $r(s)$,

$$r_0 H(t, s) = r_0 Z(t, s). \tag{1.2.6.6}$$

Now let r_0 be the transpose of the ith basis vector e^i. Then (1.2.6.6) implies that the ith rows of $H(t, s)$ and $Z(t, s)$ are equal for $0 \leq s \leq t$. The theorem follows as t is arbitrary.

1.3 Nonlinear Existence

Existence theory for integral equations is endless and we will make no attempt to cover it fully. However, we will give two classical existence theorems here and there will be another one for integrodifferential equations in Chapter 3. This is a book about Liapunov theory and we will generally state in our theorems that if a solution exists then it satisfies the stated conclusions. Our first result is no surprise and the interested reader can extend the interval of existence using a weighted norm or a well-known method of steps, so long as the solution remains in the region of definition. In the second result the reader will recognize that a similar result could be obtained by Schauder's second fixed point theorem, dated 1930. But Tonelli obtained this one by a rather unique method two years earlier. It is offered here both for its interesting construction and the fact that it gives us a complete proof without resort to a deep fixed point theorem.

When we consider

$$x(t) = f(t) + \int_0^t g(t, s, x(s))\, ds\,, \quad t \geq 0\,, \tag{1.3.1}$$

it is to be understood that $x(0) = f(0)$ and we are looking for a continuous solution $x(t)$ for $t \geq 0$. However, it may happen that $x(t)$ is specified to be a certain *initial function* on an *initial interval*, say,

$$x(t) = \phi(t) \quad \text{for} \quad 0 \leq t \leq t_0$$

(see Fig. 1.1). We are then looking for a solution of

$$x(t) = f(t) + \int_0^{t_0} g(t, s, \phi(s))\, ds + \int_{t_0}^t g(t, s, x(s))\, ds\,, \quad t \geq t_0\,.$$

Notice that at $t = t_0$ we have

$$x(t_0) = f(t_0) + \int_0^{t_0} g(t, s, \phi(s))ds$$

which may not equal $\phi(t_0)$ so the graph in Fig. 1.1 could have a discontinuity, not just the indicated cusp.

As an example, (1.3.1) may describe the population density $x(t)$. A given population is observed over a time period $[0, t_0]$ and is given by $\phi(t)$. The subsequent behavior of that density may depend greatly on $\phi(t)$.

A change of variable will reduce the problem back to one of form (1.3.1).

Let $x(t + t_0) = y(t)$, so that we have

$$\begin{aligned} x(t+t_0) &= f(t+t_0) + \int_0^{t_0} g(t+t_0, s, \phi(s))\, ds \\ &\quad + \int_{t_0}^{t_0+t} g(t_0+t, s, x(s))\, ds \\ &= f(t+t_0) + \int_0^{t_0} g(t+t_0, s, \phi(s))\, ds \\ &\quad + \int_0^{t} g(t_0+t, u+t_0, x(u+t_0))\, du \end{aligned}$$

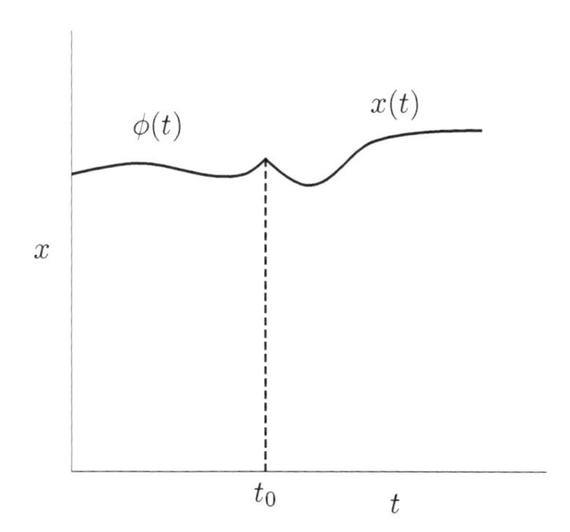

Figure 1.1: The cusp (or discontinuity).

or

$$y(t) = h(t) + \int_0^t g(t_0 + t, u + t_0, y(u))\,du$$

where

$$h(t) = f(t + t_0) + \int_0^{t_0} g(t + t_0, s, \phi(s))\,ds$$

and we want the solution for $t \geq 0$.

Thus, the initial function on $[0, t_0]$ is absorbed into the forcing function, and hence, it always suffices to consider (1.3.1) with the simple condition $x(0) = f(0)$.

Exactly the same may happen for an integrodifferential equation

$$x'(t) = A(t)x(t) + \int_0^t B(t,s)x(s)ds + f(t).$$

The natural initial condition is $x(0) = x_0$ and we seek a solution for $t \geq 0$. But there may be a given continuous initial function $\phi : [0, t_0] \to \Re^n$. Make the same change of variable as was done for the integral equation and we obtain a new equation of this same form where the initial function is absorbed into the forcing function. There is one large difference. This time we will get a continuous solution, but it will likely have a corner at t_0 just as our figure indicates. The solution does become smooth as time goes on.

We now proceed to obtain two of the most standard existence results for integral equations.

Definition 1.3.1. *Let $U \subset \Re^{n+1}$ and $G : U \to \Re^n$. We say that G satisfies a local Lipschitz condition with respect to x if for each compact subset H of U there is a constant K such that (t, x_i) in H implies*

$$\left|G(t, x_1) - G(t, x_2)\right| \leq K|x_1 - x_2|\,.$$

Frequently we need to allow t on only one side of t_0.

Let x, f, and g be n vectors and consider again (1.3.1)

$$x(t) = f(t) + \int_0^t g(t, s, x(s))\,ds\,.$$

From the proof of Theorem 1.2.1.1 and the introductory remarks of this section we know that an integrodifferential equation with initial conditions can be put into the form of (1.3.1).

Theorem 1.3.1. *Let a, b, and L be positive numbers, and for some fixed $\alpha \in (0,1)$ define $c = \alpha/L$. Suppose*

(a) *f is continuous on $[0,a]$,*

(b) *g is continuous on*

$$U = \left\{(t,s,x) : 0 \le s \le t \le a \text{ and } |x - f(t)| \le b\right\},$$

(c) *g satisfies a Lipschitz condition with respect to x on U of the form*

$$\big|g(t,s,x) - g(t,s,y)\big| \le L|x-y|$$

if (t,s,x), $(t,s,y) \in U$.

If $M = \max_U |g(t,s,x)|$, then there is a unique solution of (1.3.1) on $[0,T]$, where $T = \min[a, b/M, c]$.

Proof. Let $\mathcal{S}$ be the space of continuous functions from $[0,T] \to \Re^n$ with $\psi \in \mathcal{S}$ if

$$\|\psi - f\| \stackrel{\text{def}}{=} \max_{0 \le t \le T} |\psi(t) - f(t)| \le b\,.$$

Define an operator $A : \mathcal{S} \to \mathcal{S}$ by

$$(A\psi)(t) = f(t) + \int_0^t g(t,s,\psi(s))\,ds\,.$$

To see that $A : \mathcal{S} \to \mathcal{S}$ notice that ψ continuous implies $A(\psi)$ continuous and that

$$\begin{aligned}\|A(\psi) - f\| &= \max_{0 \le t \le T} |(A\psi)(t) - f(t)| \\ &= \max_{0 \le t \le T} \left| \int_0^t g(t,s,\psi(s))\,ds \right| \le MT \le b\,.\end{aligned}$$

To see that A is a contraction mapping, notice that if ϕ and $\psi \in \mathcal{S}$, then

$$\begin{aligned}\|A\phi - A\psi\| &= \max_{0 \le t \le T} \left| \int_0^t g(t,s,\phi(s))\,ds - \int_0^t g(t,s,\psi(s))\,ds \right| \\ &\le \max_{0 \le t \le T} \int_0^t \big|g(t,s,\phi(s)) - g(t,s,\psi(s))\big|\,ds \\ &\le \max_{0 \le t \le T} L \int_0^t |\phi(s) - \psi(s)|\,ds \\ &\le T \max_{0 \le t \le T} L|\phi(t) - \psi(t)| \\ &= TL\|\phi - \psi\| \le cL\|\phi - \psi\| = \alpha\|\phi - \psi\|\,.\end{aligned}$$

Thus, by the contraction mapping principle, there is a unique function $x \in \mathcal{S}$ with

$$(Ax)(t) = x(t) = f(t) + \int_0^t g(t,s,x(s))\,ds\,.$$

This completes the proof.

Remark 1.3.1. We can certainly get by with less than continuity of g in t and s. We need the Lipschitz condition and we need to say that if $\psi \in \mathcal{S}$ then $A\psi \in \mathcal{S}$. This will allow for weak singularities where the definition is modified to fit the particular form of $g(t,s,x)$ under consideration. One may find very general existence results throughout Miller (1971a), Corduneanu (1991), and many other places.

The interval $[0,T]$ can be improved by using a different norm, as we saw in Theorem 1.2.1. The conclusion can be strengthened to $T = \min[a, b/M]$. On an interval $[0,T]$ one may ask that

$$\|\phi\| = \max_{0\le s\le T} Ae^{-Bs}|\phi(s)|$$

for A and B positive constants.

Definition 1.3.2. *Let $\{f_n(t)\}$ be a sequence of functions from an interval $[a,b]$ to real numbers.*

(a) $\{f_n(t)\}$ is uniformly bounded *on $[a,b]$ if there exists M such that*

$$n \text{ a positive integer and } t \in [a,b]$$

imply $|f_n(t)| \le M$.

(b) $\{f_n(t)\}$ is equicontinuous *if for any $\varepsilon > 0$ there exists $\delta > 0$ such that*

$$\big[n \text{ a positive integer, } t_1 \in [a,b],\ t_2 \in [a,b], \text{ and } |t_1 - t_2| < \delta\big]$$

imply $|f_n(t_1) - f_n(t_2)| < \varepsilon$.

Theorem 1.3.2 (Ascoli-Arzela). *If $\{f_n(t)\}$ is a uniformly bounded and equicontinuous sequence of real functions on an interval $[a,b]$, then there is a subsequence that converges uniformly on $[a,b]$ to a continuous function.*

The theorem is, of course, also true for vector functions. Suppose that $\{F_n(t)\}$ is a sequence of functions from $[a,b]$ to R^p, for instance, $\mathbf{F}_n(t) = \big(f_n(t)_1, \dots, f_n(t)_p\big)$. [The sequence $\{\mathbf{F}_n(t)\}$ is uniformly bounded and equicontinuous if all the $\{f_n(t)_j\}$ are.] Pick a uniformly convergent subsequence $\{f_{kj}(t)_1\}$ using the theorem. Consider $\{f_{kj}(t)_2\}$ and use the theorem to obtain a uniformly convergent subsequence $\{f_{kjr}(t)_2\}$. Continue and conclude that $\{\mathbf{F}_{kjr\cdots s}(t)\}$ is uniformly convergent.

According to C. Corduneanu, the next result is by L. Tonelli and appeared in the Bull. Calcutta Math. Soc. in 1928. We have not seen the paper.

Theorem 1.3.3. *Let a and b be positive numbers and let $f : [0, a] \to \Re^n$ and $g : U \to \Re^n$ both be continuous, where*

$$U = \{(t, s, x) : 0 \le s \le t \le a \text{ and } |x - f(t)| \le b\}.$$

Then there is a continuous solution of (1.3.1)

$$x(t) = f(t) + \int_0^t g(t, s, x(s))\, ds$$

on $[0, T]$, where $T = \min[a, b/M]$ and $M = \max_U |g(t, s, x)|$.

Proof. We construct a sequence of continuous functions on $[0, T]$ such that

$$x_1(t) = f(t),$$

and if j is a fixed integer, $j > 1$, then $x_j(t) = f(t)$ when $t \in [0, T/j]$ and

$$x_j(t) = f(t) + \int_0^{t-(T/j)} g\big(t - (T/j), s, x_j(s)\big)\, ds$$

for $T/j \le t \le T$.

Let us examine this definition. If j is a fixed integer, $j > 1$, then

$$x_j(t) = f(t) \quad \text{on} \quad [0, T/j]$$

and

$$x_j(t) = f(t) + \int_0^{t-(T/j)} g\big(t - (T/j), s, f(s)\big)\, ds$$

for $t \in [T/j, 2T/j]$. At $t = 2T/j$, the upper limit is T/j, so the integrand was defined on $[T/j, 2T/j]$, thereby defining $x_j(t)$ on $[T/j, 2T/j]$. Now, for $t \in [2T/j, 3T/j]$ we still have

$$x_j(t) = f(t) + \int_0^{t-(T/j)} g\big(t - (T/j), s, x_j(s)\big)\, ds,$$

and the upper limit goes to $2T/j$ on that interval, so $x_j(s)$ is defined, and hence the integral is well defined. This process is continued to

$[(j-1)T/j, T]$, obtaining $x_j(t)$ on $[0,T]$ for each j. A Lipschitz condition will be avoided because $x_j(t)$, $j > 1$, is independent of all other $x_k(t)$, $j \neq k$.

Notice that

$$
\begin{aligned}
|x_j(t) - f(t)| &\le \int_0^{t-(T/j)} \big|g\big(t-(T/j), s, x_j(s)\big)\big|\, ds \\
&\le M(t-(T/j)) \le M(b/M) = b\,.
\end{aligned}
$$

This sequence $\{x_j(t)\}$ is uniformly bounded because

$$
|x_j(t)| \le |f(t)| + b \le \max_{0\le t\le T} |f(t)| + b\,.
$$

To see that $\{x_j(t)\}$ is an equicontinuous sequence, let $\varepsilon > 0$ be given, let n be an arbitrary integer, let t and v be in $[0,T]$ with $t > v$, and consider

$$
\begin{aligned}
|x_n(t) - x_n(v)| \le |f(t) - f(v)| &\\
&+ \bigg| \int_0^{v-(T/n)} \big[g\big(t-(T/n), s, x_n(s)\big) \\
&\qquad - g\big(v-(T/n), s, x_n(s)\big)\big]\, ds \bigg| \\
&+ \bigg| \int_{v-(T/n)}^{t-(T/n)} g\big(t-(T/n), s, x_n(s)\big)\, ds \bigg| \\
\le |f(t) - f(v)| + M|t-v| &\\
&+ \int_0^{v-(T/n)} \big|g\big(t-(T/n), s, x_n(s)\big) \\
&\qquad - g\big(v-(T/n), s, x_n(s)\big)\big|\, ds\,.
\end{aligned}
$$

By the uniform continuity of f, there is a $\delta_1 > 0$ such that $|t-v| < \delta_1$ implies $|f(t) - f(v)| < \varepsilon/3$.

By the uniform continuity of g on U, there is a $\delta_2 > 0$ such that $|t-v| < \delta_2$ implies

$$
T\big|g\big(t-(T/n), s, x_n(s)\big) - g\big(v-(T/n), s, x_n(s)\big)\big| < \varepsilon/3\,.
$$

Let $\delta = \min[\delta_1, \delta_2, \varepsilon/3M]$. This yields equicontinuity. The conclusion now follows from the Ascoli's-Arzela theorem.

Theorems 1.3.1 and 1.3.3 are local existence results. They guarantee a solution on an interval $[0,T]$. There is extensive discussion of existence in Burton (1983b). It is shown in detail how the solution can be continued until it approaches the boundary of the set where f and g are defined or there is a number L with $\limsup_{t\uparrow L} |x(t)| = +\infty$. This is finite escape time.

One can construct examples of equations whose solutions tend to infinity in finite time in several ways. Perhaps the easiest way is as follows. Let $g : \Re \to \Re$, $xg(x) > 0$ if $x \neq 0$, g be continuous, and let $\int_0^\infty [dx/g(x)] < \infty$. Suppose that C and C_t are continuous. If, in addition, $C(t,s) \geq c_0 > 0$ and $C_t(t,s) \geq 0$, then for $x_0 > 0$ the solution of

$$x(t) = x_0 + \int_0^t C(t,s)g(x(s))ds$$

tends to infinity in finite time. Just note that $x' \geq c_0 g(x)$, divide by $g(x)$ and integrate both sides from 0 to t. For more details, see p. 88f. in Burton (1983b) or p. 94f. in Burton (2005c).

1.3.1 Global existence

In this subsection we assume continuity of the functions in the integral equation and show that there is global existence of solutions if either the growth of the functions involved is not too fast or if there is a certain type of Liapunov function. A convexity condition will be assumed on the kernel which will be encountered again and again throughout the book. A Liapunov functional will be constructed based on that convex kernel and it will be necessary to differentiate it. The details of the differentiation are not particularly simple. To reduce the repetition of that differentiation, we give a sketch of how it is done here. References are given for the details found elsewhere in this book.

All of this work starts from a well-known global existence result for the ordinary differential equation

$$x' = f(t,x) \tag{1.3.1.1}$$

where $f : [0,\infty) \times \Re^n \to \Re^n$ is continuous. Our task is to extend that to the scalar equation

$$x(t) = a(t) - \int_0^t D(t,s)g(s,x(s))ds \tag{1.3.1.2}$$

where $D : [0,\infty) \times [0,\infty) \to \Re$, $a : [0,\infty) \to \Re$, and $g : [0,\infty) \times \Re \to \Re$ are all continuous with $xg(t,x) \geq 0$. If we extend the domain of D to $-\infty < s \leq t < \infty$ and of g to $\Re \times \Re \to \Re$, we will obtain results for the infinite delay equation

$$x(t) = b(t) - \int_{-\infty}^{t} D(t,s)g(s,x(s))ds. \tag{1.3.1.3}$$

To specify a solution of (1.3.1.3), we require a continuous initial function $\varphi : (-\infty, 0] \to R$ such that

$$a(t) := b(t) - \int_{-\infty}^{0} D(t,s)g(s,\varphi(s))ds \text{ is continuous}$$

so that (1.3.1.3) is essentially of the form of (1.3.1.2) when φ is chosen so that $\varphi(0) = a(0)$.

We come then to the question of how to rule out noncontinuable solutions of the kind mentioned above. In principle, there is a fine way of doing so. Kato and Strauss (1967) prove that it always works.

Definition 1.3.1.1. *A continuous function $V : [0,\infty) \times R^n \to [0,\infty)$ which is locally Lipschitz in x is said to be mildly unbounded if for each $T > 0$, $\lim_{|x|\to\infty} V(t,x) = \infty$ uniformly for $0 \le t \le T$.*

Notice that there is no mention of the differential equation or its solution. That will necessarily change when we advance the theory to integral equations.

If there is a mildly unbounded V which is differentiable, then we invoke the local existence theory and consider a solution $x(t)$ of (1.3.1.1) on $[t_0, \alpha)$ so that $V(t, x(t))$ is an unknown but well-defined function. The chain rule then gives

$$\frac{dV}{dt}(t,x(t)) = \sum_{i=1}^{n} \frac{\partial V}{\partial x_i}\frac{dx_i}{dt} + \frac{\partial V}{\partial t} = \operatorname{grad} V \cdot f + \frac{\partial V}{\partial t}.$$

We can also compute V' when V is only locally Lipschitz in x (cf. Yoshizawa (1966; p. 3)) and we will display such an example in a moment; in that case, one uses the upper right-hand derivative.

If V is so shrewdly chosen that it is mildly unbounded and $V' \le 0$, then there can be no $\alpha < \infty$ with $\lim_{t\to\alpha-} |x(t)| = \infty$ because $V(t,x(t)) \le V(t_0, x_0)$.

There is a converse theorem: If f is continuous and locally Lipschitz in x for each fixed t, then Kato and Strauss (1967) show that there is a mildly unbounded V with $V' \le 0$ if and only if all solutions can be continued for all future time. Their result is not constructive, but investigators have constructed suitable V for many important systems without any growth condition on f. In the example $x' = -x^3$ mentioned above, $V = x^2$ yields $V' = -2x^4 \leq 0$, showing global existence.

These remarks for (1.3.1.1) apply in large measure to (1.3.1.2). In those cases we require a functional $V(t, x(\cdot))$. More importantly, for (1.3.1.1) the only way a solution can fail to be continuable to $+\infty$ is for there to exist an α with $\lim_{t \to \alpha^-} |x(t)| = +\infty$; but for (1.3.1.2) we must take the limit supremum.

Wintner derived conditions on the growth of f to ensure that solutions of (1.3.1.1) could be continued to $+\infty$ and Conti used these to construct a suitable V. These results are most accessible in Hartman (1964; pp. 29–30) for the Wintner condition and Sansone and Conti (1964; p. 6) for V. A proof here will show how it works. The interested reader should also consult Kato and Strauss (1967) concerning the concept of mildly unbounded.

Theorem 1.3.1.1 (Conti-Wintner). *If there are continuous functions $\Gamma : [0, \infty) \to [0, \infty)$ and $W : [0, \infty) \to [1, \infty)$ with*

$$|f(t, x)| \leq \Gamma(t)W(|x|) \text{ and } \int_0^\infty \frac{ds}{W(s)} = \infty, \text{ then}$$

$$V(t, x) = \left\{ \int_0^{|x|} \frac{ds}{W(s)} + 1 \right\} \exp\left(- \int_0^t \Gamma(s)ds \right)$$

is mildly unbounded and $V'(t, x(t)) \leq 0$ along any solution of (1.3.1.1) and that solution can be continued on $[t_0, \infty)$.

Proof. Let $x(t)$ be a noncontinuable solution of (1.3.1.1) on $[t_0, \alpha)$. By examining the difference quotient we see that $|x(t)|' \leq |x'(t)|$. Thus,

$$V'(t, x(t)) \leq \frac{|x(t)|'}{W(|x(t)|)} \left[\exp\left(- \int_0^t \Gamma(s)ds \right) \right] - \Gamma(t)V(t, x(t)) \leq 0$$

when we use $|x(t)|' \leq |x'(t)| \leq \Gamma(t)W(|x(t)|)$. This means that $V(t, x(t)) \leq V(t_0, x_0)$; since V is mildly unbounded, $\lim_{t \to \alpha^-} |x(t)| \neq \infty$. This completes the proof.

Remark 1.3.1.1. Notice here and throughout the section that we only used $|f(t, x)| \leq \Gamma(t)W(|x|)$ for $0 \leq t \leq \alpha$. In many problems it turns out that the inequality fails for t greater than some α, but holds for all smaller α. In such cases we have proved that the solution can be continued to the latter α and that may be all that is needed.

The main result of this section is based on the following modified theorem of Schaefer (1955) which is discussed and proved also in Smart (1980; p. 29). Schaefer's result was for a locally convex topological vector space.

Theorem 1.3.1.2 (Schaefer). *Let $(X, \|\cdot\|)$ be a normed space, H a continuous mapping of X into X which is compact on each bounded subset of X. Then either*

(i) *the equation $x = \lambda Hx$ has a solution for $\lambda = 1$, or*

(ii) *the set of all such solutions x, for $0 < \lambda < 1$, is unbounded.*

We now turn to (1.3.1.2) and refer to Definition 1.3.1.1. There, the function V is mildly unbounded without any reference to (1.3.1.2). A typical example would be $V(t,x) = x^2/(t+1)$. Now things are going to change radically. For example, in (1.3.1.2) we will prepare to use Schaefer's theorem, introduce λ with $0 \le \lambda \le 1$, and write

$$\begin{aligned} x(t) &= \lambda\left[a(t) - \int_0^t D(t,s)g(s,x(s))ds\right] \\ &=: \lambda(Hx)(t), \end{aligned} \tag{1.3.1.2$_\lambda$}$$

retaining all of the continuity conditions.

We will see this several times and significant preparation is needed. First, we require a continuous function M with

$$\begin{aligned} &-2g(t,x)[x - \lambda a(t)] \le M(t), \\ &\qquad g(t,x) \text{ bounded for } t \ge 0 \text{ if } x \text{ is bounded .} \end{aligned} \tag{1.3.1.4}$$

But the defining property of this example is that the kernel is convex:

$$D(t,s) \ge 0, \quad D_s(t,s) \ge 0, \quad D_{st}(t,s) \le 0, \ D_t(t,0) \le 0. \tag{1.3.1.5}$$

Now, define

$$\begin{aligned} V(\lambda, t, x(\cdot)) &= \left\{\int_0^t D_s(t,s)\left(\int_s^t \lambda g(v,x(v))dv\right)^2 ds \right. \\ &\left. + D(t,0)\left(\int_0^t \lambda g(v,x(v))dv\right)^2 + 1\right\} \exp - \int_0^t M(s)ds. \end{aligned} \tag{1.3.1.6}$$

It seems very clear that this function is not necessarily mildly unbounded as it stands. But we will see that it definitely is mildly unbounded along a solution and that is all that is needed. To avoid too much repetition, we ask the reader to refer to the proof of Theorem 2.1.10 in the linear case. Here are the steps: Differentiate V; integrate by parts the term in V' obtained from differentiating the inner integral in the first integral in V; substitute $\lambda a(t) - x(t)$ from (1.3.1.2$_\lambda$) into the expression just obtained.

Since $D_{st} \le 0$ if there is a solution of $(1.3.1.2_\lambda)$ then we will have

$$V'(\lambda, t, x(\cdot)) \le \{-2\lambda g(t, x(t))[x(t) - \lambda a(t)] - M(t)\}e^{-\int_0^t M(s)ds} \quad (1.3.1.7)$$

and this is not positive by the assumptions on $M(t)$ and the fact that $xg(t,x) \ge 0$. Hence

$$V(\lambda, t, x(\cdot)) \le V(\lambda, 0, x(\cdot)) = 1, \quad (1.3.1.8)$$

a bound on V along that supposed solution, and the bound is uniform in λ.

We now show that V is mildly unbounded along that supposed solution. We have

$$(\lambda a(t) - x(t))^2 = \left(\int_0^t \lambda D(t,s) g(s, x(s))ds\right)^2$$

(from $(1.3.1.2_\lambda)$)

$$= \left(D(t,0)\int_0^t \lambda g(v, x(v))dv + \int_0^t D_s(t,s)\int_s^t \lambda g(v, x(v))dv\, ds\right)^2$$

(upon integration by parts)

$$\le 2\int_0^t D_s(t,s)ds \int_0^t D_s(t,s)\left(\int_s^t \lambda g(v,x(v))dv\right)^2 ds$$
$$+ 2D^2(t,0)\left(\int_0^t \lambda g(v, x(v))dv\right)^2$$

(by Schwarz's inequality)

$$\le 2[D(t,0) + D(t,t) - D(t,0)]V(\lambda, t, x(\cdot))e^{\int_0^t M(s)ds}$$
$$= 2D(t,t)V(\lambda, t, x(\cdot))e^{\int_0^t M(s)ds}.$$

But from (1.3.1.8) we have

$$(\lambda a(t) - x(t))^2 \le 2D(t,t)e^{\int_0^t M(s)ds}.$$

Here is the objective which we have achieved. Independent of $\lambda \in [0,1]$, for each $T > 0$ we can find $K > 0$ with $|x(t)| \le K$ if $0 \le t \le T$ and $0 \le \lambda \le 1$. Of course, there are many different Liapunov functions which could give us the same result so we formulate the end result as a definition.

Definition 1.3.1.2. *Let Q be the set of continuous solutions $\phi : [0,\infty) \to \Re^n$ of (1.3.1.2$_\lambda$) and let $V : [0,1] \times [0,\infty) \times Q \to [0,\infty)$. Then $V(\lambda, t, x(\cdot))$ is said to be mildly unbounded along any solution of (1.3.1.2$_\lambda$) if there is an $L > 0$ with $V(\lambda, 0, x(\cdot)) \leq L$ and for each $T > 0$ there is a $K > 0$ such that $V(\lambda, t, x(\cdot)) \leq L$ for $t \geq 0$ implies that $|x(t)| \leq K$ for $0 \leq t \leq T$. If the property is independent of T then V is radially unbounded along solutions.*

Note that $\lambda = 0$ yields $x = 0 \in Q$. We also note that, whereas a Liapunov function for a differential equation is mildly or radially unbounded independent of solutions, that is not true for integral equations.

The next result is adapted from Burton (1994b).

Theorem 1.3.1.3. *If either of the following conditions hold, then (1.3.1.2) has a solution on $[0,\infty)$:*

(I) There are continuous increasing functions $\Gamma : [0,\infty) \to [0,\infty)$ and $W : [0,\infty) \to [1,\infty)$ with

$$\int_0^\infty \frac{ds}{W(s)} = \infty \text{ and } |D(t,s,x)| \leq \Gamma(t)W(|x|) \text{ for } 0 \leq s \leq t. \quad (1.3.1.9)$$

(II) There is a differentiable scalar functional $V(\lambda, t, x(\cdot))$ which satisfies Definition 1.3.1.2 with $V'(\lambda, t, x(\cdot)) \leq 0$ along any solution of (1.3.1.2$_\lambda$).

Proof. Let $T > 0$ and $(X, \|\cdot\|)$ be the Banach space of continuous functions $\varphi : [0,T] \to R^n$ with the supremum norm. We will show that there is a solution $x(t)$ of (1.3.1.2) on $[0,T]$. These lemmas are for $0 \leq t \leq T$.

Lemma 1. *If H is defined by (1.3.1.2$_\lambda$) then $H : X \to X$ and H maps bounded sets into compact sets.*

Proof. If $\varphi \in X$, then $D(t,s,\varphi(s))$ is continuous and so $H\varphi$ is a continuous function of t. Let $J > 0$ be given and let $B = \{\varphi \in X \mid \|\varphi\| \leq J\}$. Now $a(t)$ is uniformly continuous on $[0,T]$ and $D(t,s,x)$ is uniformly continuous on $\Delta = \{(t,s,x) \mid 0 \leq s \leq t \leq T, |x| \leq J\}$. Thus, for each $\varepsilon > 0$ there is a $\delta > 0$ such that for $(t_i, s_i, x_i) \in \Delta$, $i = 1,2$, then $|(t_1,s_1,x_1) - (t_2,s_2,x_2)| < \delta$ implies that $|D(t_1,s_1,x_1) - D(t_2,s_2,x_2)| < \varepsilon$; a similar statement holds for $a(t)$. If $\varphi \in B$ then $0 \leq t_i \leq T$ and $|t_1 - t_2| < \delta$ imply that

$$\begin{aligned}
&|(H\varphi)(t_1) - (H\varphi)(t_2)| \leq |a(t_1) - a(t_2)| \\
&+ \left| \int_0^{t_1} \Big[D\big(t_1, s, \varphi(s)\big) - D\big(t_2, s, \varphi(s)\big) \Big] ds \right| \\
&+ \left| \int_0^{t_1} D\big(t_2, s, \varphi(s)\big) ds - \int_0^{t_2} D\big(t_2, s, \varphi(s)\big) ds \right| \\
&\leq \varepsilon + t_1 \varepsilon + |t_1 - t_2| M \leq \varepsilon(1+T) + \delta M
\end{aligned}$$

where $M = \max_{\Delta} |D(t,s,x)|$. Hence, the set $A = \{H\varphi | \varphi \in B\}$ is equicontinuous. Moreover, $\varphi \in B$ implies that $\|H\varphi\| \leq \|a\| + TM$. Thus, A is contained in a compact set by Ascoli's theorem.

Lemma 2. *H is continuous in φ.*

Proof. Let $J > 0$ be given, $\|\varphi_i\| \leq J$ for $i = 1, 2$, and for a given $\varepsilon > 0$ find the δ of uniform continuity on the region Δ of the proof of Lemma 1 for D. If $\|\varphi_1 - \varphi_2\| < \delta$, then

$$\begin{aligned} |(H\varphi_1)(t) - (H\varphi_2)(t)| &\leq \int_0^t \big|D\big(t,s,\varphi_1(s)\big) - D\big(t,s,\varphi_2(s)\big)\big| ds \\ &\leq T\varepsilon \end{aligned}$$

so $\|(H\varphi_1) - (H\varphi_2)\| \leq T\varepsilon$.

Lemma 3. *There is a $K > 0$ such that any solution of ($1.3.1.2_\lambda$) on $[0,T]$, satisfies $\|\varphi\| \leq K$.*

Proof. Let (I) hold. If φ satisfies (1.3.1.5) on $[0,T]$, then

$$|\varphi(t)| \leq \lambda\left[A(T) + \int_0^t \Gamma(T)W(|\varphi(s)|)ds\right] \text{ for } 0 \leq t \leq T$$

where $A(T) = \max_{0 \leq t \leq T} |a(t)|$. If we define $y(t)$ by

$$y(t) = \lambda\left[A(T) + 1 + \Gamma(T)\int_0^t W(|y(s)|)ds\right]$$

then $y(t) \geq |\varphi(t)|$ on $[0,T]$; clearly, $y(0) > |\varphi(0)|$ so if there is a first t_1 with $y(t_1) = |\varphi(t_1)|$, then a contradiction is clear. But the Conti-Wintner result gives a bound K on $\|y\|$ and so the lemma is true for (I). The argument when (II) holds follows directly from Definition 1.3.1.2 and $V' \leq 0$.

Interesting global existence is also found in Lakshmikantham and Leela (1969; p. 46).

1.4 Memory, singularity, limiting equations

An ordinary differential equation

$$x' = f(t, x), \quad x(t_0) = x_0,$$

with backward and forward uniqueness has a perfect memory, but a very uninteresting one. If $x(t, t_0, x_0)$ is a solution with $t_1 > t_0$, then

$$x' = f(t, x), \quad x(t_1) = x_1$$

has a unique solution, $x(t, t_1, x_1)$, for $t < t_1$ and $x(t_0, t_1, x_1) = x_0$. The solution remembers exactly where it started.

An nth order linear normalized (leading coefficient equals 1) can be written as a scalar integral equation, as we will soon show and we will see an interesting weighting system. Here is the way the weighting system works.

An integral equation

$$x(t) = a(t) - \int_0^t C(t, s)g(s, x(s))ds$$

is an equation with an obvious memory. The function x at time t depends on its whole history. The function $C(t, s)$ is a weight applied to $g(s, x(s))$. The integral sums all of those products, subtracts the sum from $a(t)$, and yields the quantity $x(t)$. In most physical problems there is a fading memory: earlier values of $x(s)$ are weighted less heavily than later values. Often the memory fades completely so that for fixed s then $C(t, s) \to 0$ as $t \to \infty$. Convexity is a natural and very prevalent example, especially as singularities are allowed. Volterra noted this in 1928 and it persists to this day. There is ample reason to focus much of our attention on it, especially as we expand the idea to complete monotonicity including singularities. Notice that we have $C(t, s) > 0$, but $C_t(t, s) < 0$ so the memory is fading: with s fixed, as $t \to \infty$ we see $C(t, s)$ decreasing. But the interesting case is generally at $s = t$. We now display cases progressing to singularities.

a) If $C(t, s) = 1 + e^{-(t-s)}$, then the weight is always at least1, but never greater than 2.

b) If $C(t, s) = e^{-(t-s)}$ then the weight is 1 at $s = t$, but tends to zero for fixed s as $t \to \infty$.

c) For $C(t, s) = C(t - s)$ and for T a fixed positive number we define

$$\begin{aligned} C(t) &= (t - T)^4, \quad 0 \le t \le T \\ &= 0, \quad t > T. \end{aligned}$$

It is readily shown that C, C', C'' are continuous on $[0, \infty)$ and that C is a convex function. We will look at two different kinds of equations.

(ci) Notice that for $t \geq T$ then

$$\begin{aligned} x &= a(t) - \int_0^t C(t-s)g(s,x(s))ds \\ &= a(t) - \int_{t-T}^t C(t-s)g(s,x(s))ds. \end{aligned}$$

The memory fades and then abruptly ends. All of this can be smoothed out so that we do not need to restrict $t \geq T$.

(cii) For $-\infty < s \leq t < \infty$, then

$$\begin{aligned} x &= a(t) - \int_{-\infty}^t C(t-s)g(s,x(s))ds \\ &= a(t) - \int_{t-T}^t C(t-s)g(s,x(s))ds. \end{aligned}$$

These are called truncated equations. We will see a unification in Section 4.4.

d) If $C(t,s) = e^{-(t-2s)}$ then at $s = t$ we have $C(t,t) = e^t$. The memory grows exponentially for the event at time t, but then fades as s is fixed and $t \to \infty$. This has meaning for many theorems in the pages to come. Frequently, we ask that $C(t,t)$ be bounded. In the work following (1.3.1.8) we showed that

$$(\lambda a(t) - x(t))^2 \leq 2D(t,t)V(\lambda, t, x(\cdot))e^{\int_0^t M(s)ds}$$

and $D(t,t)$ is our $C(t,t)$. This means that the Liapunov functional is mildly unbounded. But if $D(t,t)$, $\int_0^t M(s)ds$, and $a(t)$ are bounded, then V is radially unbounded. So here we retain the mildly unbounded property, but lose the radially unbounded property. That is critical to note in later problems. Moreover, the worst is yet to come. This is a transitional example leading us to singularities.

e) We make a major leap to $C(t,s) = (t-s)^{q-1}$ for $0 < q < 1$ in fractional differential equations. At $s = t$ the weight applied to $g(t, x(t))$ is infinite, but its effect on $x(t)$ is measurable because it is weakly singular. This is typical of heat problems; enormous smoothing occurs.

In the foregoing, we started with memory and progressed to singularities. Now we look at problems where we start with large dimensional systems and convert them to scalar integral equations. Next, we will show that a higher order ordinary differential equation can be reduced to a scalar integral equation. Finally, we will show how to pass from a standard integral equation to one with infinite delay.

The problem of Lurie concerns a control problem of $n+1$-dimension in the form

$$x' = Ax + bf(\sigma)$$
$$\sigma' = c^T x - rf(\sigma)$$

where x, c, and b are n-vectors, while σ, $f(\sigma)$, and r are scalars. Also A is an $n \times n$ constant matrix.

To explain what is being studied, we note that we begin with the linear constant coefficient system $y' = Ay$ and seek to improve the performance of the system by adding a control, $bf(\sigma)$. It is assumed that $\sigma f(\sigma) > 0$ for $\sigma \neq 0$. Now σ is determined from the feedback of the position $x(t)$ by the equation $\sigma' = c^T x(t) - rf(\sigma)$. That equation uses the information, $x(t)$, to manufacture a value $\sigma(t)$ which is then relayed back to $x' = Ax + bf(\sigma(t))$.

Modern versions of the problem insist that there are time delays. It takes a certain amount of time, T_1, to relay to σ the value of $x(t)$; thus, a better second equation is

$$\sigma' = c^T x(t - T_1) - rf(\sigma).$$

With this information we can obtain $\sigma(t)$; but it takes a certain amount of time, say T_2, to relay that to the first equation controlling x, yielding the improved first equation as

$$x' = Ax + bf(\sigma(t - T_2))$$

and, hence, the corrected system

$$x' = Ax(t) + bf(\sigma(t - T_2))$$
$$\sigma' = c^T x(t - T_1) - rf(\sigma(t)).$$

But, in either case we readily reduce the problem to a scalar integral equation. We will give the details for the original system and leave the delayed system for an exercise. The interested reader should be able to apply techniques of Chapter 2 to obtain stability results for this scalar equation.

Treating $bf(\sigma)$ as an inhomogeneous term, we use the variation of parameters formula to write

$$x(t) = e^{At}x(0) + \int_0^t e^{A(t-s)} bf(\sigma(s))ds$$

so that we obtain the scalar equation

$$\sigma' = c^T \left[e^{At}x(0) + \int_0^t e^{A(t-s)} bf(\sigma(s))ds \right] - rf(\sigma)$$

and then, upon integration and interchange of the order of integration, a scalar nonlinear integral equation is obtained of the form

$$\sigma = a(t) + \int_0^t H(t,s)f(\sigma(s))ds.$$

This problem was introduced by Lurie (1951) and enjoyed much attention. The book by Lefschetz (1965) is devoted entirely to it, while significant material on it is also found in LaSalle and Lefschetz (1961). Cao, Li, and Ho (2005), Somolinos (1977), and Burton and Somolinos (2007) contain treatments with delays. During the last ten years there has been a great resurgence of interest. A check of the online Mathematical Reviews will yield almost one hundred papers on this problem, most of which are recent. Burton and Somolinos (2007) recently treated the problem with a variety of delays and with A having zero as a characteristic root. The delayed equation is a good source of research problems.

The work just finished showed how large dimension systems can be collapsed into a simple scalar integral equation which can be studied by very elementary techniques. The work depended on the equation being in a particular form. The present example is intended to show that the same thing can be done for any linear normalized ordinary differential equation, regardless of the form.

Let f and $a_1(t), \dots, a_n(t)$ be continuous on $[0, T)$ in

$$x^{(n)} + a_1(t)x^{(n-1)} + \cdots + a_n(t)x = f(t)\,,$$

with $x(0), x'(0), \dots, x^{(n-1)}(0)$ given initial conditions, and set $x^{(n)}(t) = z(t)$. Then

$$x^{(n-1)}(t) = x^{(n-1)}(0) + \int_0^t z(s)\,ds\,,$$

$$x^{(n-2)}(t) = x^{(n-2)}(0) + tx^{(n-1)}(0) + \int_0^t (t-s)z(s)\,ds\,,$$

$$x^{(n-3)}(t) = x^{(n-3)}(0) + tx^{(n-2)}(0) + \frac{t^2}{2!}\,x^{(n-1)}(0) + \int_0^t \frac{(t-s)^2}{2}\,z(s)\,ds\,,$$

$$\vdots$$

$$x(t) = x(0) + tx'(0) + \cdots + \frac{t^{n-1}}{(n-1)!}\,x^{(n-1)}(0) + \int_0^t \frac{(t-s)^{n-1}}{(n-1)!}\,z(s)\,ds\,.$$

If we replace these values of x and its derivatives in our differential equation we have a scalar integral equation for $z(t)$.

Not only is this a compact expression, but it is a sobering admonition that, while a scalar linear first order differential equation is a thinly disguised exercise in elementary integration, a scalar linear integral equation commands our full attention and respect.

We now show one way of formulating an integral equation on the whole line. If there is a $T > 0$ such that

$$a(t+T) = a(t), \quad C(t+T, s+T) = C(t, s)$$

then the solution of

$$x(t) = a(t) - \int_0^t C(t,s)x(s)ds$$

will often have a very well-behaved solution. In fact, it will often approach a periodic function. For n a positive integer we write

$$\begin{aligned} x(t+nT) &= a(t+nT) - \int_0^{t+nT} C(t+nT, s)x(s)ds \\ &= a(t) - \int_{-nT}^t C(t+nT, s+nT)x(s+nT)ds \\ &= a(t) - \int_{-nT}^t C(t,s)x(s+nT)ds. \end{aligned}$$

If $x(t+nT)$ converges to some function $y(t)$ and if the integral of C converges in a certain way, then y satisfies

$$y(t) = a(t) - \int_{-\infty}^t C(t,s)y(s)ds,$$

called a limiting equation. That equation is just set up for fixed point theory because if ϕ is $T-$periodic, so is the mapping

$$(P\phi)(t) = a(t) - \int_{-\infty}^t C(t,s)\phi(s)ds.$$

A fixed point is a periodic solution. Under general conditions $x(t+nT)$ converges to that periodic function. Thus, we see that from the standard integral equation we come to study an equation on the whole line in a natural way. This will be of interest in Sections 4.3 and 4.4.

1.5 The Convolution Lemma

Very often we encounter an integral of the form

$$H(t) = \int_0^t f(s)g(t-s)ds$$

where either both functions are in $L^1[0,\infty)$ or $g \in L^1[0,\infty)$, while $f(t) \to 0$ as $t \to \infty$. It is shown in Rudin (1966; p. 156) that in the first case then for H interpreted as the Lebesgue integral it is true that $H(t) \in L^1[0,\infty)$. It is a deep result. The second case offers some pitfalls which we want to discuss early. Very frequently when we study fractional differential equations we consider the resolvent kernel $R(t)$ and the forcing function $f(t)$ in the relation

$$F(t) = \int_0^t R(t-s)f(s)ds.$$

We know that $R \in L^1[0,\infty)$ and that $R(t) \to 0$ as $t \to \infty$. Thus, R can play the role of either f or g in the second case mentioned above. Now, here is the pitfall. If $f \in L^1[0,\infty)$ then we would like to say that $F(t) \to 0$ as $t \to \infty$, relying on the decay of $R(t)$ as $t \to \infty$. But there are counterexamples to that statement and they follow from the fact that, while $R(t) \to 0$ as $t \to \infty$, it is also true that $R(t)$ is unbounded as $t \to 0$. We are indebted to Bo Zhang for the construction of just such counterexample for the following situation. We are also indebted to Leigh Becker for offering us several alternatives to our up coming lemma. In the fractional differential equation of Caputo type

$${}^cD^q x = -[1+f(t)]x(t), 0 < q < 1, x(0) \in \Re,$$

we invert to

$$x(t) = z(t) - \int_0^t R(t-s)f(s)x(s)ds.$$

One would like to say that this is an asymptotic contraction because

$$\int_0^t R(t-s)|f(s)|ds \to 0$$

as $t \to \infty$. That may be true under some conditions on f and q, but the counterexample of Bo Zhang shows that we must take care.

Throughout this book we will refer to the following result.

Convolution Lemma. *Let $f, g : [0, \infty) \to \Re$ with $g \in L^1[0, \infty)$. Denote the convolution by*

$$H(t) := \int_0^t f(s)g(t-s)ds.$$

(i) If $f \in L^1[0, \infty)$ and if H denotes the Lebesgue integral, then the Lebesgue integral of H on $[0, \infty)$ is finite.
(ii) If f is bounded, locally integrable, and tends to zero as $t \to \infty$, then $H(t) \to 0$ as $t \to \infty$.

Proof. Part (i) is the quoted result of Rudin. Here is a proof of Part (ii). Let $|f(t)| \leq M$ for $0 \leq t < \infty$ and let $\int_0^\infty |g(s)|ds = K$. If $0 < T < t$ then

$$\begin{aligned}\int_0^t |f(s)g(t-s)|ds &\leq \int_0^T |f(s)g(t-s)|ds + \int_T^t |f(s)g(t-s)|ds \\ &\leq M \int_0^T |g(t-s)|ds + \sup_{T \leq s \leq t} |f(s)| \int_T^t |g(t-s)|ds \\ &\leq M \int_{t-T}^t |g(s)|ds + \sup_{T \leq s \leq t} |f(s)|K.\end{aligned}$$

For a given $\epsilon > 0$ find T so that the last term is less than $\epsilon/2$. Then choose t so large that the first term satisfies the same relation.

Notes Sections 1.2.1 - 1.2.6 are entirely the work of Becker (1979), (2006). The rest of the material in Chapter 1 is either taken from the classical literature, is original, or specific credit is listed with the theorems.

Chapter 2

Integral Equations

2.1 Basic theory

Most of this section is concerned with basic classical type results for an integral equation $x(t) = a(t) - \int_0^t C(t,s)x(s)ds$ whose kernel satisfies a condition typified by

$$\sup_{t\geq 0} \int_0^t |C(t,s)|ds \leq \alpha < 1 \text{ or } \int_s^t |C(u,s)|du \leq \beta < 1.$$

Such conditions generally promote the idea that $a(t)$ and the solution $x(t)$ lie in the same space. Indeed, the variation of parameters formula shows x, a, and $\int_0^t R(t,s)a(s)ds$ all in the same space; that is what we hope the reader will focus on when reading every theorem in this section. While such results are classical and fundamental, they are not astonishing. We call them pedestrian.

Nowhere in mathematics can one find a more mysterious function than the resolvent. The scalar second order equation $x'' + tx = 0$ can be written as a scalar integral equation and Ritt (1966) has shown that its solution is arbitrarily complicated. Thus, one may infer that the resulting resolvent, $R(t,s)$, is also arbitrarily complicated. Yet, such resolvents can have precise control over vast vector spaces of unbounded and badly behaved functions. For a function ϕ from such a space it can happen that $\int_0^t R(t,s)\phi(s)ds$ is virtually an exact copy of $\phi(t)$. On the other hand, for the same $R(t,s)$ there are vector spaces of small and well-behaved functions over which R has virtually no control in that integral. The goal of the investigator is to determine conditions on $C(t,s)$ and vector spaces that go with such C resulting in precise or poor control. Such information is critical in applied problems; we may identify large perturbations with virtually no effect, while certain small perturbations can lead to disaster.

We are concerned again with the three equations

$$x(t) = a(t) - \int_0^t C(t,s)x(s)ds, \tag{2.1.1}$$

$$\begin{aligned} R(t,s) &= C(t,s) - \int_s^t C(t,u)R(u,s)du \\ &= C(t,s) - \int_s^t R(t,u)C(u,s)du, \end{aligned} \tag{2.1.2}$$

and

$$x(t) = a(t) - \int_0^t R(t,s)a(s)ds \tag{2.1.3}$$

where $a : [0,\infty) \to \Re^n$ is continuous and $C(t,s)$ is an $n \times n$ matrix of functions continuous for $0 \le s \le t < \infty$. Refer back to Theorem 1.2.1 for existence and uniqueness. We will consider many results in which knowledge of R will tell us much about x from (2.1.3). But there are other reasons for studying R. Frequently, embedded in $a(t)$ is a nonlinear function or functional of x and by passing to (2.1.3) we obtain a new integral equation requiring the properties of $R(t,s)$.

Much of this chapter will concern a mapping $P : V \to W$ where V and W are vector spaces of functions $\phi : [0,\infty) \to \Re^n$ with P defined by

$$(P\phi)(t) = \phi(t) - \int_0^t R(t,s)\phi(s)ds. \tag{2.1.4}$$

The spaces V and W will say much about the character of R.

Notation. *Throughout this book, $\mathcal{BC}$ will denote the vector space of bounded and continuous functions $\phi : [0,\infty) \to \Re^n$.*

One of the projects in this book is to show that $\int_0^t R(t,s)a(s)ds$ can make a good copy of $a(t)$. The following definition is formalizing this property. The first property states that regardless of how badly $a(t)$ is behaved, the difference between $a(t)$ and the integral is always bounded.

Definition 2.1.1. *Let P, defined by (2.1.4), map a vector space V into a vector space W.*

(i) The resolvent $R(t,s)$ is said to generate an approximate identity on V if $W \subseteq \mathcal{BC}$.

(ii) Let the resolvent $R(t,s)$ generate an approximate identity on V. Then $R(t,s)$ generates an asymptotic identity on V if $\phi \in V$ implies that for P defined by (2.1.4), then $(P\phi)(t) \to 0$ as $t \to \infty$.

(iii) The resolvent $R(t,s)$ is said to generate an L^p approximate identity on V if for P defined by (2.1.4) there is a positive integer p with $P : V \to L^p$.

There is a major result of Perron (1930) which plays a central role here.

Theorem 2.1.1. *(Perron) Let H be an $n \times n$ matrix of functions continuous from $0 \le s \le t < \infty$ to $\Re$. Then*

$$\sup_{t\ge 0} \int_0^t |H(t,s)|ds < \infty$$

if and only if $\int_0^t H(t,s)\phi(s)ds$ is bounded for every $\phi \in \mathcal{BC}$.

Theorem 2.1.2. *(The Fundamental Theorem) Every solution of (2.1.1) is in $\mathcal{BC}$ for every function $a \in \mathcal{BC}$ if and only if*

$$\sup_{t\ge 0} \int_0^t |R(t,s)|ds < \infty. \tag{2.1.5}$$

Proof. By (2.1.3) it is clear that if (2.1.5) holds then the solution is bounded. But by (2.1.3) again, if x is bounded for every bounded a then by Perron's theorem (2.1.5) holds.

The classical idea is that for $C(t,s)$ well-behaved then the solution of (2.1.1) follows $a(t)$. We will go through the ideas beginning with a repeat of the Adam and Eve idea. The idea for part (iii) below comes from Strauss (1970).

Parts (iii) and (iv) suggest that we need the kernel to have a small integral in order for R to have a small integral. Nothing could be farther from the truth. These are merely handy results. Some of the most important real-world problems have kernels with infinite integrals, but the corresponding resolvents may have very small integrals. This makes it possible to construct very effective Liapunov functionals. All of our subsequent work with fractional differential equations falls in this category, as do many of the results concerning convex kernels.

Theorem 2.1.3. *Let $C(t,s)$ be an $n \times n$ matrix of continuous functions and suppose that there is an $\alpha < 1$ with*

$$\sup_{t\ge 0} \int_0^t |C(t,s)|ds \le \alpha < 1. \tag{2.1.6}$$

(i) If $a \in \mathcal{BC}$ so is the solution x of (2.1.1); hence, (2.1.5) holds.

(ii) Suppose, in addition, that for each $T > 0$ then $\int_0^T |C(t,s)|ds \to 0$ as $t \to \infty$. If $a(t) \to 0$ as $t \to \infty$, so does $x(t)$ and $\int_0^t R(t,s)a(s)ds$. Also, $\int_0^T |R(t,s)|ds \to 0$ as $t \to \infty$.

(iii) $\int_0^t |R(t,s)|ds \leq \frac{\alpha}{1-\alpha}$.

Moreover, if (2.1.6) is replaced by the condition that there is a $\beta < 1$ with

$$\int_s^t |C(u,s)|du \leq \beta, \quad 0 \leq s \leq t < \infty$$

then

(iv) $\int_s^t |R(u,s)|du \leq \frac{\beta}{1-\beta}$ for $0 \leq s \leq t < \infty$.

Proof. For (i) we define a mapping $Q : \mathcal{BC} \to \mathcal{BC}$ by

$$(Q\phi)(t) = a(t) - \int_0^t C(t,s)\phi(s)ds.$$

By (2.1.6) it is a contraction using the operator norm on C, as in (1.2.15). There is a unique fixed point in $\mathcal{BC}$ which satisfies (2.1.1).

For (ii), we add to the mapping set, say M, the condition that for each $\phi \in M$ then $\phi(t) \to 0$ as $t \to \infty$. Then

$$|(Q\phi)(t)| \leq |a(t)| + \int_0^t |C(t,s)\phi(s)|ds.$$

We will show that the last term tends to zero as $t \to \infty$. For a given $\epsilon > 0$ and for $\phi \in M$, find T such that $|\phi(t)| < \epsilon$ if $t \geq T$ and find J with $|\phi(t)| \leq J$ for all $t \geq 0$. For this fixed T, find $\eta > T$ such that $t \geq \eta$ implies that $\int_0^T |C(t,s)|ds \leq \epsilon/J$. Then $t \geq \eta$ implies that

$$\begin{aligned} \int_0^t |C(t,s)\phi(s)|ds &\leq \int_0^T |C(t,s)\phi(s)|ds + \int_T^t |C(t,s)\phi(s)|ds \\ &\leq (J\epsilon/J) + \alpha\epsilon < 2\epsilon. \end{aligned}$$

Thus, $Q : M \to M$ and the fixed point satisfies $x(t) \to 0$ for every continuous function $a(t)$ which tends to zero. We can also write

$$x(t) = a(t) - \int_0^t R(t,s)a(s)ds$$

and so $\int_0^t R(t,s)a(s)ds \to 0$ for every continuous $a(t)$ which tends to zero. The last part of (ii) is a result of Strauss (1970) and the proof will not be given here. A more general result and application are found in Proposition 2.6.4.1, not requiring (2.1.6).

For (iii), we take the absolute values in the second form of (2.1.2) and integrate to obtain

$$\begin{aligned}\int_0^t |R(t,s)|ds &\le \int_0^t |C(t,s)|ds + \int_0^t \int_s^t |R(t,u)C(u,s)|duds \\ &\le \alpha + \int_0^t \int_0^u |C(u,s)|ds|R(t,u)|du \\ &\le \alpha + \alpha \int_0^t |R(t,u)|du,\end{aligned}$$

from which the result follows.

Finally, for (iv) we again take absolute values in the first form of (2.1.2) and have

$$\begin{aligned}\int_s^t |R(u,s)|du &\le \int_s^t |C(u,s)|du + \int_s^t \int_s^v |C(v,u)R(u,s)|dudv \\ &= \int_s^t |C(u,s)|du + \int_s^t \int_u^t |C(v,u)R(u,s)|dvdu \\ &\le \beta + \beta \int_s^t |R(u,s)|du,\end{aligned}$$

completing the proof.

Part (iii) is related to Theorem 2.6.3.1. There is an interesting alternate proof using a Liapunov functional.

Remark 2.1.1. *Notice that when (2.1.6) holds then the mapping P defined in (2.1.4) maps $\mathcal{BC}$ into itself. Thus, $R(t,s)$ generates an approximate identity on $\mathcal{BC}$. For $\phi \in \mathcal{BC}$ then $\int_0^t R(t,s)\phi(s)ds \in \mathcal{BC}$. However complicated $R(t,s)$ may be, it is still true that the integration operation yields a function in $\mathcal{BC}$. This is hardly remarkable, but we ask the reader to watch this process as we progress.*

Now, we will see an example in which the solution tends to zero so that $R(t,s)$ will generate an asymptotic identity.

Example 2.1.2. *If $r : [0, \infty) \to (0, 1]$ with $r(t) \downarrow 0$, with*

$$\sup_{t \geq 0} \int_0^t |C(t,s)| r(s)/r(t) ds \leq \alpha < 1, \tag{2.1.7}$$

and with

$$|a(t)| \leq k r(t) \tag{2.1.8}$$

for some $k > 0$, then the unique solution $x(t)$ of (2.1.1) also satisfies $|x(t)| \leq k^* r(t)$ for some $k^* > 0$. Moreover, the resolvent $R(t, s)$ in (2.1.2) generates an asymptotic identity on the space of functions $\phi : [0, \infty) \to \Re^n$ with $\sup_{t \geq 0} \left| \frac{\phi(t)}{r(t)} \right| < \infty$.

Proof. The proof is based on a weighted norm. Let $(M, |\cdot|_r)$ denote the Banach space of continuous functions $\phi : [0, \infty) \to \Re^n$ with the property that

$$|\phi|_r := \sup_{t \geq 0} \frac{|\phi(t)|}{r(t)} < \infty.$$

Define $Q : M \to M$ by $\phi \in M$ implies that

$$(Q\phi)(t) = a(t) - \int_0^t C(t,s)\phi(s) ds.$$

We have

$$\begin{aligned} |(Q\phi)(t)|/r(t) &\leq |a(t)|/r(t) + \int_0^t |C(t,s)| r(s)/r(t) |\phi(s)|/r(s) ds \\ &\leq k + |\phi|_r \int_0^t |C(t,s)| r(s)/r(t) ds \\ &\leq k + \alpha |\phi|_r \end{aligned}$$

so $Q\phi \in M$. To see that Q is a contraction in that norm we have immediately that

$$|(Q\phi)(t) - (Q\eta)(t)|/r(t) \leq \alpha |\phi - \eta|_r$$

for $\phi, \eta \in M$. Hence, there is a fixed point in M and so it has the required properties. As $x(t)$ in (2.1.3) tends to zero for $a(t)$ satisfying (2.1.8), so does $P\phi$ in (2.1.4). This completes the proof.

Equation (2.1.1) can have periodic solutions, but only under exceptionally rare conditions. Those conditions involve certain orthogonality relations which usually will not hold throughout an entire vector space of large dimension.

The problem seems to have been first studied in Burton (1984), (2005b; pp. 94-96). The periodicity depends on a special orthogonal property discussed by Lakshmikantham and Rao (1995). Recently, the quest has gained new life in connection with fractional differential equations and is discussed in Travazoei (2010), Travazoei and Haeri (2009), Kaslik and Sivasundarum (2011), and Burton and Zhang (2012). The natural solution is an asymptotically periodic solution. It will help to see this if we first change (2.1.1) so that the existence of a periodic solution can be proved. That will be our next result.

Theorem 2.1.5. *Let*

$$x(t) = a(t) - \int_{-\infty}^{t} C(t,s)x(s)ds$$

in which $a : \Re \to \Re^n$ is continuous, while $C(t,s)$ is continuous on $\Re \times \Re$, and there is a positive constant T with

$$a(t+T) = a(t) \text{ and } C(t+T, s+T) = C(t,s).$$

Suppose also that if ϕ is continuous and periodic then $\int_{-\infty}^{t} C(t,s)\phi(s)ds$ is continuous. Finally, suppose there is an $\alpha < 1$ with

$$\sup_{0 \le t \le T} \int_{-\infty}^{t} |C(t,s)|ds \le \alpha.$$

Then the equation has a T-periodic solution.

Proof. Let $(X, |\cdot|)$ be the Banach space of continuous T-periodic functions ϕ on $\Re$ into $\Re^n$ with the supremum norm and define $H : X \to X$ by $\phi \in X$ implies

$$(H\phi)(t) = a(t) - \int_{-\infty}^{t} C(t,s)\phi(s)ds.$$

A translation readily establishes that $H\phi$ is T-periodic. Next, if $\phi, \eta \in X$ then for $0 \le t \le T$ we have

$$\begin{aligned} |(H\phi)(t) - (H\eta)(t)| &\le \int_{-\infty}^{t} |C(t,s)||\phi(s) - \eta(s)|ds \\ &\le |\phi - \eta| \int_{-\infty}^{t} |C(t,s)|ds \\ &\le \alpha|\phi - \eta|. \end{aligned}$$

Hence, H is a contraction and there is a unique fixed point solving the equation and residing in X.

For (2.1.1), once we determine the proper space, then contraction mappings can provide a seemingly one-step solution of the problem. If we write (2.1.1) as

$$x(t) = a(t) - \int_{-\infty}^{t} C(t,s)x(s)ds + \int_{-\infty}^{0} C(t,s)x(s)ds$$

then

$$a(t) - \int_{-\infty}^{t} C(t,s)x(s)ds$$

suggests the periodic function just considered, while

$$\int_{-\infty}^{0} C(t,s)x(s)ds$$

can readily be expected to tend to zero for any bounded function x. It is then natural to expect a solution $x = p + q$ where p is periodic and q tends to zero. Moreover, a space of such functions with the supremum norm is a Banach space, $(Y, \|\cdot\|)$. We note that the natural mapping defined from (2.1.1) will map $Y \to Y$.

Let $\mathcal{P}_T$ be the set of continuous T-periodic functions on $\Re$ into $\Re^n$ and suppose that for $\phi \in \mathcal{P}_T$ then

$$\int_{-\infty}^{0} C(t,s)\phi(s)ds \to 0 \text{ as } t \to \infty \tag{2.1.9}$$

and is continuous. Let Q be the set of continuous functions $q : [0,\infty) \to \Re^n$ such that $q(t) \to 0$ as $t \to \infty$. For each $q \in Q$ let

$$\int_{0}^{t} C(t,s)q(s)ds \to 0 \text{ as } t \to \infty. \tag{2.1.10}$$

We will need the following lemma in the proof of the next result.

Lemma 2.1.1. *Let $(Y, \|\cdot\|)$ be the space of continuous functions $\phi : [0,\infty) \to \Re^n$ such that $\phi \in Y$ implies there is a $p \in \mathcal{P}_T$ and $q \in Q$ with $\phi = p + q$. Then $(Y, \|\cdot\|)$ is a Banach space.*

Proof. Let $\{p_n + q_n\}$ be a Cauchy sequence in $(Y, \|\cdot\|)$. Now for each $\epsilon > 0$ and each $q \in Q$ there is an $L > 0$ such that $t \geq L$ implies that $|q(t)| < \epsilon/4$. Given $\epsilon > 0$ there is an N such that for n, $m \geq N$ then

$$|p_n(t) + q_n(t) - p_m(t) - q_m(t)| < \epsilon/2.$$

Fix n, $m \geq N$; for $\epsilon/4$ find L such that $t \geq L$ implies that both $|q_n(t)| < \epsilon/4$ and $|q_m(t)| < \epsilon/4$. Then $t \geq L$ implies that

$$|p_n(t) - p_m(t)| - |q_n(t) - q_m(t)| \leq |p_n(t) + q_n(t) - p_m(t) - q_m(t)| < \epsilon/2$$

so that $t \geq L$ implies that

$$|p_n(t) - p_m(t)| < (\epsilon/2) + |q_n(t)| + |q_m(t)| < \epsilon.$$

But p_n and p_m are periodic so the end terms of the last inequality hold for all t. As this is true for every pair with $n \geq N$ and $m \geq N$, it follows that $\{p_n\}$ is a Cauchy sequence. This, in turn, shows the same for $\{q_n\}$. As both $\mathcal{P}_T$ and Q are complete in the supremum norm, Y is complete.

Theorem 2.1.6. *Let $C(t+T, s+T) = C(t,s)$, $a \in \mathcal{P}_T$, and let (2.1.9) and (2.1.10) hold. Suppose also that there is an $\alpha < 1$ with $\int_0^t |C(t,s)|ds \leq \alpha$. Finally, for each continuous and periodic function ϕ then both $\int_{-\infty}^t C(t,s)\phi(s)ds$ and $\int_{-\infty}^0 C(t,s)\phi(s)ds$ are continuous. Then (2.1.1) has a solution $x(t) = p(t) + q(t)$ where $p \in \mathcal{P}_T$ and $q \in Q$.*

Proof. Let $(Y, \|\cdot\|)$ be the Banach space of functions $\phi = p+q$ where $p \in \mathcal{P}_T$ and $q \in Q$ with the supremum norm. Define a mapping $H : Y \to Y$ by $\phi = p + q \in Y$ implies that

$$\begin{aligned}(H\phi)(t) &= a(t) - \int_0^t C(t,s)[p(s) + q(s)]ds \\ &= \left[a(t) - \int_{-\infty}^t C(t,s)p(s)ds\right] \\ &+ \left[\int_{-\infty}^0 C(t,s)p(s)ds - \int_0^t C(t,s)q(s)ds\right] \\ &=: B\phi + A\phi.\end{aligned}$$

This defines operators A and B on Y. Note that $B : Y \to \mathcal{P}_T \subset Y$ and $A : Y \to Q \subset Y$.

But from the first line of this array we see that H is a contraction with unique fixed point $\phi \in Y$ and that proves the result.

The same result holds when $a \in Y$.

We now turn to problems in which the first coordinate of C is integrated. Moreover, we change from contraction mappings to Liapunov functionals. The reader is urged to return to Section 1.1 and review the idea of a Liapunov functional. Our first result in this set is just like Theorem 1.1.3, but now it is for systems. There will be interesting contrasts in the results as we change Liapunov functionals.

Theorem 2.1.7. *Suppose that $a \in L^1[0,\infty)$ and $\int_{t-s}^{\infty} |C(u+s,s)|du$ is continuous for $0 \le s \le t < \infty$. If there is an $\alpha < 1$ with $\int_0^{\infty} |C(u+t,t)|du \le \alpha$ then the solution $x(t)$ of (2.1.1) is in $L^1[0,\infty)$ and the resolvent, $R(t,s)$, of (2.1.2) generates an L^1 approximate identity on L^1.*

Proof. Define a Liapunov functional

$$V(t) = \int_0^t \int_{t-s}^{\infty} |C(u+s,s)|du|x(s)|ds.$$

We will take the derivative of V along the unique solution of (2.1.1) and need to unite the Liapunov functional with the integral equation. The inequality accomplishing this will now be prepared. From (2.1.1) we have

$$|x(t)| \le |a(t)| + \int_0^t |C(t,s)x(s)|ds.$$

Now

$$\begin{aligned} V'(t) &\le \int_0^{\infty} |C(u+t,t)|du|x(t)| - \int_0^t |C(t,s)x(s)|ds \\ &\le \alpha|x(t)| - |x(t)| + |a(t)| = (\alpha-1)|x(t)| + |a(t)|. \end{aligned}$$

An integration from 0 to t, use of $V(t) \ge 0$, and use of $a \in L^1$ will yield $x \in L^1$. If we look at (2.1.3) and (2.1.4) we see that $P\phi \in L^1$ for each $\phi \in L^1$ and that completes the proof.

Theorems 2.1.3(ii) and 2.1.7 are nonconvolution counterparts of the Convolution Lemma in the last section of Chapter 1.

We will now add to the conditions of Theorem 2.1.7 classical conditions for boundedness of solutions of (2.1.1). Notice how a change in the Liapunov functional so that x^2 is in the integrand yields the solution in L^2.

Theorem 2.1.8. *Let (2.1.1) be a scalar equation with $\int_{t-s}^{\infty} |C(u+s,s)|du$ continuous. Suppose there exist $\alpha < 1$ and $\beta < 1$ with $\int_0^{\infty} |C(u+t,t)|du \le \alpha$ and $\sup_{t \ge 0} \int_0^t |C(t,s)|ds \le \beta$. Then $R(t,s)$ generates an L^1, an L^2, and* *an L^∞ approximate identity on the spaces L^1, L^2, and L^∞ respectively. If $\phi = \phi_1 + \phi_2 + \phi_3$ where $\phi_1 \in L^1$, $\phi_2 \in L^2$, and $\phi_3 \in L^\infty$, then $P\phi = \psi_1 + \psi_2 + \psi_3$ where $\psi_1 \in L^1$, $\psi_2 \in L^2$, and $\psi_3 \in L^\infty$.*

Proof. First we prepare the inequality which will unite the integral equation to the Liapunov functional. Notice that for any $\epsilon > 0$ there is an $M > 0$ so that by squaring both sides of (2.1.1) we can say that

$$\begin{aligned}
x^2(t) &\leq Ma^2(t) + (1+\epsilon)\left(\int_0^t C(t,s)x(s)ds\right)^2 \\
&\leq Ma^2(t) + (1+\epsilon)\int_0^t |C(t,s)|ds \int_0^t |C(t,s)|x^2(s)ds \\
&\leq Ma^2(t) + (1+\epsilon)\beta\int_0^t |C(t,s)|x^2(s)ds \\
&= Ma^2(t) + \int_0^t |C(t,s)|x^2(s)ds
\end{aligned}$$

where we choose ϵ so that $(1+\epsilon)\beta = 1$. This means that

$$-\int_0^t |C(t,s)|x^2(s)ds \leq Ma^2(t) - x^2(t)$$

which will be our fundamental uniting inequality.

Next, define a Liapunov functional by

$$V(t) = \int_0^t \int_{t-s}^\infty |C(u+s,s)|du x^2(s)ds$$

so that

$$\begin{aligned}
V'(t) &= \int_0^\infty |C(u+t,t)|du x^2(t) - \int_0^t |C(t,s)|x^2(s)ds \\
&\leq Ma^2(t) - x^2 + \int_0^\infty |C(u+t,t)|du x^2 \\
&\leq Ma^2(t) - (1-\alpha)x^2(t).
\end{aligned}$$

Hence, an integration yields

$$(1-\alpha)\int_0^t x^2(s)ds \leq M\int_0^t a^2(s)ds$$

and $x \in L^2[0,\infty)$. The L^∞ conclusion is Theorem 2.1.3, while the L^1 conclusion is Theorem 2.1.7. The results on the mappings then follow from the linearity.

The result can be taken much further.

Lemma 2.1.2. *Let (2.1.1) be a scalar equation. Suppose there is an $\alpha < 1$ with*

$$\sup_{t \geq 0} \int_0^t |C(t,s)| ds \leq \alpha.$$

Consider equation (2.1.1). There is an $M > 0$ and for each integer $n > 0$ we have

$$x^{2^n}(t) \leq M^{2^n - 1} a^{2^n}(t) + \int_0^t |C(t,s)| x^{2^n}(s) ds.$$

Proof. In (2.1.1) we square both sides to obtain

$$x^2(t) = a^2(t) - 2a(t) \int_0^t C(t,s)x(s)ds + \left(\int_0^t C(t,s)x(s)ds \right)^2.$$

Find $\epsilon > 0$ with $(1 + \epsilon)\alpha = 1$ and then find $M > 1$ with $2|a(t)||y| \leq (M-1)a^2(t) + \epsilon y^2$. Thus,

$$\begin{aligned}
x^2(t) &\leq Ma^2(t) + (1+\epsilon)\left(\int_0^t C(t,s)x(s)ds \right)^2 \\
&\leq Ma^2(t) + (1+\epsilon) \int_0^t |C(t,s)| ds \int_0^t |C(t,s)| x^2(s) ds \\
&\leq Ma^2(t) + \int_0^t |C(t,s)| x^2(s) ds
\end{aligned}$$

where we have used the Schwarz inequality. Next, suppose there is a positive integer k with

$$x^{2k}(t) \leq M^{2k-1} a^{2k}(t) + \int_0^t |C(t,s)| x^{2k}(s) ds.$$

Squaring yields

$$\begin{aligned}
x^{4k}(t) &\le M^{4k-2}a^{4k}(t) \\
&+ 2M^{2k-1}a^{2k}(t)\int_0^t |C(t,s)|x^{2k}(s)ds \\
&+ \left(\int_0^t |C(t,s)|x^{2k}(s)ds\right)^2 \\
&\le M^{4k-2}a^{4k}(t) + (M-1)\Big[M^{2k-1}a^{2k}(t)\Big]^2 \\
&+ (1+\epsilon)\left(\int_0^t |C(t,s)|x^{2k}(s)ds\right)^2 \\
&\le M^{4k-2}a^{4k}(t)(M-1+1) \\
&+ (1+\epsilon)\int_0^t |C(t,s)|ds\int_0^t |C(t,s)|x^{4k}(s)ds \\
&\le M^{4k-1}a^{4k}(t) + \int_0^t |C(t,s)|x^{4k}(s)ds.
\end{aligned}$$

That is, if $2k = 2^n$, then $4k = (2)2^n = 2^{n+1}$ which establishes the induction.

Theorem 2.1.9. *Let (2.1.1) be a scalar equation with $\int_{t-s}^{\infty} |C(u+s,s)|du$ continuous. Suppose there are constants $\alpha < 1$ and $\beta < 1$ with*

$$\int_0^t |C(t,s)|ds \le \alpha \text{ and } \int_0^\infty |C(u+t,t)|du \le \beta.$$

If there is an $n > 0$ with $a \in L^{2^n}[0,\infty)$ then the solution of (2.1.1) satisfies $x \in L^{2^n}[0,\infty)$.

Proof. Define a Liapunov functional

$$V(t) = \int_0^t \int_{t-s}^\infty |C(u+s,s)|du x^{2^n}(s)ds$$

so that

$$\begin{aligned}
V'(t) &= \int_0^\infty |C(u+t,t)|du x^{2^n}(t) - \int_0^t |C(t,s)|x^{2^n}(s)ds \\
&\le \beta x^{2^n}(t) - x^{2^n}(t) + M^{2^n-1}a^{2^n}(t).
\end{aligned}$$

Thus, an integration yields

$$0 \le V(t) \le V(0) - (1-\beta)\int_0^t x^{2^n}(s)ds + M^{2^n-1}\int_0^\infty a^{2^n}(t)dt,$$

as required.

We are now going to look at problems in which we ask much about the derivatives of C, but little about the magnitude of C. It uses what may be called a perfect Liapunov functional. It has a type of lower wedge, but the casual reader would never see it. Its derivative along the solution is accomplished without any type of inequality; it is a perfect match for the equation. When $a(t) = 0$, it can display all the classical wedges above and below the Liapunov functional and above the derivative of the functional.

Theorem 2.1.10. *Let (2.1.1) be a scalar equation where*

$$C(t,s) \ge 0, \quad C_s(t,s) \ge 0, \quad C_t(t,s) \le 0, \quad C_{st}(t,s) \le 0 \tag{2.1.11}$$

and they are all continuous. Then along the solution of (2.1.1) the functional

$$V(t) = \int_0^t C_s(t,s)\left(\int_s^t x(u)du\right)^2 ds + C(t,0)\left(\int_0^t x(s)ds\right)^2 \tag{2.1.12}$$

satisfies

$$V'(t) \le -(x(t)-a(t))^2 - x^2(t) + a^2(t). \tag{2.1.13}$$

(i) If $a \in L^2[0,\infty)$, so are x and $\int_0^t R(t,s)a(s)ds$; moreover, $V(t)$ is bounded.

(ii) If there is a constant B with

$$C(t,t) \le B \tag{2.1.14}$$

then along the solution of (2.1.1) we have

$$\left(\int_0^t R(t,s)a(s)ds\right)^2 = (a(t)-x(t))^2 \le 2BV(t) \tag{2.1.15}$$

where (2.1.15) does not require $a \in L^2$. However, if $a \in L^2$ and bounded then both $V(t)$ and x are bounded.

Proof. We have

$$V(t) = \int_0^t C_s(t,s)\Big(\int_s^t x(u)du\Big)^2 ds + C(t,0)\Big(\int_0^t x(s)ds\Big)^2$$

and differentiate to obtain

$$\begin{aligned}V'(t) &= \int_0^t C_{st}(t,s)\Big(\int_s^t x(u)du\Big)^2 ds + 2x\int_0^t C_s(t,s)\int_s^t x(u)duds \\ &\quad + C_t(t,0)\Big(\int_0^t x(s)ds\Big)^2 + 2xC(t,0)\int_0^t x(s)ds.\end{aligned}$$

We now integrate the third-to-last term by parts to obtain

$$\begin{aligned}&2x\Big[C(t,s)\int_s^t x(u)du\Big|_0^t + \int_0^t C(t,s)x(s)ds\Big] \\ &= 2x\Big[-C(t,0)\int_0^t x(u)du + \int_0^t C(t,s)x(s)ds\Big].\end{aligned}$$

Cancel terms, use the sign conditions, and use (2.1.1) in the last step of the process to unite the Liapunov functional and the equation obtaining

$$\begin{aligned}V'(t) &= \int_0^t C_{st}(t,s)\Big(\int_s^t x(u)du\Big)^2 ds + C_t(t,0)\Big(\int_0^t x(s)ds\Big)^2 \\ &\quad + 2x[a(t) - x(t)] \le 2xa(t) - 2x^2(t) \\ &= -(x(t) - a(t))^2 + a^2(t) - x^2(t).\end{aligned}$$

From this we obtain

$$0 \le V(t) \le V(0) + \int_0^t a^2(s)ds - \int_0^t x^2(s)ds - \int_0^t (a(s) - x(s))^2 ds;$$

when $a \in L^2[0,\infty)$ then $x \in L^2[0,\infty)$ and V is bounded.

To obtain (2.1.15), we begin with the integral equation, square it, integrate by parts, and use the Schwarz inequality as follows:

$$
\begin{aligned}
(x(t)-a(t))^2 &= \left(-\int_0^t C(t,s)x(s)ds\right)^2 \\
&= \left(C(t,s)\int_s^t x(u)du\Big|_0^t - \int_0^t C_s(t,s)\int_s^t x(u)duds\right)^2 \\
&= \left(-C(t,0)\int_0^t x(u)du - \int_0^t C_s(t,s)\int_s^t x(u)duds\right)^2 \\
&\le 2\Bigg[C(t,0)C(t,0)\left(\int_0^t x(u)du\right)^2 \\
&+ \int_0^t C_s(t,s)ds\int_0^t C_s(t,s)\left(\int_s^t x(u)du\right)^2 ds\Bigg] \\
&\le 2\left[C(t,0)+\int_0^t C_s(t,s)ds\right]V(t) \\
&= 2\left[C(t,0)+C(t,s)\Big|_0^t\right]V(t) \\
&= 2[C(t,0)+C(t,t)-C(t,0)]V(t) \\
&= 2C(t,t)V(t).
\end{aligned}
$$

The left side of (2.1.15) is the variation of parameters formula.

Incidentally, this shows how V was constructed. Historically, Liapunov functions have often been constructed in exactly that way and the technique goes back to Lagrange.

Remark Notice that we have shown that the Liapunov functional is mildly unbounded along a solution. If $a(t)$ is bounded then the Liapunov functional is radially unbounded along the solution. We would never see these properties if we looked at the Liapunov functional alone and not in conjunction with the integral equation.

Investigators have always relied on variations of (2.1.13) to bound the Liapunov functional and, hence, the solution. We will obtain a relation from (2.1.13) and substitute that into (2.1.12). In order to let $a(t)$ become large, we will discover three new things. First, we show how to replace $x(t)$ by $a(t)$ in the Liapunov functional. In the coming sections we show how to replace $x(t)$ by $a'(t)$ and allow $a'(t)$ to be bounded and continuous. Finally, we show how to replace $\int_0^t x(s)ds$ by $a(t)$.

We begin by replacing x in the Liapunov functional by $a(t)$. This will give us a condition to ensure that $V(t)$ is bounded. Then notice that when V is bounded and when (2.1.15) holds then x is bounded if and only if a

is bounded. We will, thereby, obtain a condition showing that $x(t)$ follows $a(t)$ regardless of the behavior of $a(t)$.

Notice that if $\int_0^t C_s(t,s)(t-s)^2 ds + C(t,0)t^2$ is bounded then there is a vector space of functions $a(t)$ satisfying (2.1.16), below. That space includes continuous functions $a = \phi + \psi$ where ϕ is bounded and $\psi \in L^2[0,\infty)$.

Theorem 2.1.11. *Let (2.1.1) be a scalar equation and let (2.1.11) and (2.1.14) hold. If, in addition, there is a constant M with*

$$\int_0^t C_s(t,s)(t-s)\int_s^t a^2(u)duds + C(t,0)t\int_0^t a^2(s)ds \le M \quad (2.1.16)$$

then $V(t)$ is bounded along the solution of (2.1.1), where V is defined in (2.1.12). Noting (2.1.15), we have that $\left(\int_0^t R(t,s)a(s)ds\right)^2 = (a(t)-x(t))^2$ is bounded so x is bounded if and only if a is bounded. Finally, when (2.1.16) holds for every $a \in \mathcal{BC}$ then $\int_0^t R(t,s)a(s)ds$ is bounded for every $a \in \mathcal{BC}$ so $\sup_{t\ge 0}\int_0^t |R(t,s)|ds < \infty$.

Proof. Focus on (2.1.13). Suppose that $V(t)$ is not bounded. Then there is a monotone increasing sequence $\{t_n\} \to \infty$ with $V(s) \le V(t_n)$ for $0 \le s \le t_n$. Let t denote any such t_n and let $0 \le s \le t$. From (2.1.13)

$$0 \le V(t) - V(s) \le \int_s^t a^2(u)du - \int_s^t x^2(u)du$$

so that

$$\int_s^t x^2(u)du \le \int_s^t a^2(u)du.$$

If we use the Schwarz inequality on both integrals of x in (2.1.12) we obtain

$$V(t) \le \int_0^t C_s(t,s)(t-s)\int_s^t a^2(u)duds + C(t,0)t\int_0^t a^2(u)du \le M$$

by (2.1.16). Hence, V is bounded and we apply (2.1.15). As $\int_0^t R(t,s)a(s)ds$ is bounded for every $a \in \mathcal{BC}$, it follows from Perron's theorem that $\sup_{t\ge 0}\int_0^t |R(t,s)|ds < \infty$.

Corollary. *If $M = M(t)$ and/or if $C(t,t)$ is not bounded, then we obtain*

$$|x(t)| \le |a(t)| + \sqrt{2C(t,t)M(t)},$$

a Liapunov growth condition.

In this example we can take $a^2(t)$ to be the sum of a bounded function and an L^1-function which could have a sequence of spikes of magnitude going to infinity. By (2.1.15) the solution will follow those spikes in a very faithful way, differing only by a fixed bounded function.

Example 2.1.1. *Let $C(t,s) = e^{-(t-s)}$ and $a^2(t) = \gamma + \mu(t)$ where γ is a fixed positive constant and $\mu \in L^1[0,\infty)$. Condition (2.1.14) becomes*

$$\int_0^t e^{-(t-s)} ds \leq 1 =: B,$$

while $C(t,0) = e^{-t} \leq 1 := K$. Then (2.1.16) is

$$V(t) \leq \int_0^t e^{-(t-s)}(t-s) \int_s^t [\gamma + \mu(u)] du ds + e^{-t} t \int_0^t [\gamma + \mu(s)] ds$$

which is bounded. Using this in (2.1.15) yields $(x(t) - a(t))^2$ bounded so $x(t)$ follows $a(t)$ on those spikes going off to infinity. Note that $C(t,s) = ke^{-(t-s)}$ works for any $k > 0$.

In the following exercise we see that very different kernels can be put together in a smooth way and the Liapunov functionals simply added in the study of integral equations. It works in the same way with nonlinearities, but this method often fails for nonlinear integrodifferential equations.

Exercise 2.1.1. Consider the equation

$$x(t) = a(t) - \int_0^t C(t,s)x(s)ds - \int_0^t D(t,s)x(s)ds.$$

Let the conditions on C from Theorem 2.1.10 hold, $\int_0^t |C(t,s)|ds \leq \alpha < 1$, and let $\int_0^\infty |D(u+t,t)|du \leq \beta < 1$. If $a \in L^2$ then prove that the solution x is also. To do this, define

$$V(t) = \int_0^t C_s(t,s)\left(\int_s^t x(u)du\right)^2 ds + C(t,0)\left(\int_0^t x(s)ds\right)^2 + \int_0^t \int_{t-s}^\infty |D(u+s,s)|du|x(s)|^2 ds$$

and show that the derivative satisfies

$$V'(t) \leq Ma^2(t) - \beta x^2(t)$$

for a certain positive constant M.

We are now going to look at some problems in which the kernel is related to a convolution kernel. The proofs are no longer particularly simple, but they do show us that with more work we can obtain more information.

In (2.1.1) suppose that

$$\int_{t-s}^{\infty} |C(u+s,s)|du \tag{2.1.17}$$

is continuous for $0 \leq s \leq t < \infty$ and that there is a number $\alpha < 1$ with

$$\int_{0}^{\infty} |C(u+t,t)|du \leq \alpha \tag{2.1.18}$$

for $0 \leq t < \infty$.

Theorem 2.1.12. *Let (2.1.17) and (2.1.18) hold, let $a(t)$ be bounded and continuous, and suppose there is a differentiable and decreasing function $\Phi : [0,\infty) \to (0,\infty)$ with $\Phi \in L^1[0,\infty)$, and*

$$\Phi(t-s) \geq \int_{t-s}^{\infty} |C(u+s,s)|du. \tag{2.1.19}$$

If $x(t)$ is the unique solution of (2.1.1) and if

$$V(t) := \int_0^t \int_{t-s}^{\infty} |C(u+s,s)|du|x(s)|ds, \tag{2.1.20}$$

then $V(t)$ is bounded. If, in addition, there is a $K > 0$ with

$$\int_{t-s}^{\infty} |C(u+s,s)|du \geq K|C(t,s)| \tag{2.1.21}$$

then $x(t)$ is bounded and $\sup_{t\geq 0} \int_0^t |R(t,s)|ds < \infty$.

Proof. Note that in (2.1.20) we have

$$V'(t) = \int_0^{\infty} |C(u+t,t)|du|x(t)| - \int_0^t |C(t,s)x(s)|ds$$

and from (2.1.1) that

$$|x(t)| - |a(t)| \leq \int_0^t |C(t,s)x(s)|ds$$

or by (2.1.18)

$$V'(t) \le \alpha|x(t)| + |a(t)| - |x(t)| =: -\delta|x(t)| + |a(t)| \tag{2.1.22}$$

for $\delta > 0$. Thus, for $0 \le s \le t < \infty$ we can write

$$\frac{dV(s)}{ds}\Phi(t-s) \le -\delta|x(s)|\Phi(t-s) + |a(s)|\Phi(t-s).$$

Suppose there is a $t > 0$ satisfying $V(t) = \max_{0 \le s \le t} V(s)$.

Then

$$\begin{aligned}\int_0^t \frac{dV(s)}{ds}\Phi(t-s)ds &= V(s)\Phi(t-s)|_0^t - \int_0^t V(s)\frac{d}{ds}\Phi(t-s)ds \\ &= V(t)\Phi(0) - \int_0^t V(s)\frac{d}{ds}\Phi(t-s)ds \\ &\ge V(t)\Phi(0) - V(t)\int_0^t \frac{d}{ds}\Phi(t-s)ds \\ &= V(t)\Phi(0) - V(t)\Phi(0) + V(t)\Phi(t) \\ &= V(t)\Phi(t).\end{aligned}$$

Hence,

$$\begin{aligned}V(t)\Phi(t) &\le -\delta\int_0^t \Phi(t-s)|x(s)|ds + \int_0^t |a(s)|\Phi(t-s)ds \\ &\quad \text{(and by (2.1.19))} \\ &\quad \le -\delta V(t) + \|a\|k\end{aligned}$$

for some $k > 0$ and $\|a\|$ the supremum of a. Thus,

$$V(t)[\Phi(t) + \delta] \le \|a\|k \tag{2.1.23}$$

and $V(t)$ is bounded.

If (2.1.21) holds, then $V(t) \ge K[|x(t)| - |a(t)|]$. As V is bounded, so is $x(t)$. But $x(t) = a(t) - \int_0^t R(t,s)a(s)ds$ is bounded for every bounded continuous $a(t)$ and so $\int_0^t R(t,s)a(s)ds$ is bounded for every bounded and continuous $a(t)$. By Perron's theorem $\sup_{t\ge 0}\int_0^t |R(t,s)|ds < \infty$.

2.1.1 Nonlinearities

We are now going to substitute $g(x)$ for x in the integral and study the equation

$$x(t) = a(t) - \int_0^t C(t,s)g(x(s))ds \tag{2.1.24}$$

where

$$xg(x) > 0, \quad x \neq 0, \tag{2.1.25}$$

with $g : \Re \to \Re$ being continuous, while $a(t)$ and $C(t,s)$ satisfy the continuity conditions with (2.1.1).

In working with our next Liapunov functional we find

$$V'(t) \leq 2a(t)g(x) - 2xg(x) \tag{2.1.26}$$

and we need to separate $a(t)g(x)$. There are many *ad hoc* ways of doing that. We could simply ask for positive constants c_i with

$$2a(t)g(x) - 2xg(x) \leq c_1|a(t)| - c_2xg(x). \tag{2.1.26*}$$

Under certain conditions there is a very exact result which we now develop.

Lemma 2.1.1.1. *Let $g(x) = -g(-x)$, g be strictly increasing, for $x \geq 0$ let $\phi(x) := \frac{d}{dx}xg^{-1}(x)$ be monotone increasing to infinity. Then*

$$2|a(t)g(x)| \leq xg(x) + \int_0^{2|a(t)|} \phi^{-1}(s)ds. \tag{2.1.27}$$

Proof. Young's inequality (Hewitt and Stromberg (1971; p. 189)) states that if $\phi : [0,\infty) \to [0,\infty)$ is continuous, strictly increasing to ∞, satisfies $\phi(0) = 0$, and if $\psi = \phi^{-1}$ then for $\Phi(x) = \int_0^x \phi(u)du$ and $\Psi(x) = \int_0^x \psi(u)du$ we have

$$2|a(t)g(x)| \leq \Phi(g(x)) + \Psi(2|a(t)|). \tag{2.1.28}$$

But

$$\Phi(g(x)) = \int_0^{g(x)} \frac{d}{ds}sg^{-1}(s)ds = g(x)g^{-1}(g(x)) = xg(x),$$

as required.

A vector extension of the next result is found in Zhang (2009).

Theorem 2.1.1.1. *If $a : [0,\infty) \to R$ is continuous, if (2.1.25) holds, and if*

$$C(t,s) \geq 0, \quad C_s(t,s) \geq 0, \quad C_t(t,s) \leq 0, \quad C_{st}(t,s) \leq 0 \tag{2.1.29}$$

then along the solution of (2.1.24) the functional

$$V(t) = \int_0^t C_s(t,s)\left(\int_s^t g(x(u))du\right)^2 ds + C(t,0)\left(\int_0^t g(x(s))ds\right)^2 \tag{2.1.30}$$

satisfies

$$V'(t) \leq 2a(t)g(x) - 2xg(x).$$

(i) If there are constants B and K with

$$\sup_{t\geq 0}\int_0^t C_s(t,s)ds = B < \infty \text{ and } \sup_{t\geq 0} C(t,0) = K < \infty \tag{2.1.31}$$

then along any solution of (2.1.24) on $[0,\infty)$ we have

$$(a(t) - x(t))^2 \leq 2(B+K)V(t). \tag{2.1.32}$$

(ii) If the conditions of Lemma 2.1.1.1 hold then along the solution of (2.1.24) we have

$$V'(t) \leq -x(t)g(x(t)) + \int_0^{2|a(t)|} \phi^{-1}(s)ds.$$

Hence, if the last term is $L^1[0,\infty)$ then so is $x(t)g(x(t)$. Moreover, V is then bounded so if (2.1.31) holds then $|a(t) - x(t)|$ is bounded.

Proof. The details are like the linear case in Theorem 2.1.10. We have

$$V(t) = \int_0^t C_s(t,s)\left(\int_s^t g(x(u))du\right)^2 ds + C(t,0)\left(\int_0^t g(x(s))ds\right)^2$$

and differentiate along any solution of (2.1.24) to obtain

$$\begin{aligned} V'(t) = \\ &\int_0^t C_{st}(t,s)\left(\int_s^t g(x(u))du\right)^2 ds + 2g(x)\int_0^t C_s(t,s)\int_s^t g(x(u))duds \\ &+ C_t(t,0)\left(\int_0^t g(x(s))ds\right)^2 + 2g(x)C(t,0)\int_0^t g(x(s))ds. \end{aligned}$$

We now integrate the third-to-last term by parts to obtain

$$\begin{aligned}
&2g(x)\left[C(t,s)\int_s^t g(x(u))du\Big|_0^t + \int_0^t C(t,s)g(x(s))ds\right] \\
&= 2g(x)\left[-C(t,0)\int_0^t g(x(u))du + \int_0^t C(t,s)g(x(s))ds\right].
\end{aligned}$$

Cancel terms, use the sign conditions, and use (2.1.24) in the last step of the process to unite the Liapunov functional and the equation obtaining

$$\begin{aligned}
V'(t) = &\int_0^t C_{st}(t,s)\left(\int_s^t g(x(u))du\right)^2 ds + C_t(t,0)\left(\int_0^t g(x(s))ds\right)^2 \\
&+ 2g(x)[a(t) - x(t)] \\
&\le 2g(x)a(t) - 2xg(x) \le -xg(x) + \int_0^{2|a(t)|} \phi^{-1}(s)ds.
\end{aligned}$$

The lower bound given in (2.1.32) may be derived as in Theorem 2.1.10. The final conclusion is now immediate.

The applied mathematician correctly claims that our conditions of a' bounded or in L^p may be difficult to establish because of uncertainties and even stochastic forces. There is a simple way around that if $\frac{d}{dx}g(x) =: g^*(x)$ is bounded. For a given function $b(t)$, seek a function $a(t)$ which satisfies one of our boundedness theorems with $|a(t) - b(t)|$ bounded. Here is a sample theorem. We take a simple condition known to imply that $a \in \mathcal{BC}$ implies that the solution of (2.1.24) is in $\mathcal{BC}$ when $g(x) = x$. Many other conditions are known.

Theorem 2.1.1.2. *Suppose that $|g^*(x)| \le 1$ and that there is an $\alpha < 1$ with $\int_0^t |C(t,s)|ds \le \alpha$. If $x(t) = a(t) - \int_0^t C(t,s)g(x(s))ds$ and $y(t) = b(t) - \int_0^t C(t,s)g(y(s))ds$ with $a - b \in BC$, so is $x - y$.*

Proof. Note that for fixed solutions x and y we have

$$\begin{aligned}
x(t) - y(t) &= a(t) - b(t) - \int_0^t C(t,s)[g(x(s)) - g(y(s))]ds \\
&= a(t) - b(t) - \int_0^t C(t,s)g^*(\xi(s))[x(s) - y(s)]ds
\end{aligned}$$

by the mean value theorem for derivatives where $\xi(s)$ is between $x(s)$ and $y(s)$. The resulting integral equation has a bounded solution.

Note the transition from Theorem 2.1.1.1 to Theorem 2.1.1.2. By contriving a Liapunov functional with higher powers of $g(x)$ we are able to pass from the requirement of $a' \in L^1$ (which allows only bounded $a(t)$) to $a' \in L^{2n}$ which allows $a(t) = (t+1)^\beta$ for $0 < \beta < 1$. It is an absolutely enormous advance, especially in view of Theorem 2.1.1.2.

We come now to a counterpart of the Adam and Eve theorem in which we need $|g(x)| \leq |x|$ and we ask that integration of the first coordinate of C be small, while the Adam and Eve result asked that integration of the second coordinate of C be small. The two results merge in the convolution case.

Theorem 2.1.1.3. *Let $a, g : \Re \to \Re$, $a(t) \in L^1[0,\infty)$, $|g(x)| \leq |x|$, with $a(t)$ and $g(x)$ continuous. Let $C : \Re \times \Re \to \Re$ be continuous, as is $\int_{t-s}^{\infty} |C(u+s,s)|du$. Suppose there is an $M < 1$ with $\int_0^\infty |C(u+t,t)|du \leq M$ and choose $k > 1$ with $Mk < 1$. For the equation*

$$x(t) = a(t) + \int_0^t C(t,s)g(x(s))ds \tag{2.1.33}$$

define

$$H(t) = k\int_0^t \int_{t-s}^{\infty} |C(u+s,s)|du|g(x(s))|ds.$$

Then there exists $\delta > 0$ such that

$$H'(t) \leq -\delta\left[|g(x(t))| + \int_0^t |C(t,s)g(x(s))|ds\right] + |a(t)| \tag{2.1.34}$$

so $|g(x)|$ and $|x|$ are $L^1[0,\infty)$.

Proof. We have

$$|g(x)| \leq |x| \leq |a(t)| + \int_0^t |C(t,s)g(x(s))|ds$$

and

$$\begin{aligned} H'(t) &= k\int_0^\infty |C(u+t,t)|du|g(x)| - k\int_0^t |C(t,s)g(x(s))|ds \\ &\leq kM|g(x(t))| - (k-1)\int_0^t |C(t,s)g(x(s))|ds + |a(t)| - |g(x(t))| \end{aligned}$$

from (2.1.33), so (2.1.34) holds. Moreover,

$$\delta|x| \leq \delta[|a(t)| + \int_0^t |C(t,s)g(x(s))|ds]$$

and so $H'(t) \leq -\delta|x| + (1+\delta)|a(t)|$ from which the conclusion follows.

We now give a corollary which relates this example to familiar aspects of Liapunov's direct method.

Corollary. *Let the assumptions of Theorem 2.1.1.3 be satisfied.*

(a) If $a(t)$ and $C(t,s)$ satisfy a local Lipschitz condition in t (uniform in s when $0 \le s \le t$) and if $C(t,s)$ is bounded for $0 \le s \le t < \infty$, then $x(t) \to 0$ as $t \to \infty$.

(b) If there is a continuous function $\Phi : [0,\infty) \to [0,\infty)$ with $|C(t,s)| \le \Phi(t-s)$ for $0 \le s \le t < \infty$ and if $a(t)$ and $\Phi(t) \to 0$ as $t \to \infty$, then $x(t) \to 0$ as $t \to \infty$.

(c) If there exists $\beta_1 > 0$ such that $|C(t,s)| \ge \beta_1 \int_{t-s}^{\infty} |C(u+s,s)|du$ for $t \ge 0$, then there exists $\gamma > 0$ with $H'(t) \le -\gamma H(t) + |a(t)|$.

(d) If there exists $\beta_2 > 0$ such that $|C(t,s)| \le \beta_2 \int_{t-s}^{\infty} |C(u+s,s)|du$ for $t \ge 0$, then along any solution $(k/\beta_2)\big[|x| - |a(t)|\big] \le H(t)$.

Proof. If the conditions in (a) hold and if $x(t) \not\to 0$, then there exists $\varepsilon > 0$ and $\{t_n\} \uparrow \infty$ with $|x(t_n)| \ge \varepsilon$. Let K be the Lipschitz constant for $a(t)$ and $C(t,s)$, $|C(t,s)| \le B$, and consider a sequence $\{s_n\}$ with $t_n \le s_n \le t_n + L$ where L is fixed, but yet to be determined. Then

$$\begin{aligned}
|x(t_n) - x(s_n)| &\le |a(t_n) - a(s_n)| \\
&\quad + \Big|\int_0^{t_n} C(t_n,s)g(x(s))ds - \int_0^{s_n} C(t_n,s)g(x(s))ds\Big| \\
&\quad + \Big|\int_0^{s_n} C(t_n,s)g(x(s))ds - \int_0^{s_n} C(s_n,s)g(x(s))ds\Big| \\
&\le K|t_n - s_n| + B\int_{t_n}^{s_n} |g(x(s))|ds + \int_0^{s_n} K|t_n - s_n|\,|g(x(s))|ds \\
&\le B\int_{t_n}^{s_n} |g(x(s))|ds + K|t_n - s_n|\left(1 + \int_0^{s_n} |g(x(s))|ds\right).
\end{aligned}$$

Now $g \in L^1$ implies that $\int_{t_n}^{s_n} |g(x(s))|ds =: \epsilon_n \to 0$ as $n \to \infty$. Also, there is a $D > 0$ such that $K(1 + \int_0^{s_n} |g(x(s)|ds) \le D$. Thus,

$$|x(t_n) - x(s_n)| \le B\varepsilon_n + D|t_n - s_n|.$$

Hence, there is an $N > 0$ and an $L > 0$ such that $n \ge N$ and $|t_n - s_n| \le L$ imply that $|x(t_n) - x(s_n)| < \varepsilon/2$; thus, $|x(t)| \ge \varepsilon/2$ for $t_n \le t \le t_n + L$, a contradiction to $x \in L^1[0,\infty)$. This proves (a).

To prove (b), note that

$$|x(t)| \le |a(t)| + \int_0^t \Phi(t-s)|x(s)|ds.$$

The integral is the convolution of an L^1-function ($|x(t)|$) with a function tending to zero; hence, the integral tends to zero.

To prove (c), we note from (2.1.34) that

$$\begin{aligned} H'(t) &\le -\delta\beta_1 \int_0^t \int_{t-s}^{\infty} |C(u+s,s)|du|g(x(s))|ds + |a(t)| \\ &\le -(\delta\beta_1/k)H(t) + |a(t)|, \end{aligned}$$

as required.

We prove (d) by noting that

$$\begin{aligned} H(t) &= k\int_0^t \int_{t-s}^{\infty} |C(u+s,s)|du|g(x(s))|ds \\ &\ge (k/\beta_2)\int_0^t |C(t,s)g(x(s))|ds \\ &\ge (k/\beta_2)\big[|x| - |a(t)|\big] \end{aligned}$$

as required.

Notes The material in Section 2.1 is taken from Burton (1996a) (2007c,d). The first mentioned reference was published by Walter de Gruyter in the *P*roceedings of the First World Congress of Nonlinear Analysts. The second reference is to the *C*arpathian Journal of Mathematics of the North University of Baia Mare, Romania. The third reference is to the *T*atra Mountains Mathematical Publications of the Slovak Academy of Sciences. It was presented at conferences in those three places.

2.2 The truncated equation

Recall that in Chapter 0 we pointed out that Barbashin and Krasovskii had constructed a large set of Liapunov functionals which are still being extended today for more sophisticated problems. In Section 2.1 we saw Liapunov functionals for integral equations with convex kernels. This section and the next will illustrate how those ideas can be extended. Solution spaces for truncated equations turn out to be far more complex than for the standard integral equation discussed in Section 2.1.

We consider the scalar equation

$$x(t) = a(t) - \int_{t-h}^{t} C(t,s)g(s,x(s))ds \tag{2.2.1}$$

where h is a positive constant,

$$|a(t)| < A \text{ for some } A > 0, \tag{2.2.2}$$

$$C(t,s) \geq 0, C_s(t,s) \geq 0, C_{st}(t,s) \leq 0, C(t,t-h) = 0, \tag{2.2.3}$$

and

$$xg(t,x) \geq 0. \tag{2.2.4}$$

Recall that in Section 1.4 we showed how truncated equations arise naturally from standard integral equations. Equation (2.2.1) can have three kinds of solutions. It might have a continuous solution on $(-\infty, \infty)$, possibly a periodic solution. It could have a continuous initial function $\phi : [t_0 - h, t_0] \to \Re$ and then have a solution $x(t)$ satisfying the equation for $t > t_0$ and $x(t) = \phi(t)$ on $[t_0 - h, t_0]$; if $\phi(t_0) = a(t_0) - \int_{t_0-h}^{t_0} C(t_0,s)g(s,\phi(s))ds$, then the solution is continuous, otherwise it is discontinuous. This is analogous to a functional differential equation having a solution with a cusp at t_0. It can be shown that for a given ϕ we may choose a different initial function arbitrarily close to ϕ so that the solution is continuous. See Burton and Furumochi (1994).

Theorem 2.2.1. *Let (2.2.3) and (2.2.4) hold and let x be any continuous solution of (2.2.1). Then the functional*

$$H(t) = \frac{1}{2}\int_{t-h}^{t} C_s(t,s)\left(\int_s^t g(v,x(v))dv\right)^2 ds \tag{2.2.5}$$

satisfies

$$(a(t) - x(t))^2 \leq 2H(t)C(t,t) \tag{2.2.6}$$

and

$$H'(t) \leq g(t,x(t))[a(t) - x(t)]. \tag{2.2.7}$$

Proof. We have

$$
\begin{aligned}
H'(t) = \quad & \frac{1}{2}\int_{t-h}^{t} C_{st}(t,s)\left(\int_s^t g(v,x(v))dv\right)^2 ds \\
& - \frac{1}{2}C_s(t,t-h)\left(\int_{t-h}^{t} g(v,x(v))dv\right)^2 \\
& + g(t,x(t))\int_{t-h}^{t} C_s(t,s)\int_s^t g(v,x(v))dv\, ds \\
& \le g(t,x(t))\int_{t-h}^{t} C_s(t,s)\int_s^t g(v,x(v))dvds.
\end{aligned}
$$

Integration of the last term by parts yields

$$
\begin{aligned}
g(t,x(t))&\left[C(t,s)\int_s^t g(v,x(v))dv\Big|_{s=t-h}^{s=t} + \int_{t-h}^{t} C(t,s)g(s,x(s))ds\right] \\
&= g(t,x(t))\int_{t-h}^{t} C(t,s)g(s,x(s))ds.
\end{aligned}
$$

Then from (2.2.1) we have

$$
\begin{aligned}
H'(t) &\le g(t,x(t))\int_{t-h}^{t} C(t,s)g(s,x(s))ds \\
&= g(t,x(t))[a(t)-x(t)].
\end{aligned}
$$

To obtain the lower bound on H we note that by the Schwarz inequality we have

$$
\begin{aligned}
&\left(\int_{t-h}^{t} C_s(t,s)\int_s^t g(v,x(v))dv\, ds\right)^2 \\
&\le \int_{t-h}^{t} C_s(t,s)ds\int_{t-h}^{t} C_s(t,s)\left(\int_s^t g(v,x(v))dv\right)^2 ds \\
&= 2H(t)C(t,t).
\end{aligned}
$$

Integration of the term on the left of this inequality by parts yields

$$
\begin{aligned}
&\left(C(t,s)\int_s^t g(v,x(v))dv\Big|_{s=t-h}^{s=t} + \int_{t-h}^{t} C(t,s)g(s,x(s))ds\right)^2 \\
&\qquad = (a(t)-x(t))^2,
\end{aligned}
$$

as required.

We are now going to pay dearly for the separation of $a(t)g(t,x)$. In Section 4.2 we will learn a simple way to avoid that cost under certain differentiability conditions. .

Theorem 2.2.2. *Let (2.2.3) and (2.2.4) hold, let $|g(t,x)| \leq |x|$, and let $C(t,t)$ be bounded.*

(i) If (2.2.2) holds then any continuous solution $x(t)$ of (2.2.1) is bounded and satisfies

$$(x(t) - a(t))^2 \leq h^2 A^2 C(t,t). \tag{2.2.8}$$

(ii) If $a \in L^2[0,\infty)$, then $g(t,x(t)) \to a(t)$ in $L^2[0,\infty)$.

Proof. From (2.2.7) we have

$$2H'(t) \leq 2g(t,x(t))[a(t)-g(t,x(t))] = -(a(t)-g(t,x(t)))^2 - g^2(t,x(t)) + a^2(t).$$

Part (ii) follows from this.

If (2.2.2) holds and if $x(t)$ is not bounded then from (2.2.6) and $C(t,t)$ bounded we have $H(t)$ unbounded. Thus, there is a sequence $\{t_n\} \uparrow \infty$ with $H(t_n) \geq H(s)$ for $t_n - h \leq s \leq t_n$. For these s we have

$$0 \leq 2(H(t_n) - H(s)) \leq -\int_s^{t_n} g^2(u,x(u))du + \int_s^{t_n} a^2(u)du$$

or

$$\int_s^{t_n} g^2(u,x(u))du \leq hA^2.$$

By the Schwarz inequality

$$\begin{aligned} 2H(t_n) &\leq \int_{t-h}^t C_s(t,s)(t-s)\int_s^t g^2(v,x(v))dvds \\ &\leq h^2A^2\int_{t-h}^t C_s(t,s)ds = h^2A^2C(t,s)\Big|_{t-h}^t \leq h^2A^2C(t,t) \end{aligned}$$

which is bounded, a contradiction. As $H(t)C(t,t)$ is bounded and a is bounded, from (2.2.6) we have x bounded so (i) is true.

Very often in problems in biology the functions are periodic because of life cycles or seasons changing. One of the most sought after properties is the existence of periodic solutions. It turns out that (2.2.8) almost gives us that free using Schaefer's fixed point theorem which was stated in Theorem 1.3.1.2.

For (2.2.1) we take $(X, \|\cdot\|)$ to be the Banach space of continuous $T-$periodic functions with the supremum norm and we suppose that for this period $T > 0$ we have

$$C(t+T, s+T) = C(t, s),\ a(t+T) = a(t),\ g(t+T, x) = g(t, x). \quad (2.2.9)$$

Define P by $\phi \in X$ implies

$$(P\phi)(t) = a(t) - \int_{t-h}^{t} C(t, s)g(s, \phi(s))ds$$

and readily show that $(P\phi)(t+T) = (P\phi)(t)$ so that $P : X \to X$.

We now show that P is continuous. Let $\phi \in X$ be fixed and let $Z = \{\psi \in X | \|\phi - \psi\| \leq 1\}$. Then

$$|(P\phi)(t) - (P\psi)(t)| \leq \int_{t-h}^{t} C(t, s)|g(s, \phi(s)) - g(s, \psi(s))|ds.$$

As $\int_{t-h}^{t} C(t, s)ds$ is periodic and continuous, it is bounded by some number L. Also, g is uniformly continous on $[0, T] \times Z$ and, as $g(t+T, x) = g(t, x)$, it is uniformly continuous on $\Re \times Z$. For a given $\epsilon > 0$ there is a $\delta > 0$ such that $(t, \psi) \in \Re \times Z$ implies that

$$|g(t, \phi(t)) - g(t, \psi(t))| < \epsilon/L.$$

This establishes the continuity.

Note that if Y is a fixed bounded subset of X and if ϕ is an arbitrary element of Y then

$$\frac{d}{dt}\int_{t-h}^{t} C(t, s)g(s, \phi(s))ds = C(t, t)g(t, \phi(t)) + \int_{t-h}^{t} C_t(t, s)g(s, \phi(s))ds$$

since $C(t, t-h) = 0$. This will be bounded over Y if

$$\int_{t-h}^{t} |C_t(t, s)|ds + C(t, t) \text{ is bounded.} \quad (2.2.10)$$

As a is uniformly continuous, Y is mapped into an equi-continuous set. This establishes the required compactness.

Theorem 2.2.3. *Let the conditions of Theorem 2.2.2 hold, as well as (2.2.9) and (2.2.10). Then (2.2.1) has a $T-$periodic solution.*

Proof. All of the conditions of Schaefer's theorem will be satisfied if we can show that there is a number M such that any solution of $x = \lambda Px$ for $0 < \lambda < 1$ satisfies $\|x\| \leq M$. But all the work of Theorem 2.2.2 holds for the equation

$$x(t) = \lambda[a(t) - \int_{t-h}^{t} C(t,s)g(s,x(s))ds]$$

and by (2.2.8) we have the required bound.

Notes The material in Section 2.2 was taken from Burton (2010c). It was first published in *T*rudy Instituta Matematiki i Mekhaniki and was presented at a conference in 2009 in Ekaterinburg, Russia called Internal Conference "Actual Problems of Stability and Control Theory". Truncated equations will be studied under much more general conditions in Chapter 4.

2.3 Infinite delay

Consider the equation

$$x(t) = a(t) - \int_{-\infty}^{t} C(t,s)g(s,x(s))ds \tag{2.3.1}$$

with $g(t, x(t))$ bounded if $x(t)$ is bounded,

$$xg(t,x) \geq 0, \tag{2.3.2}$$

$$C(t,s) \geq 0, C_s(t,s) \geq 0, C_{st}(t,s) \leq 0, C(t,t) \leq B, \tag{2.3.3}$$

$$\lim_{s\to-\infty}(t-s)C(t,s) = 0 \text{ for each fixed } t, \tag{2.3.4}$$

and

$$\int_{-\infty}^{t} [C(t,s) + \big(C_s(t,s) - C_{st}(t,s)\big)(t-s)^2]ds \text{ is continuous }. \tag{2.3.5}$$

Equation (2.3.1) can have an initial function $\phi : (-\infty, t_0] \to R$; in that case, in order for the solution to be continuous at t_0 it is required that $\phi(t_0) = a(t_0) - \int_{-\infty}^{t_0} C(t_0,s)g(s,\phi(s))ds$. If ϕ is bounded and continuous, then (2.3.5) implies that this integral exists. Also, (2.3.1) may have a solution ψ on all of R; in that case, ψ is its own initial function on any interval $(-\infty, t_0]$.

Theorem 2.3.1. *Let (2.3.2) - (2.3.5) hold. If $x(t)$ is a continuous solution of (2.3.1) on an interval $[t_0, \infty)$ with bounded continuous initial function ϕ, then for*

$$H(t) = \int_{-\infty}^{t} C_s(t,s)\left(\int_s^t g(v,x(v))dv\right)^2 ds \tag{2.3.6}$$

we have

$$H'(t) \le 2g(t,x(t))[a(t) - x(t)]$$

and

$$(a(t) - x(t))^2 \le C(t,t)H(t). \tag{2.3.8}$$

Proof. A computation yields

$$\begin{aligned}
H'(t) =& \int_{-\infty}^{t} C_{st}(t,s)\left(\int_s^t g(v.x(v))dv\right)^2 ds \\
&+ 2g(t,x(t))\int_{-\infty}^{t} C_s(t,s)\int_s^t g(v,x(v))dvds \\
\le& 2g(t,x(t))\left[C(t,s)\int_s^t g(v,x(v))dv\,\Big|_{s=-\infty}^{s=t}\right. \\
&\left.+\int_{-\infty}^{t} C(t,s)g(s,x(s))ds\right] \\
=& 2g(t,x(t))[a(t) - x(t)]
\end{aligned}$$

where we have used the boundedness of the initial function (hence of $g(t,\phi(t))$) and (2.3.4) to conclude that the next to the last integral is zero. The existence of the first integrals shown in H' follows from (2.3.5).

For the lower bound on H we set

$$\begin{aligned}
Q(t) :=& \left(\int_{-\infty}^{t} C_s(t,s)\int_s^t g(v,x(v))dvds\right)^2 \\
\le& \left(\int_{-\infty}^{t} C_s(t,s)ds\int_{-\infty}^{t} C_s(t,s)\left(\int_s^t g(v,x(v))dv\right)^2 ds\right) \\
=& C(t,t)H(t).
\end{aligned}$$

where we have taken $\lim_{s\to-\infty} C(t,s) = 0$ using (2.3.4). On the other hand,

$$\begin{aligned} Q(t) &= \left(C(t,s) \int_s^t g(v,x(v))dv \Big|_{s=-\infty}^{s=t} + \int_{-\infty}^t C(t,s)g(s,x(s))ds \right)^2 \\ &= (a(t) - x(t))^2. \end{aligned}$$

Theorem 2.3.2. *Let the conditions of Theorem 2.3.1 hold, let $|g(t,x)| \leq |x|$, let $a \in L^2[0,\infty)$, and let $x(t)$ be any continuous solution of (2.3.1) with bounded continuous initial function. Then $g(t,x(t)) \to a(t)$ in $L^2[0,\infty)$, $g(t,x(t)) \in L^2[0,\infty)$, while $a(t)$ and $C(t,t)$ bounded yields $x(t)$ bounded.*

Proof. We already have

$$\begin{aligned} H'(t) &\leq 2g(t,x(t))a(t) - 2g(t,x(t))x(t) \\ &\leq 2g(t,x(t))a(t) - 2g^2(t,x(t)) \\ &= -(g(t,x(t)) - a(t))^2 - g^2(t,x(t)) + a^2(t). \end{aligned}$$

As $H(t) \geq 0$, an integration yields the L^2 properties and that $H(t)$ is bounded. The rest of the boundedness follows from (2.3.8).

Recall from Section 1.4 that (2.3.1) could be a form of (2.2.1) if C is of a special type. But in this more general case there is much more to be proved for periodic solutions and that will be seen in Chapter 4.

Note The material in Section 2.3 was taken from Burton (2010c) with details given at the end of Section 2.2.

2.4 Liapunov theory for a neutral equation

We have given examples of Liapunov functionals for a variety of common problems. Now we sketch the details of Liapunov theory for a neutral integral equation with roots in mathematical biology. A basic logistic equation can be written as

$$x' = ax - bx^2$$

where x represents a density. It has many different derivations but a common one starts with the assumption of Malthusian growth and then decay according to the law of mass action. The problem evolves into a complex equation of the form

$$x'(t) = ax(t) + \alpha x'(t-h) - q(t,x(t),x(t-h)) \tag{2.4.1}$$

where $a > 0$, $0 \leq |\alpha| < 1$, $h > 0$, all are constant. Such equations have been studied with varying degrees of vigor since the middle of the last century (cf. Driver (1965), Gopalsamy (1992), Gopalsamy and Zhang (1988), El'sgol'ts (1966), Kuang (1993a,b), (1991)). Recently investigators have given heuristic arguments to support their use in describing biological phenomena and much of this is formalized in the final chapter of each of the books by Gopalsamy (1992) and Kuang (1993).

We are interested in a solution for $t \geq 0$, possibly arising from a given initial function, or for a solution on $(-\infty, \infty)$, possibly a periodic solution. This can be inverted as an integral equation as follows.

Write (2.4.1) as

$$(x(t) - \alpha x(t-h))' = a(x - \alpha x(t-h)) + a\alpha x(t-h) - q(t, x(t), x(t-h)),$$

multiply by e^{-at}, and group terms as

$$\left[(x(t) - \alpha x(t-h))e^{-at}\right]' = \left[a\alpha x(t-h) - q(t, x, x(t-h))\right]e^{-at}.$$

We search for a solution having the property that

$$(x(t) - \alpha x(t-h))e^{-at} \to 0 \text{ as } t \to \infty$$

so that an integration from t to infinity yields

$$-(x(t) - \alpha x(t-h))e^{-at} = \int_t^\infty \left[a\alpha x(s-h) - q(s, x(s), x(s-h))\right]e^{-as}ds$$

and, finally,

$$x(t) = \alpha x(t-h) + \int_t^\infty \left[q(s, x(s), x(s-h)) - a\alpha x(s-h)\right]e^{a(t-s)}ds.$$

A general form for such equations is

$$x(t) = f(x(t-h)) + \int_t^\infty Q(s, x(s), x(s-h))C(t-s)ds + p(t). \quad (2.4.2)$$

We call this a neutral delay integral equation of advanced type and it is a very interesting equation. It may have a solution on all of $\Re$ or it may have a solution on $[0, \infty)$ generated by an initial function φ on $[-h, 0]$. In the latter case, notice that we can not obtain a local solution: we must get the full solution on $[0, \infty)$. Thus, we will need to employ a fixed point theorem to get existence and that means that we will get a fixed point in the solution space; hence, we must know in advance the form of the solution space. We study this problem in Burton (1998b) and find that the solution will have discontinuities at $t = nh$, but the jumps will tend to zero as $n \to \infty$. (This is parallel to solutions of functional differential equations smoothing (cf. El'sgol'ts (1966)).) Equally important is the need to know in advance the growth of the solution so that functions in the solution space will have a weighted norm allowing such growth.

Sometimes we let $C = C(t,s)$. Here, the forcing function $p(t)$ can be critical. If the equation has an equilibrium point without $p(t)$, then the subsequent struggle to prove the existence of a bounded or periodic solution might simply yield that obvious equilibrium point.

We will ask that

$$|f(x) - f(y)| \leq \alpha|x - y|, \quad 0 \leq \alpha < 1, \tag{2.4.3}$$

and for some fixed k with $0 \leq k \leq 1$

$$\begin{aligned} &|Q(t,x,y) - Q(t,w,z)| \\ &\qquad \leq (k|x - w| + (1-k)|y - z|), \end{aligned} \tag{2.4.4}$$

$$|Q(t,0,0)| \leq 1, \tag{2.4.5}$$

Q, p, and C are continuous,

$$\int_0^\infty |C(-u)|du =: C_0 < \infty. \tag{2.4.6}$$

In keeping with a fundamental observation of Krasnoselskii, note that it may generate the sum of a contraction and compact map.

While the integral from t to ∞ is not common, it can be found in many places. Coddington and Levinson (1955; p. 331) write an ordinary differential equation in this way (without h) when studying an unstable manifold. Also, investigators have long written differential equations

$$x' = F(t,x)$$

as

$$x(t) = x(t_0) + \int_{t_0}^t F(s,x(s))ds.$$

Then, if it can be determined that every solution converges to zero as $t \to \infty$, it is permissible to bypass $x(t_0)$ and write

$$x(t) = -\int_t^\infty F(s,x(s))ds.$$

While we see many equations of the form of (2.4.2) (without h), such equations usually are studied only after existence of solutions has been established. That is a central question here.

For this equation, we must immediately establish existence on $[t_0, \infty)$. And because of this, the central problem for (2.4.2) is to determine the space in which solutions reside. Once that is done, the actual proof of existence follows from application of a fixed point theorem. Moreover, unlike the case just mentioned of a differential equation, the existence of a solution of (2.4.2) in a given space endows that solution with the properties of that space. Thus, proof of existence yields important qualitative properties as well.

Equation (2.4.2) holds many surprises. For functional differential equations of neutral type, it may be deduced from Driver (1965) that repeated discontinuities of a solution on $[t_0, \infty)$ must be expected as a result of a given initial function. We note that this is true for (2.4.2); but we also show that the magnitude of the jumps in the discontinuities tends to zero as $t \to \infty$. This is parallel to an interesting phenomenon for delay-differential equations. El'sgol'ts (1966; p. 7) discusses solutions of

$$x' = H(t, x(t), x(t-h))$$

in which H is in C^∞ and there is an initial function $\bar{\phi} : [-h, 0] \to R$ yielding a solution $x(t, 0, \bar{\phi})$ on $[0, \infty)$. He notes that $x^{(k)}(t, 0, \bar{\phi})$ may have a discontinuity at $(k-1)h$, but will be continuous for $t > kh$; the solution smooths with time.

Theorem 2.4.1. *Let p be bounded on $\Re$, let (2.4.3)–(2.4.6) hold, and suppose that*

$$\mu := \alpha + C_0 < 1, \tag{2.4.7}$$

where α and C_0 are in (2.4.3) and (2.4.6). Then (2.4.2) has a unique bounded continuous solution on $-\infty < t < \infty$.

Proof. Let $(\mathcal{B}, \|\cdot\|)$ be the Banach space of bounded continuous functions $\phi : \Re \to \Re$ with the supremum norm. Define $P : \mathcal{B} \to \mathcal{B}$ by $\phi \in \mathcal{B}$ implies that

$$(P\phi)(t) = f(\phi(t-h)) + \int_t^\infty Q(s, \phi(s), \phi(s-h))C(t-s)ds + p(t). \tag{2.4.8}$$

As ϕ is bounded and continuous, so is $Q(t, \phi(t), \phi(t-h))$ using (2.4.4)–(2.4.6); since C is $L^1[0, \infty)$ by (2.4.6) it follows from a theorem of Hewitt and Stromberg (1971; p. 398) that the integral is uniformly continuous. That integral is bounded because of (2.4.6). Hence, $P\phi$ is bounded and continuous.

Next, if $\phi, \psi \in \mathcal{B}$ then

$$\begin{aligned}
&|(P\phi)(t) - (P\psi)(t)| \leq \alpha\|\phi - \psi\| \\
&+ \int_t^\infty [k|\phi(s) - \psi(s)| + (1-k)|\phi(s-h) - \psi(s-h)|]|C(t-s)|ds \\
&\qquad \leq \alpha\|\phi - \psi\| + \|\phi - \psi\| \int_t^\infty |C(t-s)|ds \\
&\qquad = \mu\|\phi - \psi\|.
\end{aligned}$$

As $\mu < 1$, P is a contraction with unique fixed point. This completes the proof.

Corollary 2.4.1. *If the conditions of Theorem 2.4.1 hold and if there is a $T > 0$ such that $Q(t,x,y) = Q(t+T,x,y)$ and $p(t+T) = p(t)$, then (2.4.2) has a unique T-periodic solution.*

Proof. A change of variable shows that if $(\mathcal{B}, \|\cdot\|)$ is the Banach space of continuous T-periodic functions with the supremum norm, always denoted by

$$(\mathcal{P}_T, \|\cdot\|),$$

then $\phi \in \mathcal{P}_T$ implies that $P\phi \in \mathcal{P}_T$. Thus, the fixed point is in $\mathcal{P}_T$.

We are now going to prepare (2.4.2) for a Liapunov functional and a proof that there is a periodic solution under assumptions that allow for a large kernel. The full details are found in Burton (1998b). What we provide here is a sketch showing the Liapunov functional, the derivative along solutions, and the use of that in establishing *a priori* bounds. Specialize (2.4.2) to the equation

$$x(t) = f(x(t-h)) + \int_t^\infty [g(x(s)) + r(x(s-h))]C(t,s)ds + p(t) \quad (2.4.9)$$

in which

$$|f(x) - f(y)| \leq \alpha|x-y|,\ 0 \leq \alpha < 1,\ xg(x) > 0 \text{ if } x \neq 0, \quad (2.4.10)$$

$$g, r, C, C_s, C_{st}, p \text{ are continuous},\ p(t+T) = p(t), \quad (2.4.11)$$
$$C(t+T, s+T) = C(t,s),$$

and

$$\sup_{t\ge 0}\left[\int_t^\infty |C(t,s)|ds + \int_t^\infty (|C_s(t,s)| + |C_{st}(t,s)|)[1+(t-s)^2]ds\right] < \infty, \tag{2.4.12}$$
$$|C(t,s)||t-s| \to 0 \text{ as } s \to \infty.$$

In addition to (2.4.9) we consider the equation

$$x(t) = \lambda\left[f(x(t-h)/\lambda) + \int_t^\infty \big[g(x(s)) + r(x(s-h))\big]C(t,s)ds + p(t)\right],$$
$$0 < \lambda \le 1. \tag{2.4.13}$$

We will obtain an *a priori* bound on all T-periodic solutions of (2.4.13) for $0 < \lambda \le 1$. An extension of Schaefer's theorem can then be used to prove that (2.4.13) has a T-periodic solution.

Lemma 2.4.1. *If $x(t)$ is a continuous T-periodic solution of (2.4.13) and if V is defined by*

$$V(t) = \int_t^\infty \lambda^2 C_s(t,s)\left(\int_s^t [g(x(u)) + r(x(u-h))]du\right)^2 ds, \tag{2.4.14}$$

then

$$V'(t) = \lambda^2 \int_t^\infty C_{st}(t,s)\left(\int_s^t \big[g(x(u)) + r(x(u-h))\big]du\right)^2 ds$$
$$+ 2\lambda\big(g(x) + r(x(t-h))\big)(x(t) - \lambda f(x(t-h)/\lambda) - \lambda p(t)). \tag{2.4.15}$$

Proof. The first term is clear. In addition we have

$$2\lambda^2\big[g(x) + r(x(t-h))\big]\int_t^\infty C_s(t,s)\int_s^t \big[g(x(u)) + r(x(u-h))\big]du\, ds$$
$$= 2\lambda^2\big[g(x) + r(x(t-h))\big]\left\{C(t,s)\int_s^t \big[g(x(u)) + r(x(u-h))\big]du\Big|_t^\infty + \int_t^\infty C(t,s)\big[g(x(s)) + r(x(s-h))\big]ds\right\}$$
$$= 2\lambda\big[g(x) + r(x(t-h))\big][x - \lambda f(x(t-h)/\lambda) - \lambda p(t)],$$

as required.

No further details will be given here. In the stated reference, we use V' alone (not V) to obtain *a priori* bounds on solutions of (2.4.13). The goal is to show that a modification of V, say W, satisfies

$$W'(t) \leq \lambda(-K|xg(x)| + M), \text{ some } K > 0, \ M > 0,$$

or

$$W'(t) \geq \lambda(K|xg(x)| - M).$$

As $x \in \mathcal{P}_T$ implies that $W \in \mathcal{P}_T$ we obtain $W(T) = W(0)$ and so an integration of *either* inequality yields

$$\int_0^T |x(t)g(x(t))|dt \leq M/K.$$

This inequality is then parlayed into an *a priori* bound on the supremum of x using (2.4.13), (2.4.12), and a Schwarz inequality. A fixed point theorem then yields a T-periodic solution of (2.4.9). The details are long and very intricate. Our purpose here is to derive the nonstandard integral equation, construct a Liapunov functional for it, and indicate how it is used.

Notes The material in Section 2.4 was taken from Burton (1998b). It was first published by Elsevier in *N*onlinear Analysis: TMA with all details given in the bibliography.

2.5 Singular & fractional kernels

In this section we will construct Liapunov functionals for variations of the equation

$$x(t) = a(t) - \int_0^t C(t,s)g(s,x(s))ds. \tag{2.5.1}$$

The reader is reminded of the comment with Equation (1.2.1N) found in Chapter 1, Section 2 just before Lemma 1.2.1. We remarked there that the existence theory was readily advanced to this equation if g satisfies a Lipschitz condition. In fact, our subsequent condition (2.5.7) will demand that g be bounded by a linear function. Moreover, there is extensive existence theory available in standard texts and it would be a great distraction to continue our already lengthy discussion of such matters. Instead, our theorems will state that any solution which exists on $[0, \infty)$ will satisfy the conclusion of the theorem.

There are five main conditions which we require and we will discuss them under the heading of **critique** later. First, for each continuous function x, then $\int_0^t C(t,s)g(s,x(s))ds$ exists. Next, for each small $\epsilon > 0$ then

$$C(t,s) \geq 0, C_s(t,s) \geq 0, C_{st}(t,s) \leq 0, C_t(t,s) \leq 0 \tag{2.5.2}$$

provided that

$$0 \leq s \leq t-\epsilon,\ t < \infty; \tag{2.5.3}$$

thus, $t - s \geq \epsilon$. The kernel $(t-s)^{-p}$ for $0 < p < 1$ does satisfy that condition. The final three conditions are (2.5.7), (2.5.8), and (2.5.9).

Consider the scalar equation (2.5.1) with (2.5.2) and (2.5.3) satisfied and let both $a(t)$ and $g(t,x)$ be continuous for $t \geq 0$ and $x \in \Re$, while

$$xg(t,x) > 0 \text{ if } x \neq 0. \tag{2.5.4}$$

For an $\epsilon > 0$ and for $t \geq \epsilon$ we define a Liapunov functional by

$$V(t,\epsilon) = \int_0^{t-\epsilon} C_s(t,s)\Big(\int_s^t g(u,x(u))du\Big)^2 ds + C(t,0)\Big(\int_0^t g(u,x(u))du\Big)^2. \tag{2.5.5}$$

While $C_s(t,s)$ may be badly behaved at $s=t$, we have $s \leq t-\epsilon$ or $\epsilon \leq t-s$ so the bad point is always avoided in $V(t,\epsilon)$.

Theorem 2.5.1. *Let x be a continuous solution of (2.5.1) on $[0,\infty)$ and let (2.5.2) and (2.5.3) be satisfied. If $\epsilon > 0$ is chosen and if $V(t,\epsilon)$ is defined in (2.5.5) then for $t \geq \epsilon$ we have*

$$\begin{aligned}\frac{dV(t,\epsilon)}{dt} &\leq 2g(t,x(t))\Big[a(t) - x(t) + C(t,t-\epsilon)\int_{t-\epsilon}^t g(u,x(u))du \\ &- \int_{t-\epsilon}^t C(t,s)g(s,x(s))ds\Big] + C_s(t,t-\epsilon)\Big(\int_{t-\epsilon}^t g(u,x(u))du\Big)^2.\end{aligned} \tag{2.5.6}$$

Proof. For $t \geq \epsilon$ we have $C_t(t,0) \leq 0$ and $C_{st}(t,s) \leq 0$ when $0 \leq s \leq t-\epsilon$ so by Leibnitz's rule we have

$$\begin{aligned}V'(t,\epsilon) &\leq C_s(t,t-\epsilon)\Big(\int_{t-\epsilon}^t g(u,x(u))du\Big)^2 \\ &\quad + 2g(t,x(t))\int_0^{t-\epsilon} C_s(t,s)\int_s^t g(u,x(u))duds\end{aligned}$$

$$+ 2g(t,x(t))C(t,0)\int_0^t g(u,x(u))du.$$

Integration of the next-to-last term by parts yields

$$\begin{aligned}
&2g(t,x(t))\int_0^{t-\epsilon} C_s(t,s)\int_s^t g(u,x(u))duds \\
&= 2g(t,x(t))\left[C(t,s)\int_s^t g(u,x(u))du\Big|_0^{t-\epsilon} + \int_0^{t-\epsilon} C(t,s)g(s,x(s))ds\right] \\
&= 2g(t,x(t))\left[C(t,t-\epsilon)\int_{t-\epsilon}^t g(u,x(u))du - C(t,0)\int_0^t g(u,x(u))du\right. \\
&\left. + \int_0^{t-\epsilon} C(t,s)g(s,x(s))ds\right].
\end{aligned}$$

Thus,

$$\begin{aligned}
&V'(t,\epsilon) \le C_s(t,t-\epsilon)\left(\int_{t-\epsilon}^t g(u,x(u))du\right)^2 \\
&+ 2g(t,x(t))\left[C(t,t-\epsilon)\int_{t-\epsilon}^t g(u,x(u))du + \int_0^{t-\epsilon} C(t,s)g(s,x(s))ds\right] \\
&= C_s(t,t-\epsilon)\left(\int_{t-\epsilon}^t g(u,x(u))du\right)^2 \\
&+ 2g(t,x(t))\left[C(t,t-\epsilon)\int_{t-\epsilon}^t g(u,x(u))du - \int_{t-\epsilon}^t C(t,s)g(s,x(s))ds\right] \\
&+ 2g(t,x(t))[a(t)-x(t)],
\end{aligned}$$

as required. We have used the integral equation (2.5.1) in the last step.

Three relations will be needed for us to parlay this Liapunov functional derivative into a qualitative result for a solution of (2.5.1). First, we must be able to estimate the relation between $xg(t,x)$ and $g^2(t,x)$. Our conditions (2.5.8) and (2.5.9) below will allow such strongly singular kernels that it is not a great surprise to need $g(t,x)$ bounded by a linear function. In fact, the condition is needed mainly in order to separate $a(t)g(t,x)$. We ask that

$$xg(t,x) \ge g^2(t,x). \tag{2.5.7}$$

Next, we need some control over the magnitude of the singularity and this turns out to be enlightening. In the simplest case it asks for a locally L^1 kernel.

We suppose that there are positive constants α and β with $\alpha + \beta < 1$ so that there is an $\epsilon > 0$ with

$$\int_s^{s+\epsilon} [\epsilon C_s(u, u-\epsilon) + C(u, u-\epsilon) + |C(u,s)|]du < \alpha \tag{2.5.8}$$

for $0 \leq s < \infty$ and that

$$C(t, t-\epsilon)\epsilon + \int_{t-\epsilon}^{t} |C(t,s)|ds < \beta \tag{2.5.9}$$

for $\epsilon \leq t < \infty$. Note in (2.5.2) and (2.5.3) that we do not specify the sign of $C(s,s)$ so the absolute value is needed in (2.5.8).

Remark We will use the $\alpha+\beta < 1$ relation in obtaining the second line of (2.5.10). In fact, we can get by with $\alpha + \beta < 2$, as we actually do using ξ and M in the proof of Theorem 2.5.7. This, however, will complicate the relation $a - g \in L^2$ which we want to preserve. Theorem 2.5.7 will, thereby, contain this possible improvement of Theorem 2.5.2 by taking $D = 0$.

Critique

Conditions (2.5.8) and (2.5.9) allow for gross singularities to occur at $s = t$. We have already noticed that $C_s(u, u-\epsilon)$ occurring in the derivative of V is always well away from the singularity and (2.5.2) holds for it. While (2.5.8) and (2.5.9) occur as technical necessities in a later computation, repeated again and again in this work here, one would really like at least a theoretical rational for them. To begin with, for weak singularities such as $C(t,s) = [t-s]^{-p}$ for $0 < p < 1$, those conditions are too lenient to make any sense; for in that case $\alpha + \beta \to 0$ as $\epsilon \to 0$.

Liapunov growth rate

The next two results will yield a relation on the growth rate of the solution depending on the growth rate of $a(t)$. It will have special application to fractional differential equations in the next subsection.

Theorem 2.5.2. *Let x be a continuous solution of (2.5.1) on $[0, \infty)$ and let (2.5.2), (2.5.3), (2.5.7)–(2.5.9) hold. If, in addition, $a \in L^2[0, \infty)$ so are $g(t, x(t))$ and $g(t, x(t)) - a(t)$.*

Proof. We begin by organizing the derivative of V which we computed in (2.5.6). First, by the Schwarz inequality we have

$$C_s(t,t-\epsilon)\left(\int_{t-\epsilon}^{t} g(u,x(u))du\right)^2 \leq \epsilon C_s(t,t-\epsilon)\int_{t-\epsilon}^{t} g^2(u,x(u))du.$$

Next,

$$\begin{aligned}|2g(t,x(t))C(t,t-\epsilon)&\int_{t-\epsilon}^{t} g(u,x(u))du| \\ &\leq C(t,t-\epsilon)\int_{t-\epsilon}^{t}[g^2(t,x(t))+g^2(u,x(u))]du\end{aligned}$$

and

$$|2g(t,x(t))\int_{t-\epsilon}^{t} C(t,s)g(s,x(s))ds| \leq \int_{t-\epsilon}^{t} C(t,s)[g^2(t,x(t))+g^2(s,x(s))]ds.$$

These three relations in (2.5.6) yield

$$\begin{aligned}V'(t,\epsilon) &\leq 2g(t,x(t))[a(t)-x(t)] \\ &\quad + C(t,t-\epsilon)\epsilon g^2(t,x(t)) + g^2(t,x(t))\int_{t-\epsilon}^{t} C(t,s)ds \\ &\quad + \int_{t-\epsilon}^{t}[\epsilon C_s(t,t-\epsilon)+C(t,t-\epsilon)+C(t,s)]g^2(s,x(s))ds.\end{aligned}$$

By (2.5.7) we obtain

$$\begin{aligned}2g(t,x(t))[a(t)-x(t)] &\leq 2g(t,x(t))a(t) - 2g^2(t,x(t)) \\ &= -g^2(t,x(t)) - (g(t,x(t))-a(t))^2 + a^2(t). \qquad (2.5.10)\end{aligned}$$

Invoke (2.5.8) and (2.5.9) to find ϵ, α, and β. Using (2.5.9) we have

$$\begin{aligned}V'(t,\epsilon) &\leq a^2(t) - (g(t,x(t))-a(t))^2 - (1-\beta)g^2(t,x(t)) \\ &\quad + \int_{t-\epsilon}^{t}[\epsilon C_s(t,t-\epsilon)+C(t,t-\epsilon)+C(t,s)]g^2(s,x(s))ds. \qquad (2.5.11)\end{aligned}$$

Take $|C(t,s)|$ in this expression, integrate from ϵ to t, and work with the last term, interchanging the order of integration. We have

$$\int_{\epsilon}^{t}\int_{u-\epsilon}^{u}[\epsilon C_s(u,u-\epsilon)+C(u,u-\epsilon)+|C(u,s)|]g^2(s,x(s))dsdu$$

$$\leq \int_0^t \int_s^{s+\epsilon} [\epsilon C_s(u,u-\epsilon) + C(u,u-\epsilon) + |C(u,s)|]du g^2(s,x(s))ds$$
$$\leq \alpha \int_0^t g^2(s,x(s))ds$$

using (2.5.8). This now yields

$$V(t,\epsilon) \leq V(\epsilon,\epsilon) + \int_\epsilon^t a^2(u)du - \int_\epsilon^t (g(s,x(s)) - a(s))^2 ds$$
$$-(1-\alpha-\beta)\int_\epsilon^t g^2(u,x(u))du + \alpha\int_0^\epsilon g^2(u,x(u))du.$$

We assumed $x(t)$ exists so $V(\epsilon,\epsilon)$ is finite, while $V(t,\epsilon) \geq 0$. Put the negative terms on the left to finish the proof.

Theorem 2.5.3. *If x is a continuous solution of (2.5.1) on $[0,\infty)$, then for $t \geq \epsilon$*

$$(x(t) - a(t))^2 \leq 6\epsilon C^2(t,t-\epsilon)\int_{t-\epsilon}^t g^2(u,x(u))du$$
$$+ 6C(t,t-\epsilon)V(t,\epsilon) + 2\left(\int_{t-\epsilon}^t C(t,s)g(s,x(s))ds\right)^2.$$

Proof. In the calculation below, it will save much work to set

$$H := 2\left(\int_{t-\epsilon}^t C(t,s)g(s,x(s))ds\right)^2.$$

Starting with (2.5.1) we have

$$(x(t) - a(t))^2 = \left(\int_0^t C(t,s)g(s,x(s))ds\right)^2$$
$$\leq 2\left(\int_0^{t-\epsilon} C(t,s)g(s,x(s))ds\right)^2 + H$$
$$= 2\Big(-C(t,s)\int_s^t g(u,x(u))du\Big|_0^{t-\epsilon}$$
$$+\int_0^{t-\epsilon} C_s(t,s)\int_s^t g(u,x(u))duds\Big)^2 + H$$
$$= 2\Big(-C(t,t-\epsilon)\int_{t-\epsilon}^t g(u,x(u))du + C(t,0)\int_0^t g(u,x(u))du$$

$$
\begin{aligned}
&+ \int_0^{t-\epsilon} C_s(t,s) \int_s^t g(u,x(u))duds\Big)^2 + H \\
&\leq 6\Big(- C(t,t-\epsilon) \int_{t-\epsilon}^t g(u,x(u))du\Big)^2 \\
&+ 6\Big(C(t,0) \int_0^t g(u,x(u))du\Big)^2 \\
&+ 6 \int_0^{t-\epsilon} C_s(t,s)ds \int_0^{t-\epsilon} C_s(t,s)\Big(\int_s^t g(u,x(u))du\Big)^2 ds + H \\
&\leq 6\Big(- C(t,t-\epsilon) \int_{t-\epsilon}^t g(u,x(u))du\Big)^2 \\
&+ 6\Big[C(t,0) + \int_0^{t-\epsilon} C_s(t,s)ds\Big] V(t,\epsilon) + H. \\
&= 6C^2(t,t-\epsilon)\epsilon \int_{t-\epsilon}^t g^2(u,x(u))du \\
&+ 6C(t,t-\epsilon)V(t,\epsilon) + H.
\end{aligned}
$$

Corollary 1. *Let x be a continuous solution of (2.5.1) on $[0,\infty)$ and let (2.5.2), (2.5.3), and (2.5.7)–(2.5.9) hold. Suppose also that*

$$I(t) := \int_{t-\epsilon}^t C^2(t,s)ds$$

is continuous for $t \geq \epsilon$. Then from the last display in the proof of Theorem 2.5.2 we can deduce that

$$
\begin{aligned}
V(t,\epsilon)+(1-\alpha-\beta) \int_\epsilon^t g^2(u,x(u))du + \int_\epsilon^t (g(s,x(s)) - a(s))^2 ds \\
\leq \int_\epsilon^t a^2(u)du + \alpha \int_0^\epsilon g^2(u,x(u))du + V(\epsilon,\epsilon)
\end{aligned}
$$

and from Theorem 2.5.3 we can deduce that

$$(x(t) - a(t))^2 \leq \left[\frac{6\beta^2}{\epsilon} + 2I(t)\right] \int_{t-\epsilon}^t g^2(s,x(s))ds + \frac{6\beta}{\epsilon} V(t,\epsilon).$$

If we substitute the bounds on $V(t,\epsilon)$ and $\int_\epsilon^t g^2(u,x(u))du$ found in the first inequality into the second inequality, then we obtain a growth relation in the form of a continuous function f with

$$|x(t)| \leq f\Big(I(t), |a(t)|, \int_0^t a^2(s)ds\Big).$$

If we now look ahead to the fractional differential equation and Theorems 2.5.1.2 and 2.5.1.3 we have the corresponding corollary. In that problem we have a constant k and for $1/2 < q < 1$ then

$$\begin{aligned}\int_{t-\epsilon}^{t} C^2(t,s)ds &= \int_{t-\epsilon}^{t} k(t-s)^{2(q-1)}ds \\ &= -\frac{k(t-s)^{2q-1}}{2q-1}\Big|_{t-\epsilon}^{t} \\ &= k\frac{\epsilon^{2q-1}}{2q-1}\end{aligned}$$

which is a constant. But if we had used the corollary to Theorem 2.5.2, then the integral used there would be unbounded.

Theorem 2.5.4. *Suppose that x is a continuous solution of (2.5.1) on $[0,\infty)$, that $g(t,x(t)) \in L^2[0,\infty)$, that for each large T then*

$$\int_0^T |C(t,s)|ds \to 0$$

as $t \to \infty$, and that

$$\sup_{0\leq t} \int_0^t C^2(t,s)ds \leq M$$

for some $M > 0$. Then $x(t) \to a(t)$ as $t \to \infty$.

Proof. We have

$$\begin{aligned}|x(t)-a(t)| &\leq \int_0^T |C(t,s)||g(s,x(s))|ds \\ &\quad + \sqrt{\int_T^t C^2(t,s)ds \int_T^t g^2(s,x(s))ds} \\ &\leq \|g\|^{[0,T]}\int_0^T |C(t,s)|ds + \sqrt{M\int_T^t g^2(s,x(s))ds}.\end{aligned}$$

For a given $\epsilon > 0$, take T so large that the last term is less than $\epsilon/2$. Then we have $J > 0$ with $\|g\|^{[0,T]} \leq J$, where the notation means the supremum on $[0,T]$. Finally, if t is large enough then $\int_0^T |C(t,s)|ds \leq \epsilon/2J$.

This section concerns convex kernels but some of these will be perturbed with small kernels so we give a short treatment of those here. Let

$$x(t) = a(t) - \int_0^t D(t,s)g(s,x(s))ds \tag{2.5.12}$$

where g again satisfies (2.5.4) and D satisfies several integrability conditions to follow. Again we will be treating an assumed solution, rather than give detailed conditions under which the solution exists. Once more the problem of separating $a(t)g(t,x)$ leads to a severe condition on g.

Theorem 2.5.5. *Let (2.5.4) hold, $|g(t,x)| \le |x|$, D be continuous, $\int_0^\infty |D(u+t,t)|du \le \delta$, $\int_0^t |D(t,s)|ds \le \gamma$, and for p an even positive integer let $a \in L^p[0,\infty)$. If*

$$\delta + (p-1)\gamma - p < 0$$

and if $x(t)$ is a continuous solution of (2.5.12) on $[0,\infty)$ then

$$\int_0^\infty g^p(s,x(s))ds < \infty.$$

Proof. Define

$$V(t) = \int_0^t \int_{t-s}^\infty |D(u+s,s)|du g^p(s,x(s))ds$$

and compute the derivative. We quickly arrive at

$$V'(t) \le \delta g^p(t,x) - \int_0^t |D(t,s)|g^p(s,x(s))ds$$
$$+ pg^{p-1}(t,x(t))[a(t) - x(t) - \int_0^t D(t,s)g(s,x(s))ds].$$

That last term is identically zero since x solves (2.5.12); this will be clarified in the remark following the proof of Theorem 2.5.6. Multiplying out that last term and using the inequality (under appropriate conditions) $ab \le (a^p/p) + (b^s/s)$ we can find positive numbers q and M with q as small as we please, while M is large, and show that the last term is bounded by

$$p\Big[a(t)g^{p-1}(t,x(t)) - g^{p-1}(t,x(t))x(t)$$
$$+ \int_0^t |D(t,s)|\big|g^{p-1}(t,x(t))g(s,x(s))\big|ds\Big]$$

$$\begin{aligned}
&\le q(p-1)(g^{p-1}(t,x(t)))^{\frac{p}{p-1}} + Ma^p(t) - pg^p(t,x(t)) \\
&+ \int_0^t |D(t,s)| \left[(p-1)(g^{p-1}(t,x(t)))^{\frac{p}{p-1}} + g^p(s,x(s))\right] ds \\
&\le q(p-1)g^p(t,x(t)) + Ma^p(t) - pg^p(t,x(t)) \\
&+ (p-1)\int_0^t |D(t,s)|ds g^p(t,x(t)) + \int_0^t |D(t,s)|g^p(s,x(s))ds.
\end{aligned}$$

We now have

$$V'(t) \le \left[\delta + q(p-1) - p + (p-1)\gamma\right] g^p(t,x(t)) + Ma^p(t).$$

Choose q so small that we can find $\mu > 0$ and have

$$V'(t) \le -\mu g^p(t,x(t)) + Ma^p(t).$$

An integration yields the result.

We now consider the perturbed scalar equation

$$x(t) = a(t) - \int_0^t [C(t,s) + D(t,s)]g(s,x(s))ds \tag{2.5.13}$$

where (2.5.2), (2.5.3), and (2.5.4) hold. We also suppose that there are positive constants γ and δ with

$$\int_0^t |D(t,s)|ds \le \gamma, \ \int_0^\infty |D(u+t,t)|du \le \delta, \ \gamma + \delta < 2. \tag{2.5.14}$$

Unlike differential equations, we can add kernels and add Liapunov functionals in a completely seamless manner.

Theorem 2.5.6. *Let (2.5.2), (2.5.3), and (2.5.14) hold for (2.5.13) and let D be continuous. Then for*

$$\begin{aligned}
V(t,\epsilon) &= \int_0^t \int_{t-s}^\infty |D(u+s,s)|du g^2(s,x(s))ds \\
&+ \int_0^{t-\epsilon} C_s(t,s)\left(\int_s^t g(u,x(u))du\right)^2 ds + C(t,0)\left(\int_0^t g(u,x(u))du\right)^2
\end{aligned} \tag{2.5.15}$$

we have

$$V'(t,\epsilon) \le 2g(t,x(t))\big[a(t) - x(t) - \int_0^t D(t,s)g(s,x(s))ds$$

$$+ C(t,t-\epsilon)\int_{t-\epsilon}^{t} g(u,x(u))du]$$

$$- 2g(t,x(t))\int_{t-\epsilon}^{t} C(t,s)g(s,x(s))ds$$

$$+ C_s(t,t-\epsilon)\Big(\int_{t-\epsilon}^{t} g(u,x(u))du\Big)^2$$

$$+ \delta g^2(t,x(t)) - \int_0^t |D(t,s)|g^2(s,x(s))ds.$$

Proof. As $C_t \le 0$ and $C_{st} \le 0$ we have

$$V'(t,\epsilon) \le \int_0^\infty |D(u+t,t)|dug^2(t,x(t))$$

$$- \int_0^t |D(t,s)|g^2(s,x(s))ds$$

$$+ C_s(t,t-\epsilon)\Big(\int_{t-\epsilon}^{t} g(u,x(u))du\Big)^2$$

$$+ 2g(t,x(t))\int_0^{t-\epsilon} C_s(t,s)\int_s^t g(u,x(u))duds$$

$$+ 2g(t,x(t))C(t,0)\int_0^t g(u,x(u))du$$

(Integrate the next-to-last term by parts.)

$$\le \delta g^2(t,x(t)) - \int_0^t |D(t,s)|g^2(s,x(s))ds$$

$$+ 2g(t,x(t))C(t,0)\int_0^t g(u,x(u))du$$

$$+ C_s(t,t-\epsilon)\Big(\int_{t-\epsilon}^{t} g(u,x(u))du\Big)^2$$

$$+ 2g(t,x(t))\Big[C(t,s)\int_s^t g(u,x(u))du\Big|_0^{t-\epsilon}$$

$$+ \int_0^{t-\epsilon} C(t,s)g(s,x(s))ds\Big]$$

$$= \delta g^2(t,x(t)) + C_s(t,t-\epsilon)\Big(\int_{t-\epsilon}^{t} g(u,x(u))du\Big)^2$$

$$+ 2g(t,x(t))C(t,0)\int_0^t g(u,x(u))du - \int_0^t |D(t,s)|g^2(s,x(s))ds$$

$$+ 2g(t,x(t))\left[C(t,t-\epsilon)\int_{t-\epsilon}^{t} g(u,x(u))du - C(t,0)\int_0^t g(u,x(u))du\right.$$

$$\left.+\int_0^{t-\epsilon} C(t,s)g(s,x(s))ds\right]$$

(Write the last term as $\int_0^t C(t,s)g(s,x(s))ds - \int_{t-\epsilon}^t C(t,s)g(s,x(s))ds$ and replace the first of these by its value in (2.5.13), thereby linking the Liapunov functional to the integral equation.)

$$= \delta g^2(t,x(t)) + C_s(t,t-\epsilon)\left(\int_{t-\epsilon}^t g(u,x(u))du\right)^2$$

$$-\int_0^t |D(t,s)|g^2(s,x(s))ds + 2g(t,x)C(t,t-\epsilon)\int_{t-\epsilon}^t g(u,x(u))du$$

$$+ 2g(t,x(t))\left[a(t) - x(t) - \int_0^t D(t,s)g(s,x(s))ds\right]$$

$$- 2g(t,x(t))\int_{t-\epsilon}^t C(t,s)g(s,x(s))ds,$$

as required.

Remark When $D = 0$ then V' coincides with V' in Theorem 2.5.1. When $C = 0$, then V' coincides with V' in Theorem 2.5.5 for $p = 2$. The replacement described in the sentence before the last set of expressions in the above proof is well-motivated as it links the Liapunov functional with the integral equation. That work coincides with the rather mysterious addition of the second line in the derivation of V' in the proof of Theorem 2.5.5.

Caution! In view of this remark we would conjecture that if the conditions of Theorems 2.5.2 and 2.5.5 hold for $p = 2$, then a solution of (2.5.13) satisfies $g(t,x(t)) \in L^2$ when $a \in L^2$. But this is wrong. Both of those theorems use the x of the equation in the term $2g(t,x)x$. When we add $C(t,s) + D(t,s)$ we only have x once. Thus, instead of asking $\alpha + \beta < 2$, as in the remark following (2.5.8) and (2.5.9), followed by asking $\delta + \gamma < 2$, as in (2.5.14), we are forced to ask $\alpha + \beta + \delta + \gamma < 2$.

Take courage! It is often true. For weak singularities, such as $C(t,s) = [t-s]^{-1/2}$, the conditions $\alpha + \beta < 1$ in (2.5.8) and (2.5.9) ask entirely too little. They can be replaced by $\alpha + \beta \to 0$ as $\epsilon \to 0$. In such cases the conjecture is saved and $\delta + \gamma < 2$ is all that is needed. The perturbation D is added entirely without cost.

Theorem 2.5.7. *Let (2.5.2) – (2.5.4), (2.5.7)–(2.5.9) without $\alpha+\beta<1$, and (2.5.14) hold for (2.5.13). For α and β defined in (2.5.8) and (2.5.9) and for γ and δ defined in (2.5.14) let*

$$\gamma+\delta+\alpha+\beta<2.$$

Then $a\in L^2[0,\infty)$ implies $g(t,x(t))\in L^2[0,\infty)$.

Proof. Starting with V' in Theorem 2.5.6 we can find an ξ as small as we please and a correspondingly large M with

$$\begin{aligned}
&V'(t,\epsilon)\le 2a(t)g(t,x(t))\\
&-2g^2(t,x(t))+\int_0^t |D(t,s)|[g^2(s,x(s))+g^2(t,x(t))]ds\\
&+C(t,t-\epsilon)\int_{t-\epsilon}^t [g^2(u,x(u))+g^2(t,x(t))]du\\
&+\int_{t-\epsilon}^t C(t,s)[g^2(t,x(t))+g^2(s,x(s))]ds\\
&+\epsilon C_s(t,t-\epsilon)\int_{t-\epsilon}^t g^2(u,x(u))du\\
&+\delta g^2(t,x(t))-\int_0^t |D(t,s)|g^2(s,x(s))ds\\
&\le \xi g^2(t,x(t))+Ma^2(t)-2g^2(t,x(t))+\delta g^2(t,x(t))\\
&+g^2(t,x(t))\left[\int_0^t |D(t,s)|ds+C(t,t-\epsilon)\int_{t-\epsilon}^t ds\right.\\
&\left.+\int_{t-\epsilon}^t C(t,s)ds\right]\\
&+\int_{t-\epsilon}^t [C(t,t-\epsilon)+C(t,s)+\epsilon C_s(t,t-\epsilon)]g^2(s,x(s))ds\\
&\le g^2(t,x(t))[\xi-2+\delta+\gamma+\epsilon C(t,t-\epsilon)+\int_{t-\epsilon}^t C(t,s)ds]+Ma^2(t)\\
&+\int_{t-\epsilon}^t [C(t,t-\epsilon)+C(t,s)+\epsilon C_s(t,t-\epsilon)]g^2(s,x(s))ds.
\end{aligned}$$

We then have

$$\begin{aligned}
V'(t,\epsilon)\le\, &Ma^2(t)+g^2(t,x(t)[\xi-2+\delta+\gamma+\beta]\\
&+\int_{t-\epsilon}^t [C(t,t-\epsilon)+C(t,s)+\epsilon C_s(t,t-\epsilon)]g^2(s,x(s))ds.
\end{aligned}$$

When we integrate from ϵ to t as we did after (2.5.11) the last term is bounded by $\alpha \int_0^t g^2(s,x(s))ds$. That integration will now yield the result.

Volterra recognized two kinds of singularities: a discontinuous kernel or an infinite delay. Here, we consider the combination in the form of

$$x(t) = a(t) - \int_{-\infty}^{t} C(t,s)g(s,x(s))ds \tag{2.5.16}$$

where C satisfies (2.5.2), (2.5.3), and several integrability and limit conditions. Such equations can have three kinds of solutions. Given a continuous initial function $\phi : (-\infty, t_0] \to \Re$ we seek a solution $x(t)$ for $t > t_0$ with $x(t) = \phi(t)$ for $t \leq t_0$. If

$$\phi(t_0) = a(t_0) - \int_{-\infty}^{t_0} C(t_0,s)g(s,\phi(s))ds \tag{2.5.17}$$

then $x(t)$ is continuous on $(-\infty, t)$ for $t > t_0$; otherwise, the solution has a discontinuity at t_0. It can be shown that there is an ψ arbitrarily near ϕ with $x(t)$ continuous. This gives two kinds of solutions. The third kind occurs when x satisfies (2.5.16) on $(-\infty, \infty)$ so at any t_0, then $x(t)$ is its own initial function. Periodic solutions are central examples of this. Several existence results using fixed point theory for this kind of equation are found in Burton (1994b) and in Burton and Makay (2002). See also Zhang (2009).

Here, we again allow for C to be weakly singular and (2.5.8) and (2.5.9) hold. Moreover, we suppose that ϕ is chosen so that (2.5.17) holds and that there is a continuous solution on $[t_0, \infty)$. The work of Burton-Makay (2002) speaks extensively to that case when the kernel is nonsingular. Our Liapunov functional will have the form

$$V(t,\epsilon) = \int_{-\infty}^{t-\epsilon} C_s(t,s)\left(\int_s^t g(v,x(v))dv\right)^2 ds. \tag{2.5.18}$$

The critical condition which makes this a viable Liapunov functional is

$$\lim_{s\to-\infty} (t-s)C(t,s) = 0 \tag{2.5.19}$$

for fixed t. It is used in the integration by parts formula for the derivative and it allows us to skip one term in the usual Liapunov functional.

Theorem 2.5.8. *Let (2.5.2), (2.5.3), and (2.5.19) hold. If $x(t)$ solves (2.5.16) with a bounded and continuous initial function ϕ satisfying (2.5.17), then for $\epsilon > 0$ the derivative of V defined by (2.5.18) satisfies (2.5.6). If (2.5.8) and (2.5.9) hold and if $|g(t,x)| \leq |x|$ then $a \in L^2[0,\infty)$ implies that $g(t,x(t)) \in L^2[t_0,\infty)$.*

Proof. As $C_{st} \leq 0$ we have

$$V'(t,\epsilon) \leq C_s(t,t-\epsilon)\left(\int_{t-\epsilon}^{t} g(v,x(v))dv\right)^2 + 2g(t,x(t))\int_{-\infty}^{t-\epsilon} C_s(t,s)\int_s^t g(v,x(v))dvds.$$

If we integrate the last term by parts as we have done before and use (2.5.19) on the lower limit with the bounded initial function then we obtain

$$\begin{aligned} V'(t,\epsilon) &\leq -2g(t,x(t))\int_{t-\epsilon}^{t} C(t,s)g(s,x(s))ds \\ &+ C_s(t,t-\epsilon)\left(\int_{t-\epsilon}^{t} g(v,x(v))dv\right)^2 \\ &+ 2g(t,x(t))C(t,t-\epsilon)\int_{t-\epsilon}^{t} g(v,x(v))dv + 2g(t,x(t))[a(t)-x(t)]. \end{aligned}$$

The rest of the proof proceeds just as we have seen several times before.

We can add the perturbation D, add the Liapunov functional

$$\int_{-\infty}^{t}\int_{t-s}^{\infty} |D(u+s,s)|du g^2(s,x(s))ds,$$

and find a lower bound on the Liapunov functional as we did in Theorem 2.5.3. In summary, (2.5.1) can be perturbed with a singularity at $t=s$, it can be perturbed with $D(t,s)$, and it can be perturbed to $(-\infty,0]$ without substantially changing the solution. Equation (2.5.1) is very stable and it is a defensible practice to use a kernel satisfying (2.5.2) to model real-world problems, just as noted by Volterra in 1928.

Consider the equation

$$x(t) = a(t) - \int_{t-h}^{t} C(t,s)g(s,x(s))ds, \ h>0, \tag{2.5.20}$$

and modify (2.5.2) in a crucial way, asking that

$$C(t,s)\geq 0,\ C_{st}(t,s)\leq 0,\ C_s(t,s)\geq 0,\ C(t,t-h)=0 \tag{2.5.21}$$

when

$$0<\epsilon<h \text{ and } t-h\leq s\leq t-\epsilon. \tag{2.5.22}$$

Theorem 2.5.9. *Let (2.5.21) and (2.5.22) hold for (2.5.20) and let V be defined by*

$$V(t,\epsilon) = \int_{t-h}^{t-\epsilon} C_s(t,s)\left(\int_s^t g(v,x(v))dv\right)^2 ds.$$

Then along a continuous solution of (2.5.20) $V'(t,\epsilon)$ satisfies (2.5.6).

The proof of the theorem and the consequences should now be routine.

Notes The material in Section 2.5 was taken from Burton (2010b). It was first published by Elsevier in *N*onlinear Analysis: TMA with full details given in the bibliography.

2.5.1 Fractional Differential Equations

In this section we will deal with an integral equation with a parameter q and a kernel which is so often related to the C of Section 2.5 that we want to begin with the association

$$C(t,s) = C(t-s) := \frac{k}{\Gamma(q)}(t-s)^{q-1}, \quad 0 < q < 1, \tag{2.5.1.1}$$

where Γ is the Euler gamma function. We will be concerned with the fractional differential equation of Caputo type

$${}^cD^q x = f(t) - kg(t,x(t)), \quad x(0) = 0, \quad 0 < q < 1, \tag{2.5.1.2}$$

where k is a positive constant, $g : [0,\infty) \times \Re \to \Re$, $f : [0,\infty) \to \Re$ with both f and g continuous,

$$xg(t,x) \geq 0 \tag{2.5.1.3}$$

for $t \geq 0$ and $x \in \Re$. Later in this subsection we deal with the case of $x(0) \neq 0$.

We follow Lakshmikantham, Leela, and Vasundhara Devi (2009; p. 12) and define the Caputo fractional derivative of order q of a function x as

$${}^cD^q x(t) = \frac{1}{\Gamma(1-q)} \int_{t_0}^t (t-s)^{-q} x'(s)ds.$$

It will not be used here because we will immediately invert (2.5.1.2) as an integral equation. It is to be emphasized that this is not the Riemann-Liouville derivative, denoted by D^q. Caputo introduced his derivative to avoid the initial conditions imposed by the Riemann-Liouville derivative which were difficult to reconcile with many real-world problems.

Define

$$F(t) := \frac{1}{\Gamma(q)} \int_0^t (t-s)^{q-1} f(s)ds. \tag{2.5.1.4}$$

Often we will ask that

$$\int_0^\infty F^2(t)dt < \infty, \tag{2.5.1.5}$$

but it plays the role of $a(t)$ in the Main Remark stated before Theorem 2.5.3. This will relate the Liapunov functional to the iterated kernels of Proposition 1.2.2. Just exactly as q was allowed to decrease as we continued iterating the kernels, as p increases we will be able to include smaller values of q. This seems to be an entirely new relation.

Liapunov growth Keep in mind the Liapunov growth property which was stated just before Theorems 2.5.3 and 2.5.4. It will yield a growth rate on x for any continuous function $F(t)$,

The Caputo equation (2.5.1.2) will be inverted as

$$x(t) = F(t) - \frac{k}{\Gamma(q)} \int_0^t (t-s)^{q-1} g(s, x(s))ds. \tag{2.5.1.7}$$

We refer the reader to Lakshmikantham, Leela, and Vasundhara Devi (2009; p. 54) or to Chapter 6 of Diethelm (2004; pp. 78, 86, 103) for proofs of the inversion.

When (2.5.1.5) is required, it is still true that F can be an arbitrarily large perturbation and it is persistent. We have $x(0) = 0$ and then we apply F.

In Section 2.5 conditions (2.5.8) and (2.5.9) were fundamental. They are repeated here as (2.5.1.8) and (2.5.1.9) so that we can repeat our main theorems from that section for this problem. But our first result will show that they automatically hold for (2.5.1.7). That theory requires positive constants α, β with $\alpha + \beta < 1$. Moreover, for some $\epsilon > 0$ and for $C(t,s) = (t-s)^{q-1}$ we must have

$$\int_s^{s+\epsilon} [\epsilon C_s(u, u-\epsilon) + C(u, u-\epsilon) + |C(u,s)|]du < \alpha \tag{2.5.1.8}$$

for $\epsilon \leq s < \infty$ and

$$C(t, t-\epsilon)\epsilon + \int_{t-\epsilon}^t |C(t,s)|ds < \beta \tag{2.5.1.9}$$

for $\epsilon \leq t < \infty$. The absolute value on C was needed in the general case, but for the present problem it is not needed.

Theorem 2.5.1.1. *If $C(t,s) = (t-s)^{q-1}$ where $0 < q < 1$ then for each fixed q there is an $\epsilon > 0$ for which (2.5.1.8) and (2.5.1.9) hold for $\alpha = \beta = 1/4$.*

Proof. Note that $C_s(u, u-\epsilon) = C_s(\epsilon) = (1-q)\epsilon^{q-2}$, $C(u, u-\epsilon) = \epsilon^{q-1}$, while $|C(u,s)| = (u-s)^{q-1}$. Thus, (2.5.1.8) is

$$\begin{aligned}\int_s^{s+\epsilon} [(1-q)\epsilon^{q-1} + \epsilon^{q-1} + (u-s)^{q-1}]du &= (2-q)\epsilon^q + \frac{(u-s)^q}{q}\bigg|_s^{s+\epsilon} \\ &= (2-q)\epsilon^q + \frac{\epsilon^q}{q} \\ &= (2-q+\frac{1}{q})\epsilon^q < 1/4\end{aligned}$$

if ϵ is sufficiently small and $q \in (0,1)$ is fixed.

Next, (2.5.1.9) is

$$\begin{aligned}\epsilon^q + \int_{t-\epsilon}^t (t-s)^{q-1}ds &= \epsilon^q - \frac{(t-s)^q}{q}\bigg|_{t-\epsilon}^t \\ &= \epsilon^q + \frac{\epsilon^q}{q} \\ &= (1+\frac{1}{q})\epsilon^q < 1/4\end{aligned}$$

for ϵ small and $q \in (0,1)$, a fixed number.

The Liapunov functional defined in (2.5.5) is rewritten for this problem using (2.5.1.1) as follows. For some $\epsilon > 0$ and for $t \geq \epsilon$ let

$$\begin{aligned}V(t,\epsilon) = &\int_0^{t-\epsilon} C_s(t,s)\left(\int_s^t H(u,x(u))du\right)^2 ds \\ &+ C(t,0)\left(\int_0^t H(u,x(u))du\right)^2 \qquad (2.5.1.10)\end{aligned}$$

where

$$H(t,x(t)) = \frac{k}{\Gamma(q)}g(t,x(t)), \quad C(t,s) = (t-s)^{q-1}. \qquad (2.5.1.11)$$

With C defined in this way and for $0 \leq s \leq t-\epsilon$, $t < \infty$, and $0 < q < 1$ we have

$$C(t,s) \geq 0, C_s(t,s) \geq 0, C_{st}(t,s) \leq 0, C_t(t,0) \leq 0. \qquad (2.5.1.12)$$

As $s \leq t-\epsilon$ or $\epsilon \leq t-s$, both C_s and C_{st} are continuous.

Replace the $g(t, x)$ in (2.5.5) by $[k/\Gamma(q)]g(t, x)$. Then Theorem 2.5.1 becomes the following, where we have replaced C according to (2.5.1.1) and then factored out $[k^2/\Gamma^2(q)]$.

Theorem 2.5.1.2. *Let (2.5.1.3) and (2.5.1.12) hold. Let x be a continuous solution of (2.5.1.7) on $[0, \infty)$, let (2.5.1.10) be defined with this x, and let $\epsilon > 0$ be chosen so that (2.5.1.8) and (2.5.1.9) hold. Then*

$$\begin{aligned}\frac{dV(t,\epsilon)}{dt} &\leq \frac{k^2}{\Gamma^2(q)}\left[(1-q)\epsilon^{q-2}\left(\int_{t-\epsilon}^{t} g(u,x(u))du\right)^2\right. \\ &+ 2g(t,x(t))\left(\epsilon^{q-1}\int_{t-\epsilon}^{t} g(u,x(u))du - \int_{t-\epsilon}^{t}(t-s)^{q-1}g(s,x(s))ds\right) \\ &\left.+ 2g(t,x(t))[F(t)-x(t)]\frac{\Gamma(q)}{k}\right]. \qquad (2.5.1.13)\end{aligned}$$

Three relations will be needed for us to parlay this Liapunov functional derivative into a qualitative result for a solution of (2.5.1.7). First, we must be able to estimate the relation between $xg(t, x)$ and $g^2(t, x)$. Our conditions (2.5.1.8) and (2.5.1.9) will allow such strongly singular kernels that it is not a great surprise to need $g(t, x)$ bounded by a linear function. We ask that

$$\frac{\Gamma(q)}{k}xg(t,x) \geq g^2(t,x). \qquad (2.5.1.14)$$

Note that (2.5.1.14) defines sectors $\frac{\Gamma(q)}{k}|x| \geq |g(t,x)|$; as $q \downarrow 0$ the sector approximates the first and third quadrants. That property will vanish entirely when we study the resolvent and we do not have an interpretation of why that should occur.

Under the substitution (2.5.1.1), Theorem 2.5.2 becomes the following.

Theorem 2.5.1.3. *Let x be a continuous solution of (2.5.1.7) on $[0, \infty)$ and let (2.5.1.12) and (2.5.1.14) hold. If, in addition, $F \in L^2[0, \infty)$ so are $g(t, x(t))$ and $g(t, x(t)) - \frac{\Gamma(q)}{k}F(t)$.*

Main Remark 2 If we went back to Section 2.5 and went through the proof of this theorem under the substitution of (2.5.1.1) we would see in the last steps that we had integrated V' and obtained a relation

$$\begin{aligned}V(t,\epsilon) &\leq V(\epsilon,\epsilon) \\ &+ \left[\int_{\epsilon}^{t}\frac{\Gamma^2(q)}{k^2}F^2(u)du - \int_{\epsilon}^{t}\left(g(s,x(s)) - \frac{\Gamma(q)}{k}F(s)\right)^2 ds\right. \\ &\left.-(1-\alpha-\beta)\int_{\epsilon}^{t}g^2(u,x(u))du + \alpha\int_0^{\epsilon}g^2(u,x(u))du\right]\frac{k^2}{\Gamma^2(q)}.\end{aligned}$$

With this information, go back to the Liapunov growth property stated just before Theorems 2.5.3 and 2.5.4.

We can also find a constant $M > 0$ with an approximation to the relation

$$\int_{\epsilon}^{t}(g(s,x(s)) - \frac{\Gamma(q)}{k}F(s))^2 ds \le M\int_{\epsilon}^{t}\frac{\Gamma^2(q)}{k^2}F^2(u)du. \tag{2.5.1.15}$$

Thus, $g(t,x(t)) - \frac{\Gamma(q)}{k}F(t)$ gets small infinitely often. That relation shows that the growth of the solution is controlled by the growth of F even if F is not in L^2 as required for (2.5.1.5). But it is important to know when $F \in L^2[0,\infty)$ and the following is intended as a first step in putting us on the road to understanding that.

Proposition 2.5.1.1. *Suppose there is a continuous and decreasing function $h : [0,\infty) \to \Re$ with $|f(t)| \le h(t)$ and $h \in L^1[0,\infty)$. If $0 < q < 1/2$, then $F \in L^2[0,\infty)$.*

Proof. If $H = \int_0^\infty h(s)ds$ and if $\varepsilon > 0$ then

$$\begin{aligned}
F^2(t) &\le \frac{1}{\Gamma^2(q)}\left(\int_0^t (t-s)^{q-1}h(s)ds\right)^2 \\
&= \frac{1}{\Gamma^2(q)}\left[\int_0^{t-\varepsilon}(t-s)^{q-1}h(s)ds + \int_{t-\varepsilon}^{t}(t-s)^{q-1}h(s)ds\right]^2 \\
&\le \frac{1}{\Gamma^2(q)}\left[2\left(\int_0^{t-\varepsilon}(t-s)^{q-1}h(s)ds\right)^2\right. \\
&\left. + 2\left(\int_{t-\varepsilon}^{t}(t-s)^{q-1}h(s)ds\right)^2\right] \\
&\le \frac{1}{\Gamma^2(q)}\left[2\int_0^{t-\varepsilon}h(s)ds\int_0^{t-\varepsilon}(t-s)^{2q-2}h(s)ds\right. \\
&\left. + 2h^2(t-\varepsilon)\left(\int_{t-\varepsilon}^{t}(t-s)^{q-1}ds\right)^2\right] \\
&\le \frac{1}{\Gamma^2(q)}\left[2H\int_0^{t-\varepsilon}(t-s)^{2q-2}h(s)ds\right. \\
&\left. + 2h(0)h(t-\varepsilon)\left(\int_{t-\varepsilon}^{t}(t-s)^{q-1}ds\right)^2\right] \\
&= \frac{1}{\Gamma^2(q)}\left[2H\int_{\varepsilon}^{t}(t-s+\varepsilon)^{2q-2}h(s-\varepsilon)ds + 2h(0)h(t-\varepsilon)\frac{\varepsilon^{2q}}{q^2}\right].
\end{aligned}$$

As $2q-2 < -1$, the first term is bounded by the convolution of two L^1 functions, so it is in $L^1[\varepsilon,\infty)$. The last term is in $L^1[\varepsilon,\infty)$ since $h \in$

$L^1[0,\infty)$. It is clear that $F(t)$ is bounded for $0 \leq t \leq \varepsilon$. Thus, $F \in L^2[0,\infty)$, as required.

Remark Our results here center largely on the convexity of the kernel and the relation (2.5.1.14) which does allow $g(t,x)$ to vanish for values of x which are not zero. Keep these in mind as we move ahead. In Proposition 2.5.1.1 we are restricted to $0 < q < 1/2$. Now, we are going to use more than the convexity. In fact, our kernel is completely monotone. This will give us a Liapunov functional which yields results for $0 < q < 1$, but in the main result it will not allow $g(t,x)$ to vanish for $x \neq 0$. Thus, there is still good reason to work with the present Liapunov functional.

Review Proposition 1.2.2 requiring $q \geq 1/3$. We are approaching the problem from two different directions.

The whole problem facing us in (2.5.1.7) is that the kernel is not in $L^1[0,\infty)$. It is a great victory to discover that this equation can be mapped into an equation with an L^1 kernel. This is a deep matter and we will have to borrow from several sources in order to present it here. But that seems proper in order to maintain continuity of thought.

The Fundamental Transformation for Fractional Equations Completely Monotone Kernels

Return to (2.5.1.2) and suppose that $x(0)$ is arbitrary, that $G : [0,\infty) \times \Re \to \Re$ is a continuous function defined by

$$g(t,x) = x + G(t,x). \tag{2.5.1.16}$$

We shall frequently ask that $|G(t,x)| \leq \alpha|x|$ for some $\alpha < 1$. The reader may wish to compare such conditions with (2.5.1.14) which we needed for the Liapunov functional. Our counterpart of (2.5.1.7) is

$$x(t) = x(0) - \frac{k}{\Gamma(q)} \int_0^t (t-s)^{q-1} [x(s) + G(s,x(s)) - \frac{f(s)}{k}] ds. \tag{2.5.1.17}$$

From (2.5.1.1) we have the critical property that for any $T > 0$ then

$$\int_0^T |C(u)| du < \infty. \tag{2.5.1.18}$$

We now refer to a lengthy study detailed in Miller (1971a) extending from p. 193 to p. 222. For us it begins on p. 221 where it is noted that for $C(t)$ defined in (2.5.1.1) by $C(t,s) = C(t-s) = (k/\Gamma(q))(t-s)^{q-1}$ then

$$C(t) \text{ is completely monotone on } (0,\infty) \tag{2.5.1.19}$$

in the sense that $(-1)^m C^{(m)}(t) \geq 0$ for $m = 0, 1, 2, ...$ and $t \in (0, \infty)$. Moreover $C(t)$ satisfies the conditions of Miller's Theorem 6.2 on p. 212. That theorem states that if the resolvent equation for the kernel C is

$$R(t) = C(t) - \int_0^t C(t-s)R(s)ds \tag{2.5.1.20}$$

then that resolvent kernel, R, satisfies

$$0 \leq R(t) \leq C(t) \text{ for all } t > 0 \text{ so as } t \to \infty \text{ then } R(t) \to 0 \tag{2.5.1.21}$$

and that

$$C \notin L^1[0, \infty) \quad \Longrightarrow \int_0^\infty R(s)ds = 1. \tag{2.5.1.22}$$

Continuing on to Miller's pp. 221-224 (Theorem 7.2) we see that R is also completely monotone. The results are not simple and are gathered from an extensive list of sources.

Next, under the conditions here, it is shown in Miller (1971a; pp. 191-207] and also in Section 2.8.1 that (2.5.1.17) can be decomposed into

$$y(t) = x(0) - \int_0^t C(t-s)y(s)ds \tag{2.5.1.23}$$

and, having found $y(t)$, then the solution $x(t)$ of (2.5.1.17) solves

$$x(t) = y(t) - \int_0^t R(t-s)[G(s, x(s)) - \frac{f(s)}{k}]ds. \tag{2.5.1.24}$$

Something quite remarkable has happened. The kernel in (2.5.1.17) is not integrable on $[0, \infty)$, but in (2.5.1.24) it is replaced, not only by an integrable kernel, but the value of the integral is one and the new kernel is also completely monotone. Among the many useful properties that (2.5.1.24) has, we will see that the requirement (2.5.1.14) is not needed in (2.5.1.24). The term $\frac{k}{\Gamma(q)}$ has been absorbed. Moreover, our old $F(t)$ which gave so much difficulty because it was frequently not in L^2 is replaced by a new $F(t)$ which is even nicer than the original function f. A long line of results applies to (2.5.1.24), including the oldest and most elementary result in all of integral equation theory.

First, notice that the variation of parameters formula for (2.5.1.23) employs (2.5.1.20) and yields

$$y(t) = x(0) - \int_0^t R(t-s)x(0)ds = x(0)\left[1 - \int_0^t R(s)ds\right]. \quad (2.5.1.25)$$

Immediately we see that when (2.5.1.21) and (2.5.1.22) hold then

$$x(0) - \int_0^t R(t-s)x(0)ds \to 0, \quad (2.5.1.26)$$

while if we now define

$$F^*(t) := \int_0^t R(t-s)[f(s)/k]ds \quad (2.5.1.27)$$

then

$$f \text{ bounded implies } F^* \text{ bounded,} \quad (2.5.1.28)$$

and

$$f \in L^1[0,\infty) \implies F^* \in L^1[0,\infty), f \in L^2 \implies F^* \in L^2. \quad (2.5.1.29)$$

Conclusions (2.5.1.26) and (2.5.1.28) are obvious, while (2.5.1.29) follows from the fact that the convolution of two L^1 functions is an L^1 function. Continuing that argument, if $f \in L^2[0,\infty)$ then

$$\left(\int_0^t R(t-s)f(s)ds\right)^2 \leq \int_0^t R(t-s)ds \int_0^t R(t-s)f^2(s)ds$$

and that last integral is an L^1 function, being the convolution of two L^1 functions.

Adam and Eve revisited

We study

$$x(t) = x(0) - \frac{k}{\Gamma(q)} \int_0^t (t-s)^{q-1} g(s, x(s))ds$$

here because it is an excellent example of a singular integral equation and it has global applications in applied mathematics even for the single value $q = 1/2$. Moreover, it is nontrivial in every mathematical sense of the word.

By contrast, we studied (1.1.1)

$$x(t) = a(t) - \int_0^t C(t,s)x(s)ds$$

in Theorem 1.1.2 under the conditions

$$|a(t)| \le K, \quad \sup_{t\ge 0} \int_0^t |C(t,s)|ds \le \alpha < 1$$

because it was the oldest and simplest problem in integral equations; thus, it was a good starting place.

Through the transformation here, certain aspects of the two problems merge. Certainly, it would be equivalent in Theorem 1.1.2 to study

$$x(t) = a(t) - \int_0^t [(1/\alpha)C(t,s)][\alpha x(s)]ds \tag{*}$$

so that if we named $(1/\alpha)C(t,s) = R(t,s)$ we would have $\int_0^t |R(t,s)|ds \le 1$ as in (2.5.1.22). Continuing, if $g(t,x) = x + G(t,x)$ where $|G(t,x)| \le \alpha|x|$ with $\alpha < 1$ then this admits the function $g(t,x) = x + \alpha x = [1+\alpha]x$ which is weaker than (*).

In short, by asking $|G(t,x)| \le \alpha|x|$ where $\alpha < 1$ in (2.5.1.24) we will be asking less than was asked in the Adam and Eve theorem. Notice, too, that from (2.5.1.21) the resolvent $R(t)$ is bounded by $C(t)$ so that in the variation of parameters work in Chapter 1 where we needed to interchange the order of integration by invoking the Tonelli-Hobson test (See Corollary 1 to Theorem 1.2.6.), we can do the same with R. This also happens in Section 2.8.1 which we invoked in the decomposition (2.5.1.23) and (2.5.1.24).

Recall that we finished the work preceding (2.5.1.16) noting that we covered $0 < q < 1/2$ using our Liapunov functional. We are now going to take care of $0 < q < 1$, first using contraction mappings with arbitrary $x(0) \in \Re$ and obtaining $x \in L^\infty$, and then $x(0) = 0$ using a Liapunov functional which covers all such q if $F^* \in L^p$.

Remark This can be improved, but it would take us too far afield to give details here. It is shown in Burton and Zhang (2012b) that when $x(0) \ne 0$, then the solution $y(t)$ of (2.5.1.25) is in $L^p[0,\infty)$ if $p > 1/q$. Thus, one may fix q and find p. Then absorb y in F^* so that $F^* \in L^p$.

Theorem 2.5.1.4. *Suppose that there is an $\alpha < 1$ such that $x, y \in \Re$ implies that*

$$|G(t,x) - G(t,y)| \le \alpha|x-y|, \quad G(t,0) \in L^1[0,\infty).$$

If $f \in L^1[0,\infty)$ and bounded, while $x(0) \in \Re$, then the unique solution, x, of (2.5.1.24) tends to zero as $t \to \infty$.

Proof. Let $(X, \|\cdot\|)$ be the Banach space of continuous functions $\phi : [0, \infty) \to \Re$ tending to zero as $t \to \infty$ with the supremum norm and define a mapping $P : X \to X$ by $\phi \in X$ implies that

$$(P\phi)(t) = y(t) + \int_0^t R(t-s)(f(s)/k)ds - \int_0^t R(t-s)G(s, \phi(s))ds.$$

We already know that $y(t)$ and $F^*(t)$ both tend to 0 as $t \to \infty$. But $|G(t, \phi(t))| \leq |G(t, 0)| + \alpha|\phi(t)|$ so $R \in L^1[0, \infty)$ and $R(t) \to 0$ implies that $(P\phi)(t) \to 0$ as $t \to \infty$. Next, if $\phi, \eta \in X$ then

$$|(P\phi)(t) - (P\eta)(t)| \leq \int_0^t R(t-s)\alpha|\phi(s) - \eta(s)|ds \leq \alpha\|\phi - \eta\|.$$

Thus, P is a contraction with unique fixed point tending to zero.

There are many variations of this result using weighted norms and Schauder's theorem which do not require a Lipschitz condition and those will not be treated here. The next result illustrates common techniques. It does not stipulate the existence of a solution, but that can be proved.

Theorem 2.5.1.5. *Suppose there is a constant K, a non-negative constant $\alpha < 1$, and a bounded continuous function $\psi : [0, \infty) \to \Re$ with $\psi \in L^1[0, \infty)$ and $|G(t, x)| \leq K + [\alpha + \psi(t)]|x|$. Suppose also that $f \in L^1[0, \infty)$ and is bounded on $[0, \infty)$. If $x(0) \in \Re$ then any solution of (2.5.1.24) is bounded.*

Proof. Notice that $\int_0^t R(t-s)K ds$ is bounded and that $\int_0^t R(t-s)\psi(s)ds \to 0$ as $t \to \infty$ so there is a $T > 0$ and $\alpha_1 \in (\alpha, 1)$ with $\int_0^t R(t-s)[\psi(s)+\alpha]ds \leq \alpha_1$ if $t \geq T$. If there is a fixed unbounded solution then there is a sequence $\{t_n\} \uparrow \infty$ with $t_n > T$, $|x(t_n)| \uparrow \infty$, and $|x(t)| \leq |x(t_n)|$ if $0 \leq t \leq t_n$. We then have

$$\begin{aligned}|x(t_n)| &\leq |y(t_n)| + |F^*(t_n)| + K + |x(t_n)| \int_0^{t_n} R(t_n - s)\alpha_1 ds \\ &\leq |y(t_n)| + |F^*(t_n)| + K + \alpha_1|x(t_n)|\end{aligned}$$

from which we see that $|x(t_n)|$ is bounded since $y(t_n)$ and $F^*(t_n)$ are bounded and $\alpha_1 < 1$.

By choosing $\alpha = K = 0$ we readily find limit sets of the solution. Our next result leads us back to the search for L^p solutions. It will be clear that each term in the equation for (2.5.1.24) is in $L^1[0, \infty)$.

Corollary. *If $K = \alpha = x(0) = 0$ and if $f \in L^1[0, \infty)$, then $x \in L^1[0, \infty)$.*

Remark For the remainder of this subsection we take $x(0) = 0$ so that $y(t) = 0$. But keep in mind the statements just before Theorem 2.5.1.4 so that this can be avoided.

The fact that $\int_0^\infty R(u)du = 1$ leads us to the possibility of new Liapunov functionals for (2.5.1.24) of the form

$$V(t,\epsilon) = \int_0^t \int_{t-s+\epsilon}^\infty R(u)du|G(s,x(s))|^p ds, \tag{2.5.1.30}$$

where p is some positive integer, in case

$$x(0) = 0 \text{ and } |G(t,x)| \le \alpha|x|, \alpha < 1. \tag{2.5.1.31}$$

Lemma 2.5.1.1. *Let x solve (2.5.1.24) where R solves (2.5.1.20) so that (2.5.1.21) and (2.5.1.22) hold. Then for $\epsilon > 0$*

$$\sup_{0\le s\le t<\infty} \int_s^t |R(u+\epsilon-s) - R(u-s)|du \to 0$$

as $\epsilon \to 0$.

Proof. We can write the integral as

$$\int_0^{t-s} [R(v) - R(v+\epsilon)]dv \le \int_0^t [R(v) - R(v+\epsilon)]dv$$

by a change of variable and dropping the absolute values since R is positive and decreasing by the complete monotonicity. The supremum occurs as $t \to \infty$. For $t > 2\epsilon$ we have

$$\begin{aligned}
\int_0^t [R(v) - R(v+\epsilon)]dv &= \int_0^t R(v)dv - \int_0^t R(v+\epsilon)dv \\
&= \int_0^t R(v)dv - \int_\epsilon^{\epsilon+t} R(v)dv \\
&= \int_0^\epsilon R(v)dv + \int_\epsilon^t R(v)dv - \int_\epsilon^t R(v)dv - \int_t^{t+\epsilon} R(v)dv \\
&\le \frac{k}{\Gamma(q)} \int_0^\epsilon v^{q-1}dv - \int_t^{t+\epsilon} R(v)dv \\
&= \frac{k}{\Gamma(q)} \frac{v^q}{q}\bigg|_0^\epsilon - \int_t^{t+\epsilon} R(v)dv \\
&= \frac{k}{\Gamma(q)} \frac{\epsilon^q}{q} - \int_t^{t+\epsilon} R(v)dv.
\end{aligned}$$

As $\epsilon \to 0$ the first term tends to zero. As the integral of R converges, the last term tends to zero as $t \to \infty$ for any fixed ϵ.

The first inequality in the proof of the next theorem can be iterated and the result is actually true for $F^* \in L^{2^p}$ and p is any positive integer. The proof will show another way in which the Liapunov functional and the integral equation are united without using the chain rule.

Theorem 2.5.1.6. *Let x solve (2.5.1.24) so that (2.5.1.21) and (2.5.1.22) hold. Let (2.5.1.31) be satisfied. If $F^* \in L^2[0,\infty)$, so is $G(t,x(t))$.*

Proof. More work is required to unite the Liapunov functional with the integral equation. In (2.5.1.30) we now have $p = 2$ and we will contrive a derivative of V including the term $\int_0^t R(t-s)G^2(s,x(s))ds$. We must work with (2.5.1.24) to obtain that term. Here are the details. For any $\epsilon > 0$ we can find $M > 0$ so that

$$\begin{aligned} x^2(t) &\leq M(F^*)^2(t) + (1+\epsilon)\left(\int_0^t R(t-s)G(s,x(s))ds\right)^2 \\ &\leq M(F^*)^2(t) + (1+\epsilon)\int_0^t R(t-s)ds \int_0^t R(t-s)G^2(s,x(s))ds \end{aligned}$$

or

$$x^2(t) \leq M(F^*)^2(t) + (1+\epsilon)\int_0^t R(t-s)G^2(s,x(s))ds. \qquad (2.5.1.32)$$

If we had $F^* \in L^4$ we would have taken $p = 4$ and we would have iterated the above process and obtained $\int_0^t R(t-s)G^4(s,x(s))ds$. For a general p, see the proof of Lemma 2.1.2.

Let V be defined by (2.5.1.30) with $p = 2$ and obtain

$$\begin{aligned} V'(t,\epsilon) &= \int_\epsilon^\infty R(u)du G^2(t,x(t)) - \int_0^t R(t+\epsilon-s)G^2(s,x(s))ds \\ &\quad \text{(prepare to use (2.5.1.32))} \\ &\leq G^2(t,x(t)) - \int_0^t R(t-s)G^2(s,x(s))ds \\ &\quad + \int_0^t [R(t-s) - R(t+\epsilon-s)]G^2(s,x(s))ds \\ &\quad \text{(use (2.5.1.32) and note that the sense of } V' \text{ is unchanged)} \\ &\leq G^2(t,x(t)) + \frac{1}{1+\epsilon}[M(F^*)^2(t) - x^2(t)] \\ &\quad + \int_0^t [R(t-s) - R(t+\epsilon-s)]G^2(s,x(s))ds, \end{aligned}$$

as seen in the above display. But $x^2(t) \geq G^2(t,x(t))/\alpha^2$ where $\alpha < 1$, as seen in (2.5.1.31). This yields

$$V'(t,\epsilon) \leq G^2(t,x(t)) + \frac{M}{1+\epsilon}(F^*)^2(t) - \frac{1}{\alpha^2(1+\epsilon)}G^2(t,x(t)) + \int_0^t [R(t-s) - R(t+\epsilon-s)]G^2(s,x(s))ds.$$

But $\alpha < 1$ so we can take ϵ so small that $1 - \frac{1}{\alpha^2(1+\epsilon)} = -\mu$, for some $\mu > 0$.

If we integrate from 0 to t the last term is

$$\int_0^t \int_0^u [R(u-s) - R(u+\epsilon-s)]G^2(s,x(s))dsdu = \int_0^t \int_s^t [R(u-s) - R(u+\epsilon-s)]duG^2(s,x(s))ds$$

by the Hobson-Tonelli test. For a sufficiently small ϵ by Lemma 2.5.1.1 we have $\int_s^t [R(u-s) - R(u+\epsilon-s)]du < \mu/2$. This yields

$$0 \leq V(t,\epsilon) \leq V(0,\epsilon) - \frac{\mu}{2}\int_0^t G^2(s,x(s))ds + \frac{M}{1+\epsilon}\int_0^t (F^*)^2(s)ds.$$

Notes The material in Section 2.5.1 was taken from Burton (2011). It was first published by Elsevier in *N*onlinear Analysis: TMA with full details given in the bibliography.

2.6 Liapunov Functionals for Resolvents

We can watch how Liapunov functions have evolved. With small changes at each step the Liapunov function for

$$x'(t) = -x(t)$$

is applied to

$$x'(t) = -x(t) + x(t-1),$$

then to

$$x'(t) = -x(t) + \int_{t-1}^t a(s)x(s)ds,$$

and then to

$$x'(t) = -x(t) + \int_0^t a(t,s)x(s)ds,$$

with many nonlinearities added along the way. At the end of Section 3.5 we will see the Liapunov functional for

$$x'(t) = A(t)x(t) + \int_0^t B(t,u)x(u)du$$

being changed minutely to yield properties of the resolvent

$$\frac{\partial}{\partial t}Z(t,s) = A(t)Z(t,s) + \int_s^t B(t,u)Z(u,s)du, \quad Z(s,s) = I.$$

It is so simple because the main change is only the lower limit on the integral. Turning to integral equations, we have seen the old Liapunov functional constructed by Krasovskii for delay equations being modified to cover integral equations of the form

$$x(t) = a(t) - \int_0^t C(t,s)x(s)ds$$

and we have deduced properties of the resolvent, $R(t,s)$, by studying that result together with the variation of parameters formula

$$x(t) = a(t) - \int_0^t R(t,s)a(s)ds.$$

Thus, the properties of the resolvent were always intimately connected with properties of $a(t)$. We had enormous flexibility. No connection between $a(t)$ and $C(t,s)$ was required.

But the resolvent equation involves only $C(t,s)$, totally independent of any forcing function $a(t)$. Surely, we can get closer to the basic properties of $R(t,s)$ by studying

$$R(t,s) = C(t,s) - \int_s^t C(t,u)R(u,s)du.$$

Here, the forcing function $C(t,s)$ is the same as the kernel. All of our flexibility has vanished. Thus, the resolvent equation seems so different than the resolvent of the integrodifferential equation that we can hardly believe that the old Liapunov functionals can be advanced to it. But they can. And that is the substance of this section. The necessary changes are surprisingly small and the reader will anticipate most of the results as we move along.

In fact, the Liapunov functionals in this case are surprisingly superior to those in the former cases. The ones we obtain here are positive definite with negative definite derivatives (with respect to the quantity $C(t,s)-R(t,s)$). Moreover, we are now in a position to use all the theory developed in Section 8.3 of Burton (2005c) to attack the problem of showing that $C(t,s)-R(t,s)$ tends to zero. While $x(t)$ has so frequently followed $a(t)$ in previous work, here we see that $R(t,s)$ follows $C(t,s)$.

2.6.1 Introduction

We return to the three scalar equations

$$x(t) = a(t) - \int_0^t C(t,s)x(s)ds, \tag{2.6.1}$$

$$R(t,s) = C(t,s) - \int_s^t C(t,u)R(u,s)du, \tag{2.6.2}$$

and variation of parameters formula

$$x(t) = a(t) - \int_0^t R(t,s)a(s)ds. \tag{2.6.3}$$

We have seen many results supporting the idea that if C is a "nice" kernel, then $x(t)$ follows $a(t)$. Here, we notice that it is also true that $R(t,s)$ follows $C(t,s)$. In fact, for fixed s then $C(t,s)$ is an excellent $L^1[s,\infty)$ approximation to $R(t,s)$ and this should be useful in many problems. The first step in studying this is to note the fundamental property that $\int_s^t |R(u,s)|^p du \leq K\int_s^t |C(u,s)|^p du$ for some positive integer p. The real surprise is that this relation can be obtained either from properties of the derivatives of C with respect to both t and s or from integrals of C with respect to both t and s.

In Section 2.5 we worked with $C(t,s)$ being weakly singular and studied the resulting resolvent, $R(t-s)$, using it in a Liapunov functional. Here we ask more of C.

For much of the work here it will more than suffice to ask that a and C be continuous on $[0,\infty)$ and $[0,\infty)\times[0,\infty)$, respectively. However, there will be times when we will indicate certain partial derivatives of C and it will be assumed that these are continuous. The main requirement would be that the Liapunov functional defined in (2.6.10) can be differentiated by Leibnitz's rule.

Our objective is to construct Liapunov functionals for (2.6.2) from which we can deduce properties of $R(t,s)$ so that we will be able to obtain properties of the solution, $x(t)$, from the integral in (2.6.3).

It is clear that we need to know properties of the integral of R with respect to s. And it is difficult because s is the initial time. Thus, it will often be convenient to use our fundamental Theorems 2.1.2 and 2.1.3 which we state here for ready reference.

Theorem 2.6.1.1. *The solution of (2.6.1) is bounded for every bounded continuous function $a(t)$ if and only if*

$$\sup_{t\geq 0}\int_0^t |R(t,s)|ds < \infty. \tag{2.6.4a}$$

Theorem 2.6.1.2. *If there is a number $\alpha < 1$ such that*

$$\sup_{0\leq t<\infty}\int_0^t |C(t,s)|ds \leq \alpha \tag{2.6.5a}$$

then each solution of (2.6.1) is bounded for every bounded and continuous function a. (See Theorem 2.1.3.)

Thus, (2.6.5a) implies (2.6.4a) and that will be used repeatedly. We will, however, focus on cases in which (2.6.5a) does not hold.

There is a long line of results in integral equation theory in which the objective is to show that the solution is in L^p for some p and there are important problems in which the objective is to study the integral of the solution, $x(t)$, of (2.6.1), as may be seen for example in Feller (1941). This section begins with the idea that if we are willing to work with $\int_0^t x(s)ds$ instead of just $x(t)$, then we can work with the integral of R with respect to t rather than with respect to s.

The work will focus on four sets of conditions.

Set #1 asks that there is a positive number α such that

$$C(t,t) - \int_0^t |C_s(t,s)|ds \geq \alpha$$

and

$$C(t,t) + \int_0^t |C_s(t,s)|ds$$

is bounded. This will yield (2.6.4a) and its proof is parallel to the classical proof of Theorem 2.6.1.2. It will also yield $|R(t,s)|$ bounded, a central result in this discussion.

Set #2 asks that there exist an $\alpha < 1$ with

$$\sup_{t \geq 0} \int_0^\infty |C(t+v,t)|dv \leq \alpha \tag{2.6.5b}$$

and yields

$$\sup_{0 \leq s \leq t < \infty} \int_s^t |R(v,s)|dv < \infty, \tag{2.6.4b}$$

so it is a counterpart to (2.6.5a) and (2.6.4a) with corresponding applications, both alone and with (2.6.5a).

Set #3 abandons the integral conditions, asking instead that

$$C(t,s) \geq 0, \quad C_s(t,s) \geq 0, \quad C_t(t,s) \leq 0, \quad C_{st}(t,s) \leq 0.$$

Surprisingly, the derivative conditions yield virtually the same as our integral conditions.

Set #4 allows us to "cleanse" the kernel of additive functions of t, frequently constants, asking that $\int_0^\infty |C_t(u+t,t)|du - C(t,t) \leq -\alpha$ for some positive number α, yielding results parallel to those with Set #2.

MOTIVATION

We now introduce some ideas which will be developed in the coming subsections. We will be concerned with cases in which (2.6.4a) holds, or $|R(t,s)|$ is bounded, or $\int_s^t |R(u,s)|^p du$ is bounded. These simple results direct our later investigations. Here are some of the consequences of the properties which we will prove.

Theorem 2.6.1.3. *If there is a number M with $|R(t,s)| \leq M$ for $0 \leq s \leq t < \infty$ and if $a \in L^1[0,\infty)$, then the solution x of (2.6.1) satisfies $|x(t) - a(t)| \leq K$ for some $K > 0$ and all $t \geq 0$.*

This is an immediate consequence of (2.6.3) since

$$|x(t) - a(t)| \leq \int_0^t |R(t,s)||a(s)|ds \leq M \int_0^\infty |a(s)|ds.$$

Thus, we seek conditions under which $|R(t,s)| \leq M$.

We will see that there is a certain duality between a and x, as there is between R and C.

Theorem 2.6.1.4. *If (2.6.4a) holds, if $\int_s^t |R(u,s)|du$ is bounded for $0 \leq s \leq t < \infty$, and if $a \in L^{2^n}[0,\infty)$ for some positive integer n, then $x \in L^{2^n}[0,\infty)$, where x solves (2.6.1).*

Proof. There is an M with $\int_0^t |R(t,s)|ds \leq M$. If $a \in L^2$ then

$$\begin{aligned}
x^2(t) &\leq 2\left(a^2(t) + \left(\int_0^t R(t,s)a(s)ds\right)^2\right) \\
&\leq 2\left(a^2(t) + \int_0^t |R(t,s)|ds \int_0^t |R(t,s)|a^2(s)ds\right) \\
&\leq 2\left(a^2(t) + M\int_0^t |R(t,s)|a^2(s)ds\right).
\end{aligned}$$

An integration and change of order of integration yields

$$(1/2)\int_0^t x^2(s)ds \leq \int_0^t a^2(s)ds + M\int_0^t \int_s^t |R(u,s)|du a^2(s)ds.$$

This gives the result for $n = 1$. An induction with repeated squaring and inequalities as above yields the first conclusion.

As (2.6.5a) implies (2.6.4a) we seek conditions to ensure that $\int_s^t |R(u,s)|du$ is bounded.

Theorem 2.6.1.5. *If $a \in L^1[0,\infty)$ and if $\int_s^t R^2(u,s)du$ is bounded for $0 \leq s \leq u \leq t < \infty$, then $\int_0^t (x(s) - a(s))^2 ds$ is bounded.*

Proof. There is an M with $\int_0^\infty |a(u)|du \leq M$ and $\int_s^t R^2(u,s)du \leq M$ so from (2.6.3) we have

$$\begin{aligned}
(x(t) - a(t))^2 &= \left(\int_0^t R(t,s)a(s)ds\right)^2 \\
&\leq \int_0^t |a(s)|ds \int_0^t |a(s)|R^2(t,s)ds \\
&\leq M\int_0^t |a(s)|R^2(t,s)ds.
\end{aligned}$$

Upon integration and change of order of integration we have

$$\begin{aligned}\int_0^t (x(s)-a(s))^2 ds &\le M\int_0^t\int_0^u |a(s)|R^2(u,s)dsdu\\ &= M\int_0^t\int_s^t R^2(u,s)du|a(s)|ds\\ &\le M^2\int_0^t |a(s)|ds \le M^3.\end{aligned}$$

Thus, we seek conditions under which $\int_s^t R^2(u,s)du$ is bounded. See, for example, Theorem 2.6.5.2 for sufficient conditions for this to hold.

This is readily extended. In the same way we can write

$$\begin{aligned}(x(t)-a(t))^4 &\le M^2\Big(\int_0^t |a(s)|R^2(t,s)ds\Big)^2\\ &\le M^3\int_0^t |a(s)|R^4(t,s)ds.\end{aligned}$$

Inductively we can obtain the statement that if $a\in L^1[0,\infty)$ and if $\int_s^t R^{2^n}(u,s)du$ is bounded then $\int_0^t (x(s)-a(s))^{2^n}ds$ is bounded.

From the above considerations we see that $a(t)$ is an approximation to $x(t)$. But then we look at (2.6.1) and (2.6.3) with the thought that $C(t,s)$ is surely an approximation to $R(t,s)$. Indeed it is. Two of our results give conditions under which

$$\int_s^t |R(u,s)|du \le K\int_s^t |C(u,s)|du, \quad K>0,$$

and

$$\int_s^t R^2(u,s)du \le \int_s^t C^2(u,s)du.$$

That leads us to the following results.

Theorem 2.6.1.6. *If there are positive constants M_1 and M_2 with*

$$\int_s^t |R(v,s)|dv \le M_1 \text{ and } \int_s^t |C(u,s)|du \le M_2$$

for $0\le s\le t<\infty$, then

$$\int_s^t |R(u,s)-C(u,s)|du \le M_1M_2.$$

Proof. Recall that (2.6.2) may also be written as

$$R(t,s) = C(t,s) - \int_s^t R(t,u)C(u,s)du.$$

Thus,

$$\begin{aligned}\int_s^t |R(u,s) - C(u,s)|du &\leq \int_s^t \int_s^v |R(v,u)C(u,s)|dudv \\ &= \int_s^t \int_u^t |R(v,u)|dv|C(u,s)|du \\ &\leq M_1 M_2.\end{aligned}$$

See, for example, Theorem 2.6.3.1 for sufficient conditions for this result to hold.

That relation can be far better than the obvious one of $M_1 + M_2$. If, for example, $M_2 < 1$ then $M_1 M_2 < M_1$; R and C converge to each other faster than they converge to zero in L^1. There are problems in which we would like to use $C(t,s)$ as an approximation to $R(t,s)$ so we strive to improve the relationship.

Theorem 2.6.1.7. *Suppose there are positive constants M_1, M_2, and M_3 with*

$$\int_s^t R^2(u,s)du \leq M_1, \quad 0 \leq s \leq t < \infty,$$

$$\int_u^t |C(v,u)|dv \leq M_2, \, 0 \leq u \leq t < \infty,$$

and

$$\int_s^t |C(t,u)|du \leq M_3, \quad 0 \leq s \leq t < \infty.$$

Then $\int_s^t (R(u,s) - C(u,s))^2 du \leq M_1 M_2 M_3$ for $0 \leq s \leq t < \infty$.

Proof. We have from (2.6.2) that

$$\begin{aligned}\Big(R(t,s)-C(t,s)\Big)^2 &= \Big(\int_s^t C(t,u)R(u,s)du\Big)^2\\ &\le \int_s^t |C(t,u)|du \int_s^t |C(t,u)|R^2(u,s)du\end{aligned}$$

so

$$\begin{aligned}\int_s^t (R(u,s)-C(u,s))^2du &\le M_3\int_s^t\int_s^v |C(v,u)|R^2(u,s)dudv\\ &\le M_3\int_s^t\int_u^t |C(v,u)|R^2(u,s)dvdu\\ &\le M_3\int_s^t R^2(u,s)\int_u^t |C(v,u)|dvdu\\ &\le M_3M_2M_1.\end{aligned}$$

This completes the proof.

Thus, we seek conditions under which $\int_s^t R^2(u,s)du$ is bounded.

In Liapunov theory for differential equations we almost always find the derivative of the Liapunov function negative definite so that a solution is integrable in some sense. That integrability is then parlayed into a supremum bound because the Liapunov function is positive definite. The same happens with Liapunov theory for resolvents, but it is so much more difficult to see. To begin with we will look at a Liapunov functional for (2.6.2) of the form

$$V(t)=\int_s^t\int_{t-w}^{\infty} |C(u+w,w)|du|R(w,s)|dw$$

and obtain conditions to ensure that

$$V(t)\le \int_s^t |R(u,s)|du \le K\int_s^t |C(u,s)|du$$

for some fixed constant K. If that second integral is bounded, so is the first. But how can we parlay that into boundedness of $R(t,s)$? In fact, that Liapunov functional is positive definite with respect to $(R(t,s)-C(t,s))$. The next result is used in Theorem 2.6.3.2.

While we are considering C continuous here, the reader will recall from Proposition 1.2.1 and nearby material that if C is weakly singular so is R with singularities at the same place and of the same type. We also recall that after the corollary to Theorem 2.5.1.5 we introduced an ϵ in a Liapunov functional just like this one to study the singular case. The same can be done here.

Theorem 2.6.1.8. *If there is a positive constant β with*

$$\int_{t-w}^{\infty} |C(u+w,w)|du \geq \beta|C(t,w)|$$

then

$$\beta|R(t,s) - C(t,s)| \leq V(t) := \int_s^t \int_{t-w}^{\infty} |C(u+w,w)|du|R(w,s)|dw$$

along the solution $R(t,s)$ of (2.6.2).

Proof. We see immediately that

$$\begin{aligned} V(t) &\geq \beta \int_s^t |C(t,w)||R(w,s)|dw \\ &\geq \beta|R(t,s) - C(t,s)| \end{aligned}$$

where the last line follows from (2.6.2).

In the same vein we have a similar result when the Liapunov functional contains a quadratic term.

Theorem 2.6.1.9. *If there are positive constants β and M with*

$$\int_{t-w}^{\infty} |C(u+w,w)|du \geq \beta|C(t,w)|$$

and

$$\int_s^t |C(t,u)|du \leq M$$

then

$$(\beta/M)(R(t,s) - C(t,s))^2 \leq V(t) := \int_s^t \int_{t-w}^{\infty} |C(u+w,w)|duR^2(w,s)dw.$$

Proof. We have

$$
\begin{aligned}
(R(t,s)-C(t,s))^2 &\leq \left(\int_s^t |C(t,w)R(w,s)|dw\right)^2 \\
&\leq \int_s^t |C(t,w)|dw \int_s^t |C(t,w)|R^2(w,s)dw \\
&\leq M\int_s^t |C(t,w)|R^2(w,s)dw \\
&= (M/\beta)\int_s^t \beta|C(t,w)|R^2(w,s)dw \\
&\leq (M/\beta)\int_s^t \int_{t-w}^\infty |C(u+w,w)|duR^2(w,s)dw \\
&= (M/\beta)V(t).
\end{aligned}
$$

2.6.2 Another Variation of Parameters Formula

We now obtain two related consequences of (2.6.1) and (2.6.3). Integration of (2.6.1) by parts yields

$$x(t) = a(t) - C(t,t)\int_0^t x(u)du + \int_0^t C_s(t,s)\int_0^s x(u)duds.$$

When $C_s(t,s) := \frac{\partial C(t,s)}{\partial s}$ is continuous we define $y(t) = \int_0^t x(s)ds$ obtaining

$$y'(t) = a(t) - C(t,t)y(t) + \int_0^t C_s(t,s)y(s)ds,\ y(0)=0. \tag{2.6.6}$$

Next, integrate the resolvent equation for (2.6.1) which is (2.6.3) and obtain

$$\int_0^t x(s)ds = \int_0^t a(u)du - \int_0^t \int_0^u R(u,s)a(s)dsdu$$

or

$$\int_0^t x(s)ds = \int_0^t a(u)du - \int_0^t \int_s^t R(u,s)dua(s)ds, \tag{2.6.7}$$

which critically integrates R with respect to its first component. In terms of y we have

$$y(t) = \int_0^t a(u)du - \int_0^t \int_s^t R(u,s)dua(s)ds. \tag{2.6.8}$$

Notice from (2.6.6) that if

$$|a(t)| + |C(t,t)| + |y(t)| + \int_0^t |C_s(t,s)|ds$$

is bounded, so is $x(t)$. Notice also that neither $x(t)$ nor $\int_0^t x(s)ds$ depend on being able to differentiate C. Our equation (2.6.6) is merely a help in showing $x(t)$ bounded from $y(t)$ bounded.

It may happen that $C(t,s)$ has an additive constant which is cleansed in (2.6.6). Thus, for example, if $C_s(t,s) \geq 0$ then

$$\int_0^t |C_s(t,s)|ds = \int_0^t C_s(t,s)ds = C(t,s)\Big|_0^t = C(t,t) - C(t,0).$$

It may then happen that $C(t,t) - C(t,0) \geq \alpha > 0$ and that can be most useful.

The next result may be compared to Theorem 3.2.10 in Chapter 3.

Theorem 2.6.2.1. *Suppose there is an $\alpha > 0$ such that*

$$C(t,t) - \int_0^t |C_s(t,s)|ds \geq \alpha$$

on $[0,\infty)$. Then every solution of (2.6.6) is bounded for every bounded and continuous function a. In particular, then, if in addition we have

$$C(t,t) + \int_0^t |C_s(t,s)|ds$$

bounded then (2.6.4a) holds.

Proof. Let a be bounded and continuous and let y solve (2.6.6) on $[0,\infty)$. Use the Razumikhin function $V(t) = |y(t)|$ and obtain

$$V'_{(2.6.6)}(t) \leq |a(t)| - |C(t,t)||y(t)| + \int_0^t |C_s(t,s)||y(s)|ds.$$

Let $\|a\| = M$ and find $J > 0$ with $\alpha J > M$. If, by way of contradiction, $y(t)$ is not bounded then there is a $t_0 > 0$ with $|y(t)| < J$ on $[0,t_0)$, but $|y(t_0)| = J$ so that $V'(t_0) \geq 0$. But $C(t_0,t_0) > 0$ and

$$V'_{(2.6.6)}(t_0) \leq M - C(t_0,t_0)J + J\int_0^{t_0} |C_s(t_0,s)|ds \leq M - \alpha J < 0,$$

a contradiction. Now from (2.6.6) we have that $y'(t) = x(t)$ which is bounded for every bounded and continuous $a(t)$. The result now follows from Theorem 2.6.1.1.

Thus, Theorems 2.6.1.2 and 2.6.2.1 give us two simple ways of showing (2.6.4a).

We can follow the idea in the proof of Theorem 2.6.2.1 to show that $R(t,s)$ is bounded. We refer the reader to Theorem 2.6.1.3 for an application.

Theorem 2.6.2.2. *Suppose there is an $\alpha > 0$ and an $M > 0$ with $C(t,t) - \int_s^t |C_u(t,u)|du \geq \alpha$ and $|C(t,s)| \leq M$ for $0 \leq s \leq t < \infty$. Choose $J > 0$ so that $\alpha J > M$. Then the solution of (2.6.2) satisfies*

$$|R(t,s)| \leq M + 2C(t,t)J$$

for $0 \leq s \leq t < \infty$. If, in addition, $a \in L^1[0,\infty)$ then in (2.6.3) $x - a$ is bounded.

Proof. Integrate (2.6.2) by parts obtaining

$$R(t,s) = C(t,s) - C(t,t)\int_s^t R(v,s)dv + \int_s^t C_u(t,u)\int_s^u R(v,s)dvdu$$

and write $H(t,s) = \int_s^t R(v,s)dv$ so that this last equation can be written as

$$H_t(t,s) = C(t,s) - C(t,t)H(t,s) + \int_s^t C_u(t,u)H(u,s)du. \qquad (2.6.9)$$

Fix s and define a Razumikhin function by

$$V(t) = |H(t,s)|$$

so that the derivative of V along a solution of the last equation on $[s,\infty)$ satisfies

$$V'(t) \leq |C(t,s)| - C(t,t)|H(t,s)| + \int_s^t |C_u(t,u)||H(u,s)|du.$$

Let $|H(t,s)| < J$ on an interval $[s,t_0)$ and suppose that $|H(t_0,s)| = J$ so that $V'(t_0) \geq 0$. Then

$$V'(t_0) \leq M - |C(t_0,t_0)|J + J\int_s^{t_0} |C_u(t_0,u)|du \leq M - \alpha J < 0,$$

a contradiction. Thus, $|H(t,s)| \leq J$ for all $t \geq s$. Notice that $C(t,t) \geq \int_s^t |C_u(t,u)|du$ and so we have

$$|R(t,s)| = |H_t(t,s)| \leq M + 2JC(t,t),$$

as required. Referring now to (2.6.3) the final conclusion follows readily.

2.6.3 A first Liapunov Functional

One of our goals now is to find properties of $\int_s^t R(u,s)du$ for use in (2.6.7). Notice in (i) below that for fixed s, then $C(t,s)$ becomes an $L^1[s,\infty)$ approximation to $R(t,s)$, while (2.6.12) can be used directly in (2.6.7).

Theorem 2.6.3.1. *Let $\int_{t-u}^{\infty} |C(u+v,u)|dv$ be continuous for $0 \le u \le t < \infty$. If (2.6.5b) holds then for $V(t)$ defined by*

$$V(t) = \int_s^t \int_{t-u}^{\infty} |C(u+v,u)|dv|R(u,s)|du \tag{2.6.10}$$

the derivative of V along the unique solution of (2.6.2) satisfies

$$V'(t) \le -(1-\alpha)|R(t,s)| + |C(t,s)| \tag{2.6.11}$$

and for $0 \le s \le t < \infty$ we have

$$\int_s^t |R(v,s)|dv \le \frac{1}{1-\alpha}\int_s^t |C(v,s)|dv. \tag{2.6.12}$$

Thus, if there is a positive constant M with $\int_s^t |C(v,s)|dv \le M$ for $0 \le s \le t < \infty$ then (2.6.4b) holds and also, as in Theorem 2.6.1.6,

$$(i) \quad \int_s^t |R(u,s) - C(u,s)|du \le \frac{M^2}{1-\alpha}.$$

If, in addition, $a \in L^1[0,\infty)$ then

$$(ii) \quad (x-a) \in L^1[0,\infty), \quad x \in L^1[0,\infty).$$

Proof. To prove the result we have from (2.6.10) that

$$\begin{aligned} V'(t) &= \int_0^{\infty} |C(v+t,t)|dv|R(t,s)| - \int_s^t |C(t,u)||R(u,s)|du \\ &\le \alpha|R(t,s)| - \int_s^t |C(t,u)||R(u,s)|du \end{aligned}$$

so that

$$\begin{aligned} V'(t) &\le \alpha|R(t,s)| + |C(t,s)| - |R(t,s)| \\ &= -(1-\alpha)|R(t,s)| + |C(t,s)| \end{aligned}$$

from which (2.6.11) and (2.6.12) follow. Now (i) is just Theorem 2.6.1.6.

Finally, by changing the order of integration we have

$$\int_0^t |x(s) - a(s)|ds \le \int_0^t \int_s^t |R(u,s)|du|a(s)|ds$$
$$\le \frac{1}{1-\alpha} \int_0^t \int_s^t |C(u,s)|du|a(s)|ds \le \frac{1}{1-\alpha} M \int_0^\infty |a(s)|ds.$$

This completes the proof.

Notice that (ii) is a counterpart of the Adam and Eve theorem. It integrates the first coordinate of C and yields $x \in L^1$, whereas the Adam and Eve theorem integrates the second coordinate of C and yields x bounded.

As discussed following (2.6.4a) and (2.6.5a), when we ask that $\int_s^t |C(t,u)|du \le \alpha < 1$ we will have every solution of (2.6.1) bounded for every bounded and continuous $a(t)$ and we also will have (2.6.4a). But asking this condition does not seem to yield conditions on x for $a \in L^1$. We seem to need to add that $\int_s^t |C(u,s)|du$ is bounded in order to get the needed condition that $\int_s^t |R(u,s)|du$ be bounded.

We saw in Theorem 2.6.1.8 that $\beta|R(t,s) - C(t,s)| \le V(t)$. We also saw in Theorem 2.6.3.1 that

$$V'(t) \le -(1-\alpha)|R(t,s)| + |C(t,s)|.$$

If we add the condition that $\int_s^t |C(u,s|du$ is bounded then the next result will put us in a position to use the theory developed in Section 8.3 of Burton (2005c) to show that $R(t,s)$ converges to $C(t,s)$ pointwise for fixed s.

Theorem 2.6.3.2. *Let (2.6.5b) hold and define V by (2.6.10). Suppose that there are $\beta > 0$, $J > 0$ and $K > 0$ with*

$$\int_s^t (1+\beta)|C(u,s)|du \le J, \; 0 \le s \le t < \infty,$$

$$K|R(t,s) - C(t,s)| \le V(t),$$

and

$$V'(t) \le -\beta|R(t,s)| + |C(t,s)|.$$

Then for

$$W(t) = [1 + V(t)]e^{-\int_s^t (1+\beta)|C(u,s)|du}$$

we have

$$[1 + K|R(t,s) - C(t,s)|]e^{-J} \le W(t)$$

and

$$W'(t) \le -\beta|R(t,s) - C(t,s)|e^{-J}.$$

Proof. Now

$$\begin{aligned} V'(t) &\leq -\beta|R(t,s)| + |C(t,s)| \\ &\leq -\beta|R(t,s) - C(t,s)| + (\beta+1)|C(t,s)|. \end{aligned}$$

Clearly the lower bound on W holds. A calculation verifies the stated derivative of W.

Refer to Theorem 2.6.1.3 for consequences of the next result.

Theorem 2.6.3.3. *Let (2.6.5b) hold, let $\int_{t-u}^{\infty} |C(u+v,u)|dv$ be continuous, let $|C(t,s)| \leq M$ for some $M > 0$ and $0 \leq s \leq t < \infty$, and suppose there is a differentiable function $\Phi : [0,\infty) \downarrow (0,\infty)$ with $\Phi \in L^1[0,\infty)$ and*

$$\Phi(t-u) \geq \int_{t-u}^{\infty} |C(u+v,u)|dv$$

for $0 \leq u \leq t < \infty$. If, in addition, there is a $K > 0$ with

$$\int_{t-u}^{\infty} |C(u+v,u)|dv \geq K|C(t,u)|$$

then $|R(t,s)|$ is bounded.

Proof. As (2.6.5b) holds, so does (2.6.11) for (2.6.10). Fix $s \geq 0$. We first show that $V(t)$, as defined in (2.6.10), is bounded. Suppose that t is chosen so that $V(t) = \max_{s \leq u \leq t} V(u)$. Now for such u we have from (2.6.11) that

$$\frac{dV(u)}{du} \leq -(1-\alpha)|R(u,s)| + |C(u,s)|$$

where $\alpha < 1$ and then

$$\frac{dV(u)}{du}\Phi(t-u) \leq -(1-\alpha)|R(u,s)|\Phi(t-u) + M\Phi(t-u).$$

Thus,

$$\int_s^t \frac{dV(u)}{du}\Phi(t-u)du \leq -(1-\alpha)\int_s^t |R(u,s)|\Phi(t-u)du + M\int_s^t \Phi(t-u)du$$

and then since $V(s) = 0$ we have

$$\begin{aligned}
\int_s^t \frac{dV(u)}{du}\Phi(t-u)du &= V(u)\Phi(t-u)\Big|_s^t - \int_s^t V(u)\frac{d\Phi(t-u)}{du}du \\
&= V(t)\Phi(0) - \int_s^t V(u)\frac{d\Phi(t-u)}{du}du \\
&\geq V(t)\Phi(0) - V(t)\int_s^t \frac{d\Phi(t-u)}{du}du \\
&\quad (\text{since } V(t) \text{ is the maximum and } \frac{d\Phi(t-u)}{du} \geq 0) \\
&= V(t)\Phi(0) - V(t)[\Phi(0) - \Phi(t-s)] \\
&= V(t)\Phi(t-s) \geq 0.
\end{aligned}$$

Hence,

$$(1-\alpha)\int_s^t |R(u,s)|\Phi(t-u)du \leq M\int_0^\infty \Phi(u)du.$$

Then

$$\begin{aligned}
(1-\alpha)V(t) &= (1-\alpha)\int_s^t |R(u,s)|\int_{t-u}^\infty |C(u+v,u)|dvdu \\
&\leq (1-\alpha)\int_s^t |R(u,s)|\Phi(t-u)du \\
&\leq M\int_0^\infty \Phi(u)du
\end{aligned}$$

or $V(t)$ is bounded. But

$$\begin{aligned}
V(t) &= \int_s^t |R(u,s)|\int_{t-u}^\infty |C(u+v,u)|dvdu \\
&\geq K\int_s^t |C(t,u)||R(u,s)|du \\
&\geq K[|R(t,s)| - |C(t,s)|
\end{aligned}$$

so that boundedness of $V(t)$ and $|C(t,s)|$ yields the boundedness of $R(t,s)$.

Refer to Theorems 2.6.1.5 and 2.6.1.7 for consequences of the next result. Integral bounds on R^2 such as obtained below can also be used in the alternate form of (2.6.2) with the Schwarz inequality to obtain a pointwise bound on $R(t,s)$ of the type used in Theorem 2.6.1.3.

Theorem 2.6.3.4. *If (2.6.5a) and (2.6.5b) hold and if* $\int_{t-u}^{\infty} |C(u+v,u)|dv$ *is continuous then there is a constant* M *with*

$$\int_s^t R^2(u,s)du \le \frac{M}{1-\alpha}\int_s^t C^2(u,s)du.$$

Proof. Find an $\epsilon > 0$ such that $(1+\epsilon)\alpha = 1$ and then find $M > 0$ such that

$$\begin{aligned} R^2(t,s) &\le MC^2(t,s) + (1+\epsilon)\left(\int_s^t |C(t,u)R(u,s)|du\right)^2 \\ &\le MC^2(t,s) + (1+\epsilon)\int_s^t |C(t,u)|du \int_s^t |C(t,u)|R^2(u,s)du \\ &\le MC^2(t,s) + \int_s^t |C(t,u)|R^2(u,s)du. \end{aligned}$$

Next, define

$$V(t) = \int_s^t \int_{t-u}^{\infty} |C(u+v,u)|dvR^2(u,s)du$$

so that

$$\begin{aligned} V'(t) &= \int_0^{\infty} |C(t+v,t)|dvR^2(t,s) - \int_s^t |C(t,u)|R^2(u,s)du \\ &\le \alpha R^2(t,s) + MC^2(t,s) - R^2(t,s) \\ &= -(1-\alpha)R^2(t,s) + MC^2(t,s). \end{aligned}$$

An integration yields the result.

This result is a special case of the next one, but they are worth separating for the following reason. Two things can now be seen. First, if we refer to Theorem 2.6.1.9 we see that we can add conditions to ensure that

$$K(R(t,s) - C(t,s))^2 \le V(t)$$

for some positive constant K. We could then follow Theorem 2.6.3.2 and construct a new Liapunov functional, W, with the classical properties of Liapunov functions for differential equations.

Theorem 2.6.3.5. *Let (2.6.5a) and (2.6.5b) hold. Then*

$$\sup_{0\le s\le t<\infty}\int_s^t C^{2^k}(u,s)du<\infty$$

implies that

$$\sup_{0\le s\le t<\infty}\int_s^t R^{2^k}(u,s)du<\infty$$

for any positive integer k.

Proof. This is proved by squaring an inequality in the previous proof to obtain

$$\begin{aligned}R^4(t,s)&\le M^2C^4(t,s)+(1+\epsilon)\Big(\int_s^t|C(t,u)|R^2(u,s)du\Big)^2\\&\le M^2C^4(t,s)+(1+\epsilon)\int_s^t|C(t,u)|du\int_s^t|C(t,u)|R^4(u,s)du\\&\le M^2C^4(t,s)+\int_s^t|C(t,u)|R^4(u,s)du.\end{aligned}$$

Then define

$$V(t)=\int_s^t\int_{t-u}^\infty|C(u+v,u)|dvR^4(u,s)du$$

so that

$$\begin{aligned}V'(t)&=\int_0^\infty|C(t+v,t)|dvR^4(t,s)-\int_s^t|C(t,u)|R^4(u,s)du\\&\le\alpha R^4(t,s)+M^2C^4(t,s)-R^4(t,s)\\&=-(1-\alpha)R^4(t,s)+M^2C^4(t,s)\end{aligned}$$

from which the result follows for $k=2$. An induction is clear.

In the result below we do not treat the case of $k=0$, but one may verify that we can conclude also that $x\in L^1[0,\infty)$ under weaker conditions than (2.6.5a). That is, as (2.6.5b) holds, so does (2.6.12). Then ask that $\int_s^t |C(v,s)|dv$ is bounded and refer to (2.6.12) again. Thus, in

$$x(t) = a(t) - \int_0^t R(t,s)a(s)ds$$

we take the absolute value, integrate, and interchange the order of integration and obtain

$$\int_0^t |x(s)|ds \leq \int_0^t |a(s)|ds + \int_0^t \int_s^t |R(u,s)|du|a(s)|ds$$

which is bounded if $a \in L^1[0,\infty)$.

Theorem 2.6.3.6. *Let (2.6.5a) and (2.6.5b) hold with*

$$\sup_{0\leq s\leq t<\infty} \int_s^t C^{2^k}(u,s)du < \infty$$

for some positive integer k. If $a \in L^1$ and $a \in L^{2^k}$ then $x \in L^{2^k}$.

Proof. If we square (2.6.3), use the Schwarz inequality, and interchange the order of integration we have

$$\begin{aligned} x^2(t) &\leq 2\left(a^2(t) + \left(\int_0^t R(t,s)a(s)ds\right)^2\right) \\ &\leq 2\left(a^2(t) + \int_0^t |a(s)|ds \int_0^t |a(s)|R^2(t,s)ds\right). \end{aligned}$$

Then by the integrability of a there is an $M > 0$ with

$$\begin{aligned} (1/2)\int_0^t x^2(s)ds &\leq \int_0^t a^2(s)ds + M\int_0^t \int_0^u |a(s)|R^2(u,s)dsdu \\ &= \int_0^t a^2(s)ds + M\int_0^t \int_s^t R^2(u,s)du|a(s)|ds \end{aligned}$$

which is bounded by Theorem 2.6.3.5 when we take $k = 1$. An induction is clear.

We will see in Section 2.9 that essentially the same results can be obtained by asking conditions on the derivatives of C (convexity) as were obtained from the integrals of C.

2.6.4 A sum of kernels and functionals

We begin with Theorem 2.6.3.4 which asked for $\alpha < 1$ with

$$\sup_{0\leq t<\infty}\int_0^t |C(t,s)|ds \leq \alpha, \quad \sup_{t\geq 0}\int_0^\infty |C(t+v,t)|dv \leq \alpha, \tag{2.6.13}$$

yielding an $M > 0$ with

$$\int_s^t R^2(u,s)du \leq \frac{M}{1-\alpha}\int_s^t C^2(u,s)du. \tag{2.6.14}$$

Notice that if C is of convolution type then this would say that $C \in L^2[0,\infty)$ implies $R \in L^2[0,\infty)$. There would then come to mind the very useful classical results that

$$a \in L^1[0,\infty) \implies \int_0^t R^2(t-s)a(s)ds \in L^1[0,\infty)$$

and

$$a(t) \to 0 \implies \int_0^t R^2(t-s)a(s)ds \to 0$$

as $t \to \infty$. The proofs of such relations are associated with part (ii) of Theorem 2.1.3 stating that if $\int_0^t |C(t,s)|ds \leq \alpha < 1$ then for $T > 0$

$$\int_0^T |C(t,s)|ds \to 0 \implies \int_0^T |R(t,s)|ds \to 0 \tag{2.6.15}$$

as $t \to \infty$.

All of this involves the small kernel assumption (2.6.13) yielding (2.6.14). It is a very curious fact that (2.6.13) can be replaced by convexity with the result that (2.6.14) is retained. Even more interesting is a fact, shown here, for the integral equation

$$x(t) = a(t) - \int_0^t [D(t,s) + C(t,x)]x(s)ds \tag{2.6.16}$$

with $D : [0,\infty) \times [0,\infty)$ continuous and with resolvent equation

$$H(t,s) = D(t,s) + C(t,s) - \int_s^t [D(t,u) + C(t,u)]H(u,s)du. \tag{2.6.17}$$

We show that if C is convex

$$C(t,s) \geq 0,\ C_s(t,s) \geq 0,\ C_t(t,s) \leq 0,\ C_{st}(t,s) \leq 0 \tag{2.6.18}$$

and

$$\int_0^t |D(t,s)|ds \le \alpha, \quad \int_0^\infty |D(u+t,t)|du \le \beta, \quad 0 \le t < \infty \tag{2.6.19}$$

with

$$\alpha + \beta < 2 \tag{2.6.20}$$

then there is a $K > 0$ with

$$\int_s^t H^2(u,s)du \le K \int_s^t [C^2(u,s) + D^2(u,s)]du. \tag{2.6.21}$$

This is a superior form of (2.6.12) and (2.6.14) since either α or β can exceed 1 so long as (2.6.20) holds. Moreover, with additional conditions we will have

$$\int_0^T H(t,s)ds \to 0$$

as $t \to \infty$, capturing the aforementioned classical convolution type results.

Finally, we will find that for a fixed s then $H(t,s)$ converges to $D(t,s) + C(t,s)$ both pointwise and in L^2.

Theorem 2.6.4.1. *Let C satisfy (2.6.18) and D satisfy (2.6.19) and (2.6.20). Then:*

(i) There is a $K > 0$ with

$$\int_s^t H^2(u,s)du \le K \int_s^t [C^2(u,s) + D^2(u,s)]du := L(t,s). \tag{2.6.22}$$

(ii) If, in addition,

$$L(t,s),\ C(t,s) + D(t,s),\ C_t(t,s) + D_t(t,s), \tag{2.6.23}$$

$$\int_s^t [C^2(t,u) + D^2(t,u)]du,\ \int_s^t [C_t^2(t,u) + D_t^2(t,u)]du$$

are all bounded, then $\sup_{0 \le s \le t < \infty}[|H(t,s)| + |H_t(t,s)|] < \infty$ and for fixed $s \ge 0$

$$\lim_{t\to\infty} H(t,s) = 0. \tag{2.6.24}$$

(iii) If, in addition,

$$|B_s(t,s)| + |B(s,s)| + \int_s^t |B_s(u,s)|du \tag{2.6.25}$$

is also bounded, then for each $T > 0$

$$\lim_{t\to\infty} \int_0^T |H(t,s)|ds = 0. \tag{2.6.26}$$

Note. The conditions on C and D are independent in the sense that we can set $C \equiv 0$ and have the result concerning the resolvent for D alone, or we can set $D \equiv 0$ and have the conditions for the resolvent for C alone.

Proof. We begin by defining a Liapunov functional for (2.6.17) in the form

$$V(t) = \int_s^t \int_{t-v}^\infty |D(u+v,v)|duH^2(v,s)dv + C(t,s)\left(\int_s^t H(u,s)du\right)^2 + \int_s^t C_v(t,v)\left(\int_v^t H(u,s)du\right)^2 dv.$$

Taking into account that $C_{vt}(t,v) \le 0$ and $C_t(t,s) \le 0$ we have

$$V'(t) \le \int_0^\infty |D(u+t,t)|duH^2(t,s) - \int_s^t |D(t,v)|H^2(v,s)dv + 2H(t,s)C(t,s)\int_s^t H(u,s)du + 2H(t,s)\int_s^t C_v(t,v)\int_v^t H(u,s)dudv.$$

If we integrate the last term by parts we have

$$2H(t,s)\left[C(t,v)\int_v^t H(u,s)du\Big|_s^t + \int_s^t C(t,v)H(v,s)dv\right]$$
$$= 2H(t,s)\left[-C(t,s)\int_s^t H(u,s)du + \int_s^t C(t,v)H(v,s)dv\right].$$

Canceling terms and taking (2.6.17) into account we have

$$\begin{aligned}
V' &\le \beta H^2(t,s) - \int_s^t |D(t,v)|H^2(v,s)dv \\
&\quad + 2H(t,s)\left[C(t,s) + D(t,s) - H(t,s) - \int_s^t D(t,u)H(u,s)du\right] \\
&\le \beta H^2(t,s) - \int_s^t |D(t,v)|H^2(v,s)dv \\
&\quad + 2H(t,s)[C(t,s) + D(t,s) - H(t,s)] \\
&\quad + \int_s^t |D(t,u)|\big(H^2(u,s) + H^2(t,s)\big)du \\
&\le (\alpha+\beta)H^2(t,s) + 2H(t,s)[C(t,s) + D(t,s) - H(t,s)] \\
&\le (\alpha+\beta)H^2(t,s) + M\big(C^2(t,s) + D^2(t,s)\big) - \gamma H^2(t,s)
\end{aligned}$$

where γ can be chosen so that $\alpha + \beta < \gamma < 2$ and then for $\eta = \gamma - (\alpha + \beta)$ we have

$$V'(t) \leq -\eta H^2(t,s) + M\big(C^2(t,s) + D^2(t,s)\big)$$

and (2.6.22) holds.

Notice that if (2.6.23) holds then H is bounded, as may be seen using the Schwarz inequality on the last term of H. Having H bounded, we have H_t bounded, as may be seen using the Schwarz inequality on the last term of H_t. As $L(t,s)$ is bounded, it follows that $\lim_{t\to\infty} H(t,s) = 0$ for fixed s.

Remark In Theorem 2.6.4.4 we will see a string of constants $M_1 M_2 M_3 K$. They occur as a result of the last inequality in the computation of V'. With care and with more assumptions, they can be reduced. See Section 2.9 where we offer parallel work with C alone and not $C + D$.

Next, rewrite the resolvent as $H(t,s) = B(t,s) - \int_s^t H(t,u)B(u,s)du$ and use (2.6.25) to show that H_s is bounded. We are now prepared to prove the following proposition which will complete the proof.

For application of the next proposition, see Theorems 2.9.4, 2.9.5, and 2.9.6, as well as subsequent results in this section.

Proposition 2.6.4.1. *Let $H(t,s)$ be continuous for $0 \leq s \leq t < \infty$, and assume $\lim_{t\to\infty} H(t,s) = 0$ for each fixed $s \geq 0$. Assume also that $H(t,s)$ is globally Lipschitz in s, i.e., there is an $M > 0$ such that $0 \leq s_1 \leq s_2 \leq t < \infty$ implies $|H(t,s_1) - H(t,s_2)| \leq M|s_1 - s_2|$. Then for each $T > 0$ we have $\lim_{t\to\infty} \int_0^T |H(t,s)|ds = 0$.*

Proof. Assume by way of contradiction there is a $T > 0$, there is an $\epsilon > 0$, and a sequence $\{t_n\} \uparrow \infty$ such that $\int_0^T |H(t_n,s)|ds > \epsilon$ for each n. Now, if $|H(t_n,s)| \leq \epsilon/T$ for each $s \in [0,T]$ then $\int_0^T |H(t_n,s)|ds \leq \epsilon$, so it follows that for each n there is a point $s_n \in [0,T]$ such that $|H(t_n,s_n)| > \epsilon/T$. The sequence $\{s_n\}$ is contained in the compact interval $[0,T]$, so there must be some convergent subsequence $s_{n_k} \to s^*$. For the given $\epsilon > 0$, find $N > 0$ such that $n_k > N$ implies that $|s_{n_k} - s^*| < \epsilon/(2MT)$. Thus we have

$$\begin{aligned} \frac{\epsilon}{T} < |H(t_{n_k}, s_{n_k})| &\leq |H(t_{n_k}, s_{n_k}) - H(t_{n_k}, s^*)| + |H(t_{n_k}, s^*)| \\ &\leq M|s_{n_k} - s^*| + |H(t_{n_k}, s^*)|, \end{aligned}$$

so $n_k > N \implies |H(t_{n_k}, s^*)| > \epsilon/(2T)$, contradicting $H(t,s^*) \to 0$.

Remark. There is a list of conclusions which can be drawn from the theorem.

(i) The same conclusions follow from the derivative conditions (2.6.18), or from the integral conditions (2.6.19) and (2.6.20), or certain linear combinations of the two.

(ii) The integral $\int_s^t [C(t,u) + D(t,u)]H(u,s)du$ constructs an L^2 and pointwise copy of $C(t,s) + D(t,s)$ in t for fixed s.

(iii) A parallel Liapunov functional can be constructed for (2.6.16) under the conditions of (2.6.18)–(2.6.20) which will yield $x \in L^2$ for $a \in L^2$ and show that $x(t)$ converges to zero, under the conditions which sent $H(t,s)$ to zero. The integral $\int_s^t H(t,u)a(u)du$ constructs both an L^2 and an asymptotic pointwise copy of every $a \in L^2$.

Now there is a fundamental result which can be obtained from the fact that $H(t,s) \to 0$ as $t \to \infty$. It was proved by Strauss (1970) which we quoted in Theorem 2.1.3(ii) that if there is an $\alpha < 1$ with $\int_0^t |C(t,s) + D(t,s)|ds \le \alpha$ then $\int_0^t |H(t,s)|ds \le \frac{\alpha}{1-\alpha}$, and if $\lim_{t\to\infty} \int_0^T |C(t,s) + D(t,s)|ds \to 0$ for each $T > 0$ then the corresponding result also holds for $H(t,s)$.

We will see that the conclusion holds with far less than Strauss's condition and it hinges on our new conclusion that $H(t,s) \to 0$ as $t \to \infty$ for fixed s, along with Proposition 2.6.4.1.

Proposition 2.6.4.2. *If $\phi \in L^1[0,\infty)$, if ϕ is continuous or in $L^2[0,\infty)$, if $H(t,s)$ is bounded and continuous, and if for each $T > 0$ it follows that $\int_0^T |H(t,s)|ds \to 0$ as $t \to \infty$, then $\int_0^t |H(t,s)|\, \phi(s)ds \to 0$ as $t \to \infty$.*

Proof. Let $|H(t,s)| \le M$ and let $\epsilon > 0$ be given. Find $T > 0$ so that $\int_T^\infty |\phi(s)|ds < \epsilon/2M$. Then for ϕ continuous we have

$$\int_0^t |H(t,s)\phi(s)|\, ds \le \|\phi\|^{[0,T]} \int_0^T |H(t,s)|\, ds + M \int_T^\infty |\phi(s)|\, ds < \epsilon$$

for large t. On the other hand, if $\phi \in L^2[0,\infty)$ then we have

$$\begin{aligned}\int_0^t |H(t,s)\phi(s)|ds &\le \int_0^T |H(t,s)||\phi(s)|ds + M \int_T^\infty |\phi(s)|ds \\ &\le \sqrt{\int_0^T H^2(t,s)ds \int_0^\infty \phi^2(s)ds} + M \int_T^\infty |\phi(s)|ds.\end{aligned}$$

As $\int_0^T H^2(t,s)ds \le \|H\| \int_0^T |H(t,s)|ds \to 0$, the result follows as before.

Let

$$B(t, s) := C(t, s) + D(t, s) \tag{2.6.27}$$

where C satisfies (2.6.18) and D satisfies (2.6.19) and (2.6.20) so that the conclusions of Theorem 2.6.4.1 hold. We will see in Section 2.8.1 that the perturbed equation

$$x(t) = a(t) - \int_0^t B(t, s)[x(s) + G(s, x(s))]ds \tag{2.6.28}$$

can be decomposed into

$$y(t) = a(t) - \int_0^t B(t, s)y(s)ds \tag{2.6.29}$$

and

$$x(t) = y(t) - \int_0^t H(t, s)G(s, x(s))ds. \tag{2.6.30}$$

Theorem 2.6.4.2. *Let H satisfy the conditions of Proposition 2.6.4.1, let $y \in L^2[0, \infty)$ be bounded, let*

$$|G(t, x)| \leq \phi(t)|x|, \ \phi \in L^1[0, \infty),$$

and let ϕ be continuous or in $L^2[0, \infty)$. If $x(t)$ is any solution of (2.6.28) then x is bounded and $x(t) - y(t) \to 0$ as $t \to \infty$.

Proof. If $x(t)$ is not bounded, then there is a sequence $\{t_n\} \uparrow \infty$ such that $|x(t)| \leq |x(t_n)|$ for $0 \leq t \leq t_n$ and $|x(t_n)| \uparrow \infty$. Thus,

$$|x(t_n)| \leq \|y\| + \int_0^{t_n} |H(t_n, s)|\phi(s)ds|x(t_n)|.$$

For large n we have $\int_0^{t_n} |H(t_n, s)|\phi(s)ds < 1/2$ by Proposition 2.6.4.2 and this contradicts $|x(t_n)| \uparrow \infty$. It then follows that $\phi(t)|x(t)| \in L^1[0, \infty)$ and so

$$\int_0^t H(t, s)\phi(s)|x(s)|ds \to 0$$

and the result is proved.

Theorem 2.6.4.3. *Suppose there are positive constants M and K such that $\int_0^t B^2(t,u)du \leq M$ and for fixed $s \geq 0$ we have*

$$\int_s^t H^2(u,s)du \leq K \int_s^t B^2(u,s)du \leq M$$

for $s \leq t$. If, in addition, for all large fixed $T > s$ we have $\int_s^T B^2(t,u)du \to 0$ as $t \to \infty$, then

$$H(t,s) - B(t,s) \to 0$$

as $t \to \infty$.

Proof. Notice that with s fixed then $H^2(t,s) \in L^1[s,\infty)$. From (2.6.17) we have for $s < T < t$ that

$$\begin{aligned}
&\frac{1}{2}\big(H(t,s) - B(t,s)\big)^2 \\
&= \frac{1}{2}\left(\int_s^t B(t,u)H(u,s)du\right)^2 \\
&\leq \left(\int_s^T B(t,u)H(u,s)du\right)^2 + \left(\int_T^t B(t,u)H(u,s)du\right)^2 \\
&\leq \int_s^T H^2(u,s)du \int_s^T B^2(t,u)du + \int_T^t B^2(t,u)du \int_T^t H^2(u,s)du \\
&\leq M \int_s^T B^2(t,u)du + M \int_T^t H^2(u,s)du.
\end{aligned}$$

For a given $\epsilon > 0$, take T so large that $\int_T^\infty H^2(u,s)du < \frac{\epsilon}{2M}$. Then take t so large that $M\int_s^T B^2(t,u)du < \epsilon/2$. This will complete the proof.

We see that the totally unknown function, $H(t,s)$, converges pointwise to the clearly visible $B(t,s)$.

The result can be very useful in conjunction with Proposition 2.6.4.1 since if $B(t,s) \to 0$ as $t \to \infty$ for fixed s, then the same is true for $H(t,s)$ and the consequence of Proposition 2.6.4.1 is very important for work with (2.6.28) in Theorem 2.6.4.2.

Our work here is most emphatically not for the convolution case, but for purpose of illustration note that in the convolution case $H(t,s) = H(t-s)$ and for $s = 0$ we see that $H(t) \to B(t)$. In the next result we would have $\int_0^\infty \big(H(t) - B(t)\big)^2 dt < \infty$. Under the conditions of both theorems H converges to B both pointwise and in $L^2[0,\infty)$.

One by one, we transfer the asymptotic properties of B to H. One of the ultimate goals is to relate H and B so closely with integral inequalities that the unknown function H can, in effect, be replaced by the known function B in establishing long-term qualitative properties of y in (2.6.29) using its variation of parameters formula $y(t) = a(t) - \int_0^t H(t,s)a(s)ds$.

Theorem 2.6.4.4. *Suppose that there are positive constants K, M_1, M_2, and M_3 with*

$$\int_s^t H^2(u,s)du \leq K \int_s^t B^2(u,s)du,$$

$$\int_s^t |B(v,s)|dv \leq M_2,$$

and

$$\int_s^t |B(t,u)|du \leq M_3$$

all for $0 \leq s \leq t < \infty$. Then

$$\int_s^t \big(H(u,s) - B(u,s)\big)^2 du \leq M_2 M_3 K \int_s^t B^2(u,s)du.$$

If, in addition,

$$\int_s^\infty B^2(u,s)du \leq M_1$$

for any fixed $s \geq 0$, then

$$\int_s^\infty \big(H(u,s) - B(u,s)\big)^2 du \leq M_1 M_2 M_3 K.$$

Proof. From (2.6.17) we have

$$\begin{aligned}\big(H(t,s) - B(t,s)\big)^2 &= \left(\int_s^t B(t,u)H(u,s)du\right)^2 \\ &\leq \int_s^t |B(t,u)|du \int_s^t |B(t,u)|H^2(u,s)du\end{aligned}$$

so

$$\begin{aligned}\int_s^t \left(H(u,s) - B(u,s)\right)^2 du &\le M_3 \int_s^t \int_s^v |B(v,u)| H^2(u,s) du dv \\ &= M_3 \int_s^t \int_u^t |B(v,u)| H^2(u,s) dv du \\ &= M_3 \int_s^t H^2(u,s) \int_u^t |B(v,u)| dv du \\ &\le M_2 M_3 \int_s^t H^2(u,s) du \\ &\le M_2 M_3 K \int_s^t B^2(u,s) du.\end{aligned}$$

This completes the proof.

Application of Theorem 2.6.4.4 Obviously, we long to replace the unknown function H by the known function B. Consider (2.6.29) with variation of parameters formula

$$y(t) = a(t) - \int_0^t H(t,s) a(s) ds$$

which we compare with

$$Y(t) = a(t) - \int_0^t B(t,s) a(s) ds.$$

If we square $y - Y$, use the Schwarz inequality taking $\sqrt{|a(t)|}\sqrt{|a(t)|}$, integrate both sides from 0 to t, interchange order of integration, then we conclude that $a \in L^1[0,\infty)$ implies $y - Y \in L^2[0,\infty)$. Even on a finite interval we have readily measurable errors of $y - Y$. Without the squaring or the second integration, we easily show that if $a \in L^1[0,\infty)$, if $B - H$ is bounded, and if $\int_0^T |B(t,s) - H(t,s)| ds \to 0$ as $t \to \infty$ for fixed $T > 0$, then $y(t) \to Y(t)$ pointwise as $t \to \infty$.

Notes The material in Section 2.6.4 is from Burton and Dwiggins (2010). It was first published in *M*athemtica Bohemica by the Academy of Science of the Czech Republic and was presented at Equadiff 12 in Brno. Full details are given in the bibliography.

2.6.5 Large Kernels

So much of our work, both in constructing Liapunov functionals and integrating their derivatives, has involved assumptions of a finite integral of C

on $[0, \infty)$ with respect to one or both coordinates. If there is a large additive function of t, possibly a constant, then that is impossible. We solved half of that problem in Section 2.6.2, but we can get around it entirely by differentiating (2.6.2).

Thus, we consider (2.6.2) and suppose that $C_t(t,s)$ is continuous and write

$$R_t(t,s) = C_t(t,s) - C(t,t)R(t,s) - \int_s^t C_t(t,u)R(u,s)du. \tag{2.6.31}$$

At times it will be clearer to interchange the notation C_t with C_1.

Theorem 2.6.5.1. *Let $\int_{t-w}^{\infty} |C_1(u+w,w)|du$ be continuous. Suppose that there is an $\alpha > 0$ with*

$$C(t,t) - \int_0^{\infty} |C_1(u+t,t)|du \geq \alpha.$$

Then the derivative of

$$V(t) = |R(t,s)| + \int_s^t \int_{t-w}^{\infty} |C_1(u+w,w)|du|R(w,s)|dw$$

along a solution of (2.6.31) on $[s,\infty)$ satisfies

$$V'(t) \leq -\alpha|R(t,s)| + |C_t(t,s)|.$$

Thus, if, in addition, we have

$$\sup_{0\leq s\leq t<\infty} \int_s^t |C_1(u,s)|du < \infty$$

and $|C(s,s)|$ is bounded then

$$\sup_{0\leq s\leq t<\infty} \int_s^t |R(u,s)|du < \infty$$

and $R(t,s)$ is bounded.

Proof. For this function V, fix s and find

$$\begin{aligned} V'(t) &\leq |C_t(t,s)| - C(t,t)|R(t,s)| + \int_s^t |C_t(t,u)R(u,s)|du \\ &\quad + \int_0^{\infty} |C_1(u+t,t)|du|R(t,s)| - \int_s^t |C_t(t,w)||R(w,s)|dw \\ &\leq |C_t(t,s)| - \alpha|R(t,s)|. \end{aligned}$$

But $V(s) = |R(s,s)| = |C(s,s)|$ is bounded so an integration yields the second conclusion. The boundedness of R follows from boundedness of V and $V(t) \geq |R(t,s)|$.

Theorem 2.6.5.2. *Suppose that $\int_{t-w}^{\infty} |C_1(u+w,w)|du$ is continuous and that there is an $\alpha > 0$ with*

$$2C(t,t) - \int_s^t |C_t(t,u)|du - \int_0^{\infty} |C_1(u+t,t)|du \geq 2\alpha.$$

If

$$\sup_{0 \leq s \leq t < \infty} \int_s^t C_1^2(u,s)du < \infty$$

and $|C(s,s)|$ is bounded then

$$\sup_{0 \leq s \leq t < \infty} \int_s^t R^2(u,s)du < \infty.$$

Also, $R^2(t,s)$ is bounded.

Proof. For fixed s we let $q(t) = R(t,s)$ and define

$$V(t) = q^2(t) + \int_s^t \int_{t-w}^{\infty} |C_t(u+w,w)|duq^2(w)dw$$

so that along a solution of (2.6.31) we have

$$\begin{aligned}
V'(t) &= 2q(t)C_t(t,s) - 2C(t,t)q^2(t) - 2q(t)\int_s^t C_t(t,u)q(u)du \\
&+ \int_0^{\infty} |C_1(u+t,t)|duq^2(t) - \int_s^t |C_t(t,w)|q^2(w)dw \\
&\leq 2q(t)C_t(t,s) - 2C(t,t)q^2(t) + \int_s^t |C_t(t,u)|(q^2(t) + q^2(u))du \\
&+ \int_0^{\infty} |C_1(u+t,t)|duq^2(t) - \int_s^t |C_t(t,w)|q^2(w)dw \\
&= 2q(t)C_t(t,s) + [-2C(t,t) + \int_s^t |C_t(t,u)|du \\
&+ \int_0^{\infty} |C_t(u+t,t)|du]q^2(t) \\
&\leq 2q(t)C_1(t,s) - 2\alpha q^2(t) \\
&\leq MC_t^2(t,s) - \alpha q^2(t)
\end{aligned}$$

for some $M > 0$. The conclusion follows from this.

Square both sides of (2.6.2) to see that both sides are integrable with respect to t. We see then that

$$C(t,s) - \int_s^t R(t,u)C(u,s)du$$

is bounded and square integrable. That integral is constructing a copy of $C(t,s)$.

2.7 Comparing $C(t,s)b(s)$ with $R(t,s)b(s)$

There is an extensive theory concerning this comparison found in Burton and Dwiggins (2010) and (2011) and in Burton (2011a). Selected parts are given here.

For the basic linear equation we have the variation of parameters formula

$$x(t) = a(t) - \int_0^t R(t,s)a(s)ds$$

and it can happen that $a(t)$ is actually a nonlinear function of t and $x(t)$. Sometimes we want precisely the properties of $R(t,s)$ because in a nonlinear variation of parameters formula, found in Subsection 2.8.1, we obtain a new integral equation and its kernel is $R(t,s)$. But there are other times when it is not the case that we want the properties of $R(t,s)$, but rather the properties of

$$\int_0^t R(t,s)a(s)ds.$$

Moreover, when C is singular we can not study the resolvent directly by Liapunov theory, but are forced to study

$$R_1(t,s) = \int_s^t C(t,u)C(u,s)du - \int_s^t C(t,u)R_1(u,s)du$$

as we saw in (1.2.19), Theorem 1.2.6, and Proposition 1.2.1, presuming that $\int_s^t C(t,u)C(u,s)du$ is continuous.

Let $b : [0, \infty) \to \Re$ be continuous, write

$$R(t,s) = C(t,s) - \int_s^t C(t,u)R(u,s)du \tag{2.7.1}$$

as

$$R(t,s)b(s) = C(t,s)b(s) - \int_s^t C(t,u)R(u,s)dub(s),$$

and then form

$$\begin{aligned}\int_0^t R(t,s)b(s)ds &= \int_0^t C(t,s)b(s)ds - \int_0^t \int_s^t C(t,u)R(u,s)dub(s)ds\\ &= \int_0^t C(t,s)b(s)ds - \int_0^t C(t,u)\int_0^u R(u,s)b(s)dsdu\end{aligned}$$

which we will write as

$$f(t) = S(t) - \int_0^t C(t,u)f(u)du \tag{2.7.2}$$

with

$$f(t) = \int_0^t R(t,s)b(s)ds, \;\; S(t) = \int_0^t C(t,s)b(s)ds. \tag{2.7.3}$$

Remark There is application beyond the present discussion. If C is weakly singular, then S is continuous and our singular theory of Section 2.5 applies.

If C is continuous then we define

$$V(t) = \int_0^t C_s(t,s)\left(\int_s^t f(u)du\right)^2 ds + C(t,0)\left(\int_0^t f(u)du\right)^2 \tag{2.7.4}$$

and obtain under the convexity assumption on C, that

$$V'(t) \le -(S(t) - f(t))^2 - f^2(t) + S^2(t), \tag{2.7.5}$$

as seen in earlier work parallel to Theorem 2.7.1, from which we also conclude that when

$$C(t,t) \le B \tag{2.7.6}$$

then

$$(S(t) - f(t))^2 \le 2BV(t). \tag{2.7.7}$$

Thus, (2.7.5) and (2.7.7) yield the following result.

Theorem 2.7.1. *If C is convex and if S and f are defined in (2.7.3), then*

$$\int_0^t (S(u) - f(u))^2 du + \int_0^t f^2(u)du \leq \int_0^t S^2(u)du.$$

If, in addition, there is a $B > 0$ with $C(t,t) \leq B$, then

$$\frac{1}{2B}(S(t) - f(t))^2 + \int_0^t (S(u) - f(u))^2 du + \int_0^t f^2(u)du \leq \int_0^t S^2(u)du. \tag{2.7.8}$$

This theorem gives us, in a simple way, a relation between C and R. In the next result we remind the reader of how such results are used. After that we show that it can be much more powerful.

Theorem 2.7.2. *Let C be convex and suppose that $\int_0^t C(t,s)h(s)ds \to 0$ as $t \to \infty$ for every continuous function $h : [0,\infty) \to \Re$ with $h \in L^2[0,\infty)$. If $b : [0,\infty) \to \Re$ is continuous and if $S(t) \in L^2[0,\infty)$ and $S(t) \to 0$ as $t \to \infty$, then the same is true for f; here, b, S, and f satisfy (2.7.3).*

Proof. As C is convex, (2.7.5) holds. As $S \in L^2[0,\infty)$, the same is true for f from (2.7.5). As $f \in L^2[0,\infty)$ it qualifies as the h of the theorem and so $\int_0^t C(t,s)f(s)ds \to 0$ as $t \to \infty$. From (2.7.2) we then see that $f(t) - S(t) \to 0$ as $t \to \infty$. As $S(t) \to 0$ so does f. This completes the proof.

2.8 A Nonlinear Application

Virtually every result in this entire chapter can be used directly to obtain qualitative properties of solutions of the nonlinear scalar equation

$$x(t) = a(t) - \int_0^t C(t,s)[x(s) + G(s,x(s))]ds \tag{2.8.1}$$

having the linear part

$$y(t) = a(t) - \int_0^t C(t,s)y(s)ds \tag{2.8.2}$$

with two forms of the resolvent equation

$$\begin{aligned} R(t,s) &= C(t,s) - \int_s^t R(t,u)C(u,s)du \\ &= C(t,s) - \int_s^t C(t,u)R(u,s)du \end{aligned} \tag{2.8.3}$$

generating the variation of parameters formula

$$y(t) = a(t) - \int_0^t R(t,s)a(s)ds. \tag{2.8.4}$$

It is assumed that $a : [0,\infty) \to \Re$, $C : [0,\infty) \times [0,\infty) \to \Re$, and $G : [0,\infty) \times \Re \to \Re$ are all continuous. This will ensure that (2.8.1) has a solution, while (2.8.2) and (2.8.3) have unique solutions.

In the next subsection we will show that we can write (2.8.1) with another variation of parameters formula as

$$x(t) = a(t) - \int_0^t R(t,u)a(u)du - \int_0^t R(t,u)G(u,x(u))du$$

or

$$x(t) = y(t) - \int_0^t R(t,s)G(s,x(s))ds \tag{2.8.5}$$

where y is the solution of (2.8.2).

We can study (2.8.2) and (2.8.3) separately and apply our conclusions to (2.8.5) which will then be much simpler than (2.8.1). Our goal is to give conditions on a, C, and G to ensure that the solution of (2.8.1) is either bounded, is in $L^p[0,\infty)$, or tends to zero as $t \to \infty$. We ask a variant of

$$|G(t,x)| \le k_1|x| + h(t)|x|^{1/p} \tag{2.8.6}$$

where $p > 1$, $q > 1$, $k_1 > 0$, with $(1/p) + (1/q) = 1$, and $h \in L^q[0,\infty)$.

Linear preliminaries

We will now review a few of the results which we will need here, including those which yield

$$y \in L^q[0,\infty) \text{ for some } q = 1, 2, 4, ..., 2^r, ..., \infty. \tag{2.8.7}$$

We have mentioned several times that a classical idea is that the solution, y, follows the forcing function, $a(t)$. For example, if $a \in L^p$, so is y. Looking at (2.8.5) we see that we are striving for a nonlinear counterpart: if $y \in L^q$, so is x.

The next two results will be fundamental reminders. The first is Theorem 2.1.3(iii) and is discussed at some length by Islam and Neugebauer (2008).

Lemma 2.8.1. *If there is an* $\alpha \in (0,1)$ *with*

$$\int_0^t |C(t,s)|ds \leq \alpha, \ 0 \leq t < \infty, \tag{2.8.8}$$

then

$$\int_0^t |R(t,s)|ds \leq \frac{\alpha}{1-\alpha}. \tag{2.8.9}$$

Take the absolute values in the first choice in (2.8.3), integrate from 0 to t with respect to s, and finish by interchanging the order of integration.

There is a symmetric result obtained by using the second choice in (2.8.3) and integrating with respect to t; this will yield the first part of (2.8.11) below. It was also seen in Theorem 2.6.3.1 and (2.6.12).

Lemma 2.8.2. *If there is a* $\beta \in (0,1)$ *with*

$$\int_s^t |C(u,s)|du \leq \beta, \ 0 \leq s \leq t < \infty \tag{2.8.10}$$

then

$$\int_0^\infty |C(v+t,t)|dv \leq \beta \tag{2.8.10*}$$

and

$$\int_s^t |R(u,s)|du \leq \frac{\beta}{1-\beta} \text{ and } \int_0^\infty |R(v+t,t)|dv \leq \frac{\beta}{1-\beta}. \tag{2.8.11}$$

Moreover, if (2.8.10) holds, if $a \in L^1[0,\infty)$, *and if* $\int_{t-u}^\infty |C(v+u,u)|dv$ *is continuous, then* $y \in L^1[0,\infty)$.

Proof. We have indicated how to get the first part of (2.8.11). From (2.8.10) we have $\int_s^\infty |C(u,s)|du \leq \beta$ so $\int_0^\infty |C(u+s,s)|du \leq \beta$. The last inequality holds for s replaced by t.

In the same way if (2.8.10) holds and we have the first part of (2.8.11) then

$$\int_0^\infty |R(v+t,t)|dv = \int_t^\infty |R(w,t)|dw \leq \frac{\beta}{1-\beta}.$$

To prove the last sentence of the lemma, define

$$V(t) = \int_0^t \int_{t-u}^\infty |C(v+u,u)|dv|y(u)|du$$

so that along the solution of (2.8.2) we have

$$\begin{aligned} V'(t) &= \int_0^\infty |C(v+t,t)|dv|y(t)| - \int_0^t |C(t,u)y(u)|du \\ &\leq \beta|y(t)| - |y(t)| + |a(t)|. \end{aligned}$$

An integration yields the result since $V(t) \geq 0$ and $\beta < 1$. This completes the proof.

Corollary 2.8.1. *Let (2.8.10) hold. If $a \in L^1[0,\infty)$ then*

$$\int_0^t R(t,s)a(s)ds \in L^1[0,\infty). \tag{2.8.12}$$

Proof. Note that

$$-\int_0^t R(t,s)a(s)ds = y(t) - a(t) \in L^1[0,\infty).$$

There is a companion to that last result which will be needed, found in Theorem 2.1.3(ii).

Proposition 2.8.1. *If (2.8.8) holds, if $a(t) \to 0$ as $t \to \infty$, and if for each $T > 0$ then*

$$\int_0^T |C(t,s)|ds \to 0 \text{ as } t \to \infty, \tag{2.8.13}$$

then both

$$\int_0^t R(t,s)a(s)ds \text{ and } \int_0^T |R(t,s)|ds \to 0 \text{ as } t \to \infty. \tag{2.8.14}$$

Finally, we remark that there are numerous conditions on C which are known to imply that there is a positive constant K such that

$$|R(t,s)| \leq K,\ 0 \leq s \leq t < \infty. \tag{2.8.15}$$

See Section 2.6, particularly Theorems 2.6.3.2, 2.6.3.3, 2.6.4.1, 2.6.5.1.

The nonlinear problem

We begin with two parallel results.

Theorem 2.8.1. *Let (2.8.8) and (2.8.13) hold and suppose there is a continuous function $h : [0, \infty) \to [0, \infty)$ with*

$$|G(t, x)| \leq h(t)|x|, \; h(t) \to 0 \text{ as } t \to \infty. \tag{2.8.16}$$

If the solution of (2.8.2), y, is bounded so is x, the solution of (2.8.5).

Proof. Conditions (2.8.8) and (2.8.13) imply (2.8.14). This means that there is a $\gamma < 1$ and $T > 0$ with

$$\int_0^t |R(t, s)|h(s)ds \leq \gamma, \; t \geq T.$$

Let $|y(t)| \leq Y$ for some $Y > 0$. If x is not bounded, then there is a $q > T$ with $|x(t)| \leq |x(q)|$ for $0 \leq t < q$ which yields

$$|x(q)| \leq Y + |x(q)| \int_0^q |R(q, s)|h(s)ds \leq Y + \gamma|x(q)|$$

so $|x(q)| \leq Y/(1 - \gamma)$ will be the bound on x for $t \geq q$. As x is continuous, the proof is complete.

Lemma 2.8.3. *Let (2.8.10) hold so that (2.8.12) also holds and let $y \in L^1[0, \infty)$. If*

$$|G(t, x)| \leq h(t)|x|, \; h(t) \in L^1[0, \infty), \tag{2.8.17}$$

and if the solution of (2.8.5) is bounded, then it is also in $L^1[0, \infty)$.

Proof. If x is bounded, then $|G(t, x)| \leq h(t)|x| \in L^1[0, \infty)$ so

$$\int_0^t R(t, s)G(s, x(s))ds \in L^1[0, \infty)$$

and the result follows.

Theorem 2.8.2. *Let (2.8.10), (2.8.15), and (2.8.17) hold and let $y(t)$ be bounded and in $L^1[0, \infty)$. Suppose that there is an $M > 0$ with*

$$|G(t, x_1) - G(t, x_2)| \leq Mh(t)|x_1 - x_2|. \tag{2.8.18}$$

Then the solution of (2.8.5) is bounded and, hence, is in $L^1[0, \infty)$.

Proof. Let $(X, |\cdot|_h)$ be the Banach space of bounded continuous functions $\phi : [0, \infty) \to \Re$ where

$$|\phi|_h := \sup_{t \geq 0} |\phi(t)| e^{-\int_0^t (MK+1)h(s)ds}.$$

Then define $P : X \to X$ by $\phi \in X$ implies that

$$(P\phi)(t) = y(t) - \int_0^t R(t,s)G(s,\phi(s))ds$$

so

$$|(P\phi)(t)| \leq |y(t)| + \left| \int_0^t R(t,s)G(s,\phi(s))ds \right|$$

is bounded, as in the proof of Lemma 2.8.3. To see that P is a contraction, if $\phi, \eta \in X$ then

$$\begin{aligned}
&|(P\phi)(t) - (P\eta)(t)| e^{-\int_0^t (MK+1)h(s)ds} \\
&\leq \int_0^t |R(t,s)| e^{-\int_0^t (MK+1)h(s)ds} Mh(s)|\phi(s) - \eta(s)| ds \\
&\leq \int_0^t MK e^{-\int_s^t (MK+1)h(u)du} e^{-\int_0^s (MK+1)h(u)du} h(s)|\phi(s) - \eta(s)| ds \\
&\leq \frac{MK}{MK+1} |\phi - \eta|_h e^{-\int_s^t (MK+1)h(u)du} \Big|_0^t \\
&\leq \frac{MK}{MK+1} |\phi - \eta|_h,
\end{aligned}$$

completing the proof.

By changing (2.8.17) we can avoid (2.8.18). The reader may wish to review Lemma 2.8.2 and (2.8.10), (2.8.10*), and (2.8.11). If (2.8.10) holds then we can define

$$V(t) = \int_0^t \int_{t-u}^\infty |R(v+u,u)| dv |G(u,x(u))| du \tag{2.8.19}$$

and find that

$$V'(t) = \int_0^\infty |R(v+t,t)| dv |G(t,x)| - \int_0^t |R(t,u)G(u,x(u))| du$$

so that

$$V'(t) \leq \frac{\beta}{1-\beta} |G(t,x)| - |x| + |y|. \tag{2.8.20}$$

Theorem 2.8.3. *Let (2.8.10) hold, let $p > 1$, $q > 1$, $(1/p) + (1/q) = 1$, and suppose there is a function $\lambda : [0, \infty) \to [0, \infty)$, $\lambda \in L^q[0, \infty)$, and with $k_1 > 0$ so that*

$$|G(t, x)| \leq k_1|x| + \lambda(t)|x|^{1/p} \tag{2.8.21}$$

and

$$k_2 := k_1 \frac{\beta}{1 - \beta} < 1. \tag{2.8.22}$$

If $y \in L^1[0, \infty)$, so is x.

Proof. From (2.8.20) and (2.8.21) we have

$$\begin{aligned} V'(t) &\leq \frac{\beta}{1 - \beta}[k_1|x| + \lambda(t)|x|^{1/p}] - |x| + |y| \\ &= k_2|x| + \frac{\beta}{1 - \beta}\lambda(t)|x|^{1/p} - |x| + |y| \\ &= -(1 - k_2)|x| + |y| + \frac{\beta}{1 - \beta}\lambda(t)|x|^{1/p}. \end{aligned}$$

An integration yields

$$(1-k_2)\int_0^t |x(s)|ds \leq \int_0^t |y(s)|ds + \frac{\beta}{1 - \beta}\left(\int_0^t \lambda^q(s)ds\right)^{1/q}\left(\int_0^t |x(s)|ds\right)^{1/p}.$$

There are then positive constants U and L with

$$\int_0^t |x(s)|ds \leq U + L\left(\int_0^t |x(s)|ds\right)^{1/p},$$

placing a bound on $\int_0^t |x(s)|ds$. This completes the proof.

We have proved a collection of results showing that if $a \in L^{2^n}$ or if $a' \in L^{2^n}$ then $y \in L^{2^n}$. Again, it would be a distraction to repeat those here. We will assume that $y \in L^{2^n}$ and give conditions to ensure that $x \in L^{2^n}$.

Lemma 2.8.4. *Let (2.8.8) hold and let $\epsilon > 0$ be given. Then there is a positive constant M so that if*

$$\gamma := \frac{\alpha(1 + \epsilon)}{1 - \alpha} \tag{2.8.23}$$

then

$$x^2 \leq My^2 + \gamma \int_0^t |R(t, s)|G^2(s, x(s))ds. \tag{2.8.24}$$

Proof. If we square both sides of (2.8.5) we can readily find M with

$$\begin{aligned} x^2 &\leq My^2 + (1+\epsilon)\left(\int_0^t R(t,s)G(s,x(s))ds\right)^2 \\ &\leq My^2 + (1+\epsilon)\int_0^t |R(t,s)|ds \int_0^t |R(t,s)|G^2(s,x(s))ds \\ &\leq My^2 + \frac{\alpha(1+\epsilon)}{1-\alpha}\int_0^t |R(t,s)|G^2(s,x(s))ds \\ &= My^2 + \gamma\int_0^t |R(t,s)|G^2(s,x(s))ds, \end{aligned}$$

as required.

Assume that (2.8.10) holds and for γ defined in (2.8.23) define the Liapunov functional

$$V(t) = \gamma\int_0^t \int_{t-u}^{\infty} |R(u+v,u)|dvG^2(u,x(u))du. \tag{2.8.25}$$

and review (2.8.11) to see that $\int_0^\infty |R(v+t,t)|dv \leq \beta/(1-\beta)$.

Thus, the derivative of V satisfies

$$V'(t) = \gamma\int_0^\infty |R(v+t,t)|dvG^2(t,x(t)) - \gamma\int_0^t |R(t,u)|G^2(u,x(u))du$$

so that

$$V'(t) \leq \frac{\gamma\beta}{1-\beta}G^2(t,x(t)) + My^2(t) - x^2(t). \tag{2.8.26}$$

Theorem 2.8.4. *Let (2.8.8) and (2.8.10) hold, let $p > 1$, $q > 1$, $(1/p) + (1/q) = 1$, and let $y \in L^2[0,\infty)$. Suppose there is a function $\lambda \in L^q[0,\infty)$ and an $\eta > 0$ with*

$$\frac{\gamma\eta\beta}{1-\beta} =: \mu < 1 \tag{2.8.27}$$

and

$$G^2(t,x) \leq \eta x^2 + \lambda(t)|x|^{2/p}. \tag{2.8.28}$$

Then $x \in L^2[0,\infty)$. If, in addition, y and $C(t,s)$ are bounded, so is x.

Proof. From (2.8.26) and (2.8.27) we have

$$V'(t) \leq \mu x^2(t) + \frac{\mu\lambda(t)}{\eta}|x(t)|^{2/p} + My^2(t) - x^2(t) \tag{2.8.29}$$

so that

$$(1-\mu)\int_0^t x^2(s)ds \leq \frac{\mu}{\eta}\left(\int_0^t \lambda^q(s)ds\right)^{1/q}\left(\int_0^t x^2(s)ds\right)^{1/p} + M\int_0^t y^2(s)ds.$$

This places a bound on $\int_0^t x^2(s)ds$. If C is bounded, use (2.8.11) in (2.8.3) to see that R is bounded. If, in addition, y is bounded then use the last relation in (2.8.5) to see that x is bounded.

We could continue to square (2.8.5) and (2.8.28) and obtain $x \in L^{2^n}$ when $y \in L^{2^n}$.

We have been working with small C. Now we turn to possibly large C. Differentiate (2.8.3) to obtain

$$R_t(t,s) = C_t(t,s) - C(t,t)R(t,s) - \int_s^t C_t(t,u)R(u,s)du$$

and let $q(t) = R(t,s)$ for fixed s and write

$$q' = C_t(t,s) - C(t,t)q(t) - \int_s^t C_t(t,u)q(u)du.$$

Theorem 2.8.5. *Let $\int_{t-u}^{\infty}|C_1(v+u,u)|dv$ be continuous and suppose there is a constant $\beta > 0$ with*

$$C(t,t) - \int_0^{\infty}|C_1(v+t,t)|dv \geq \beta. \tag{2.8.30}$$

Then

$$|R(t,s)| + \beta\int_s^t |R(u,s)|du \leq |C(s,s)| + \int_s^t |C_1(u,s)|du. \tag{2.8.31}$$

Proof. Use

$$V(t) = |q(t)| + \int_s^t \int_{t-u}^{\infty} |C_1(v+u,u)|dv|q(u)|du,$$

note that $V(s) = |C(s,s)|$, and obtain

$$\begin{aligned} V'(t) &\le |C_t(t,s)| - C(t,t)|q(t)| + \int_s^t |C_t(t,u)q(u)|du \\ &\quad + \int_0^{\infty} |C_1(v+t,t)|dv|q(t)| - \int_s^t |C_1(t,u)q(u)|du \\ &\le |C_t(t,s)| - \beta|q(t)| = |C_t(t,s)| - \beta|R(t,s)| \end{aligned}$$

or

$$|R(t,s)| \le V(t) \le V(s) + \int_s^t |C_t(u,s)|du - \beta \int_s^t |R(u,s)|du,$$

yielding the result.

Corollary 2.8.2. *Let the conditions of Theorem 2.8.5 hold and suppose there exists $M > 0$ with*

$$|C(s,s)| + \int_s^t |C_1(u,s)|du \le M. \tag{2.8.32}$$

Then

$$|R(t,s)| + \beta \int_s^t |R(u,s)|du \le M. \tag{2.8.33}$$

If, in addition, $a \in L^1[0,\infty)$ then the unique solution y of (2.8.2) is in $L^1[0,\infty)$; moreover, if $a \in L^1$ and bounded, so is y.

Proof. Since $|R(t,s)|$ is bounded and $a \in L^1$ and bounded, y is bounded by (2.8.4). If $\int_0^t |a(u)|du \le A$ then

$$\begin{aligned} \int_0^t |y(u)|du &\le \int_0^t |a(u)|du + \int_0^t \int_0^u |R(u,s)a(s)|dsdu \\ &= \int_0^t |a(u)|du + \int_0^t \int_s^t |R(u,s)|du|a(s)|ds \\ &\le A + (M/\beta)A, \end{aligned}$$

as required.

Corollary 2.8.3. *Let the conditions of Theorem 2.8.5 and (2.8.32) hold. Suppose there are constants $p > 1$, $q > 1$, with $(1/p) + (1/q) = 1$ and a function $h \in L^q[0, \infty)$ with*

$$|G(t,x)| \le h(t)|x|^{1/p}.$$

Then the solution of (2.8.5) satisfies $x \in L^1[0, \infty)$. If, in addition, $a(t)$ is bounded then x is bounded.

Proof. Notice that

$$\int_0^\infty |R(v+t,t)|dv = \int_t^\infty |R(u,t)|du \le M/\beta$$

by (2.8.33). Thus, from the derivative of

$$V(t) = \int_0^t \int_{t-u}^\infty |R(v+u,u)|dv|G(u,x(u))|du$$

we have

$$\begin{aligned} V'(t) &\le (M/\beta)|G(t,x)| + |y| - |x| \\ &\le (M/\beta)h(t)|x|^{1/p} + |y| - |x|. \end{aligned}$$

An integration and use of Hölder's inequality will give $x \in L^1$ since Corollary 2.8.2 yields $y \in L^1$. But Corollary 2.8.2 also gives y bounded and R bounded so boundedness of x will now readily follow from (2.8.5).

We have given two ways of showing $\int_0^\infty |R(v+t,t)|dv$ bounded and there are many ways to show $y \in L^1$. Every result in Sections 2.6.3, 2.6.4, and 2.6.5 with R^{2n} integrable can be used with Hölder's inequality in the second choice in (2.8.3) to yield R bounded. We now offer a result which allows flexibility for other methods.

Theorem 2.8.6. *Let $y \in L^1[0, \infty)$ satisfy (2.8.2) and suppose that there is a constant μ with $\int_0^\infty |R(v+t,t)|dv \le \mu$. If there is a constant $k > 0$ with $|G(t,x)| \le k|x|$ and $k\mu < 1$ then the solution x of (2.8.5) is in $L^1[0, \infty)$. If, in addition, y is bounded and $R(t,s)$ is bounded then x is bounded.*

Proof. Define

$$V(t) = \int_0^t \int_{t-u}^{\infty} |R(v+u,u)|dv|G(u,x(u))|du$$

and obtain

$$\begin{aligned} V'(t) &= \int_0^{\infty} |R(v+t,t)|dv|G(t,x)| - \int_0^t |R(t,u)G(u,x(u))|du \\ &\le \mu k|x| - |x| + |y|. \end{aligned}$$

This yields $x \in L^1[0,\infty)$. Next, using these conditions in (2.8.5) yields x bounded.

Remark. Our work is stated for scalar equations, but much of it is valid also for systems. Miller's work in the next subsection is definitely for systems so one needs to be careful about commuting terms. If we go back to the early parts of the chapter we see that some of our Liapunov functionals used here also work for systems. That is true for the contraction result, as well. It fails in all the cases where we square the equation. It would be very interesting to advance Miller's variation of parameters formula to the case of $G(t,s,x(s))$.

2.8.1 The Nonlinear Variation of Parameters

Here, we follow the presentation of Miller (1971a; pp. 190-2). Given a linear system

$$X(t) = f(t) - \int_0^t C(t,s)X(s)ds, \tag{2.8.34}$$

we can write

$$X(t) = f(t) - \int_0^t R(t,u)f(u)du \tag{2.8.35}$$

and we suppose that properties of X and R are known. Then consider

$$\begin{aligned} x(t) &= f(t) - \int_0^t C(t,s)[x(s) + G(s,x(s))]ds \\ &=: F(t) - \int_0^t C(t,s)x(s)ds \end{aligned} \tag{2.8.36}$$

where

$$F(t) := f(t) - \int_0^t C(t,s)G(s,x(s))ds.$$

Thus, we apply (2.6.3) to (2.8.36) and obtain

$$\begin{aligned}
x(t) &= F(t) - \int_0^t R(t,s)F(s)ds \\
&= \left[f(t) - \int_0^t C(t,s)G(s,x(s))ds\right] - \int_0^t R(t,s)f(s)ds \\
&+ \int_0^t R(t,s)\int_0^s C(s,u)G(u,x(u))duds \\
&= \left[f(t) - \int_0^t R(t,s)f(s)ds\right] - \int_0^t C(t,s)G(s,x(s))ds \\
&+ \int_0^t \left[\int_u^t R(t,s)C(s,u)ds\right]G(u,x(u))du \\
&= \left[f(t) - \int_0^t R(t,s)f(s)ds\right] \\
&- \int_0^t \left[C(t,u) - \int_u^t R(t,s)C(s,u)ds\right]G(u,x(u))du.
\end{aligned}$$

Using (2.6.2) and (2.8.35) yields

$$\begin{aligned}
x(t) &= X(t) - \int_0^t R(t,u)G(u,x(u))du \\
&= f(t) - \int_0^t R(t,u)[f(u) + G(u,x(u))]du
\end{aligned} \tag{2.8.37}$$

which Miller calls the variation of constants form of (2.8.36).

2.9 Strong convergence to the limit set

In Chapter 3, Section 3.7.2, we study uniform asymptotic stability in which two things are specified. First, we stipulate that the limit set of solutions is the zero function (solution). Every solution starting near the zero function will converge to the zero function as $t \to \infty$. This is pointwise convergence. Next, we see that there is specified an amount of time that it may take the solution to get within a fixed neighborhood of the zero function. This is very strong stability and it implies many important properties. In this section we will investigate a parallel property for integral equations. First, we will specify a function to which the solution will converge pointwise as $t \to \infty$; that function will be denoted by $X(t) = a(t) - \int_0^t C(t,s)a(s)ds$. This is a function which can be exactly identified and calculated from the given integral equation. Next, we will show that the solution x and X will satisfy a specific relation of the form $x - X \in L^p$ where p is 1 or 2. Finally, we show the pointwise convergence in the form of $x(t) - X(t) \to 0$ as $t \to \infty$. This means that we establish very strong convergence of the solution to its limit set.

To grasp the context of such a result, let us review how we arrive at our present study. We introduce students to calculus and their first hurdle is to learn to integrate a variety of given functions. Next, we introduce them to differential equations in which the objective is to integrate unknown functions and deduce their properties. The conjecture here is that for select integral equations we may return to integrating known functions, namely, $\int_0^t C(t,s)a(s)ds$, to deduce properties of the unknown solution, $x(t)$.

Consider the scalar integral equation

$$x(t) = a(t) - \int_0^t C(t,s)x(s)ds \tag{2.9.1}$$

where $a(t)$ is continuous and $C(t,s)$ is convex:

$$C(t,s) \geq 0, C_s(t,s) \geq 0, C_{st}(t,s) \leq 0, C_t(t,s) \leq 0. \tag{2.9.2}$$

The resolvent equation is

$$\begin{aligned} R(t,s) =& C(t,s) - \int_s^t C(t,u)R(u,s)du \\ =& C(t,s) - \int_s^t R(t,u)C(u,s)du \end{aligned} \tag{2.9.3}$$

and the variation-of-parameters formula is

$$x(t) = a(t) - \int_0^t R(t,s)a(s)ds. \tag{2.9.4}$$

In Theorem 2.6.4.1 if $D = 0$ then we can proceed exactly as we did in the proof of Theorem 2.1.10 to obtain (2.1.13) and see that the derivative of

$$V(t) = \int_s^t C_v(t,v)\left(\int_v^t R(u,s)du\right)^2 dv + C(t,s)\left(\int_s^t R(u,s)du\right)^2 \tag{2.9.5}$$

along the solution of (2.9.3) satisfies

$$V'(t) \leq -(C(t,s) - R(t,s))^2 - R^2(t,s) + C^2(t,s). \tag{2.9.6}$$

Moreover, if there is a positive constant B with

$$C(t,t) \le B \tag{2.9.7}$$

then

$$(C(t,s) - R(t,s))^2 \le 2BV(t). \tag{2.9.8}$$

From these properties we can obtain the following result.

Theorem 2.9.1. *Let (2.9.2) and (2.9.7) hold so that (2.9.6) also holds. Then for* $0 \le s \le t < \infty$,

$$\frac{1}{2B}(C(t,s) - R(t,s))^2 + \int_s^t (C(u,s) - R(u,s))^2 du + \int_s^t R^2(u,s)du \le \int_s^t C^2(u,s)du. \tag{2.9.9}$$

If, in addition,

$$\sup_{0 \le s \le t < \infty} \int_s^t C^2(u,s)du < \infty, \tag{2.9.10}$$

then for fixed s,

$$\int_s^\infty (R(u,s) - C(u,s))^2 du < \infty. \tag{2.9.11}$$

If, in addition,

$$\sup_{0 \le s \le t < \infty} \int_s^t [C^2(t,u) + C_t^2(t,u)]du < \infty \tag{2.9.12}$$

then for fixed s

$$|R(t,s) - C(t,s| \to 0 \text{ as } t \to \infty. \tag{2.9.13}$$

Proof. Relation (2.9.9) is obtained by integrating (2.9.6) from s to t, observing that $V(s) = 0$, and then using (2.9.8). Next, if (2.9.10) holds then (2.9.11) is immediate from (2.9.9); moreover,

$$(C(t,s) - R(t,s))^2 \text{ is bounded for } 0 \le s \le t < \infty. \tag{2.9.14}$$

If we can show that this quantity has a bounded derivative with respect to t, then the fact that $\int_s^t (C(u,s) - R(u,s))^2 du$ converges for fixed s will imply that the integrand tends to zero as $u \to \infty$ for fixed s. To that end, we note that from (2.9.3) we have

$$R_t(t,s) - C_t(t,s) = -C(t,t)R(t,s) - \int_s^t C_t(t,u)R(u,s)du \qquad (2.9.15)$$

and

$$|R(t,s)| \leq |C(t,s)| + \sqrt{\int_s^t C^2(t,u)du \int_s^t R^2(u,s)du.} \qquad (2.9.16)$$

Since $C(t,s) \geq 0$ and $C_t(t,s) \leq 0$, for fixed s it follows that $C(t,s)$ is bounded. From (2.9.12) and (2.9.9) we now have that $|R(t,s)|$ is bounded. It follows from (2.9.15) that

$$|R_t(t,s) - C_t(t,s)| \leq |C(t,t)R(t,s)| + \sqrt{\int_s^t C_t^2(t,u)du \int_s^t R^2(u,s)du}$$

which is bounded. This, together with (2.9.14) yields a bounded t-derivative for $(C(t,s)-R(t,s))^2$ and that will show that $C(t,s)-R(t,s) \to 0$ as $t \to \infty$ for fixed s.

As R converges to C we now show that in (2.9.4) we can replace the totally unknown function R by the clearly given function C and have an excellent approximation to x.

Theorem 2.9.2. *Suppose that all conditions of the previous theorem hold and let $a \in L^1[0,\infty)$. If*

$$X(t) = a(t) - \int_0^t C(t,s)a(s)ds \qquad (2.9.17)$$

and if x solves (2.9.1), then

$$|X(t) - x(t)| \to 0 \text{ as } t \to \infty \text{ and } \int_0^\infty (X(t) - x(t))^2 dt < \infty. \qquad (2.9.18)$$

Proof. Set $\int_0^\infty |a(s)|ds = A$ and let $\int_s^t (C(u,s) - R(u,s))^2 du \le L$ for positive constants A and L. By the Schwarz inequality we have

$$(x(t) - X(t))^2 \le A \int_0^t [C(t,s) - R(t,s)]^2 |a(s)|ds.$$

Integration and interchange of the order of integration yields

$$\begin{aligned}
\int_0^t (x(u) - X(u))^2 du &\le A \int_0^t \int_0^u [C(u,s) - R(u,s)]^2 |a(s)| ds du \\
&= A \int_0^t \int_s^t [C(u,s) - R(u,s)]^2 du |a(s)| ds \\
&\le AL \int_0^t |a(s)| ds \\
&\le A^2 L.
\end{aligned}$$

We also have that $C_t(t,s) - R_t(t,s)$ is bounded so if we notice that $C(t,t) = R(t,t)$ then

$$(x(t) - X(t))' = [C(t,t) - R(t,t)]a(t) + \int_0^t [C_t(t,s) - R_t(t,s)]a(s)ds$$

is bounded. Since $\int_0^t (x(u) - X(u))^2 du$ converges, the integrand tends to zero.

The work here has been rather indirect for we wanted to derive important properties of the resolvent. However, we can start with (2.9.1), (2.9.2), and (2.9.7), define a Liapunov functional

$$W(t) = \int_0^t C_s(t,s) \left(\int_s^t x(u)du \right)^2 ds + C(t,0) \left(\int_0^t x(u)du \right)^2 \quad (2.9.19)$$

and find that the derivative of W along the solution of (2.9.1) satisfies

$$W'(t) \le 2x(t)[a(t) - x(t)] = -(a(t) - x(t))^2 - x^2(t) + a^2(t) \quad (2.9.20)$$

and that

$$(x(t) - a(t))^2 \le 2BW(t). \quad (2.9.21)$$

Theorem 2.9.3. *If (2.9.2) and (2.9.7) hold, then*

$$\frac{1}{2B}(a(t) - x(t))^2 + \int_0^t (a(s) - x(s))^2 ds + \int_0^t x^2(s)ds \le \int_0^t a^2(u)du. \quad (2.9.22)$$

If, in addition, $a \in L^2[0,\infty)$ then $\int_0^\infty (x(t) - a(t))^2 dt < \infty$, while $a(t)$ bounded implies $x(t)$ bounded. If, in addition, $\int_0^t C_t^2(t,s)ds$ is bounded, then $|x(t) - a(t)| \to 0$ as $t \to \infty$.

The conditions in the last sentence will show that $(a(t) - x(t))'$ is bounded so convergence of $\int_0^t (a(s) - x(s))^2 ds$ will show that the integrand tends to zero as $s \to \infty$.

These simple Liapunov functionals show us quite precisely both the solution of (2.9.1) for large t and the properties of the resolvent.

We are interested in showing that $x(t) - X(t) \to 0$ as $t \to \infty$ for $a \in L^1[0,\infty)$. We will do this in three ways, illustrating use of the condition that $\int_0^T |C(t,s)|ds \to 0$ as $t \to \infty$ for each $T > 0$ implies that $\int_0^T |R(t,s)|ds \to 0$ as $t \to \infty$ for each $T > 0$. Recall that we gave conditions under which this last relation is true in Theorem 2.1.3(ii) and in Proposition 2.6.4.1.

Theorem 2.9.4. *If there is a $K > 0$ with*

$$\int_s^t |R(u,s) - C(u,s)|du \leq K, \quad 0 \leq s \leq t < \infty \tag{2.9.23}$$

and if $a \in L^1[0,\infty)$, then $x - X \in L^1[0,\infty)$ where x and X are defined in (2.9.1) and (2.9.17). If (2.9.23) holds, if $a(t) \to 0$ as $t \to \infty$, and if $C(t,s) = C(t-s)$, then $x(t) - X(t) \to 0$ as $t \to \infty$.

Proof. We have

$$|x(t) - X(t)| \leq \int_0^t |R(t,s) - C(t,s)||a(s)|ds$$

so

$$\begin{aligned}\int_0^t |x(u) - X(u)|du &\leq \int_0^t \int_0^u |R(u,s) - C(u,s)||a(s)|dsdu \\ &= \int_0^t \int_s^t |R(u,s) - C(u,s)|du|a(s)|ds \\ &\leq K \int_0^t |a(s)|ds < \infty.\end{aligned}$$

On the other hand, if $a(t) \to 0$ as $t \to \infty$ and if $C(t,s) = C(t-s)$ then $R(t,s) = R(t-s)$ so

$$\int_0^t |R(t-s) - C(t-s)|ds = \int_0^t |R(s) - C(s)|ds$$

while

$$\int_s^t |R(u-s) - C(u-s)|du = \int_0^{t-s} |R(v) - C(v)|dv < \infty$$

so $R(v) - C(v)$ is an L^1-function and, hence,

$$|x(t) - X(t)| = \int_0^t |R(t-s) - C(t-s)||a(s)|ds$$

tends to zero as $t \to \infty$.

Theorem 2.9.5. *Suppose there is an $\alpha < 1$ with $\int_0^t |C(t,s)|ds \leq \alpha$, that $a(t) \to 0$ as $t \to \infty$ or that $a \in L^1[0,\infty)$, and that $\int_0^T |C(t,s)|ds \to 0$ as $t \to \infty$ for all $T > 0$ so that the same is true for R. Then $x(t) - X(t) \to 0$ as $t \to \infty$.*

Proof. As $\int_0^t |C(t,s)|ds \leq \alpha < 1$, then $\int_0^t |R(t,s)|ds \leq \alpha/(1-\alpha)$ and so

$$\int_0^t |R(t,s) - C(t,s)|ds \leq \alpha(2-\alpha)/(1-\alpha).$$

Take T large and note first that when $a(t) \to 0$ then

$$\begin{aligned}
|x(t) - X(t)| &\leq \int_0^t |R(t,s) - C(t,s)||a(s)|ds \\
&= \int_0^T |R(t,s) - C(t,s)||a(s)|ds + \int_T^t |R(t,s) - C(t,s)||a(s)|ds \\
&\leq \|a\|^{[0,T]} \int_0^T |R(t,s) - C(t,s)|ds + \|a\|^{[T,\infty)} \alpha(2-\alpha)/(1-\alpha).
\end{aligned}$$

First, take T large to make the last term small. Then take t large to make the previous term small.

When $a \in L^1$, replace that last term by $\int_T^\infty |a(s)|ds[\alpha(2-\alpha)/(1-\alpha)]$.

In Theorem 2.6.3.1 we obtained $x - a \in L^1$ and there are other conditions under which this is true. The property of $x - a \in L^1$ really says little about the exact limit set of $x(t)$. If we can say that $x(t) - X(t) \to 0$ then that does specify the limit set precisely, while $x - X \in L^p$ will say that the limit set is approached in a timely fashion.

Theorem 2.9.6. *Let x and a be defined in (2.9.1) and X be defined in (2.9.17). If $x - a \in L^1[0,\infty)$, if $\int_0^T |C(t,s)|ds \to 0$ as $t \to \infty$ for each $T > 0$, and if $C(t,s)$ is bounded, then $x(t) - X(t) \to 0$ as $t \to \infty$.*

Proof. For $T > 0$ to be determined we have

$$\begin{aligned}|x(t) - X(t)| &\leq \int_0^t |C(t,s)||x(s) - a(s)|ds \\ &= \int_0^T |C(t,s)||x(s) - a(s)|ds + \int_T^t |C(t,s)||x(s) - a(s)|ds \\ &\leq \|x - a\|^{[0,T]} \int_0^T |C(t,s)|ds + \|C\| \int_T^t |x(s) - a(s)|ds.\end{aligned}$$

For a given $\epsilon > 0$ we can take T so large that the last term is bounded by $\epsilon/2$; then take t so large that the next-to-last term is bounded by $\epsilon/2$.

We now consider a nonlinear problem

$$x(t) = a(t) - \int_0^t C(t,s)g(s, x(s))ds \tag{2.9.24}$$

in which $g : [0, \infty) \times \Re \to \Re$ is continuous, (2.9.2) is satisfied, and

$$xg(t,x) > 0 \text{ if } x \neq 0. \tag{2.9.25}$$

Define a Liapunov functional

$$\begin{aligned}Z(t) = \int_0^t C_s(t,s)\left(\int_s^t g(u,x(u))du\right)^2 ds \\ + C(t,0)\left(\int_0^t g(u,x(u))du\right)^2,\end{aligned} \tag{2.9.26}$$

and readily arrive at the relations

$$Z'(t) \leq 2g(t,x(t))[a(t) - x(t)] \tag{2.9.27}$$

and that

$$(x(t) - a(t))^2 \leq 2C(t,t)Z(t). \tag{2.9.28}$$

along solutions of (2.9.24).

There are several ways to obtain a counterpart of (2.9.20). We used Young's inequality in Theorem 2.1.1.1 and we will use a different strategy in Chapter 4, Theorem 4.2.1. But in this linear context we ask that

$$|g(t,x)| \leq |x|. \tag{2.9.29}$$

This permits us to say that $Z'(t) \leq 2g(t,x)[a(t) - g(t,x)]$ from which we obtain

$$Z'(t) \leq -(a(t) - g(t,x(t))^2 - g^2(t,x(t)) + a^2(t). \tag{2.9.30}$$

Theorem 2.9.7. *Let (2), (2.9.25), and (2.9.29) hold. If $a \in L^2[0,\infty)$, so are $(a(t) - g(t,x(t)))$ and $g(t,x(t))$, while $a(t)$ and $C(t,t)$ bounded yield $x(t)$ bounded.*

Proof. An integration of (2.9.30) from 0 to t yields the integrability properties so that $Z(t)$ is bounded. That and $C(t,t)$ bounded in (2.9.28) yield $(x(t) - a(t))^2$ bounded, yielding $x(t)$ bounded when $a(t)$ is bounded.

While it was simple to show $(x(t) - a(t))'$ bounded, it is not so easy to show $(g(t,x(t)) - a(t))'$ bounded. There is another way to reach the conclusion that $|x(t) - a(t)| \to 0$ as $t \to \infty$.

Theorem 2.9.8. *Let (2.9.2), (2.9.25), and (2.9.29) hold with $a \in L^2[0,\infty)$ so that $\int_0^\infty g^2(s,x(s))ds =: L < \infty$. Let*

$$\sup_{0 \le t < \infty} \int_0^t C^2(t,s)ds =: M < \infty$$

and suppose that for each $T > 0$ then $\lim_{t\to\infty} \int_0^T C^2(t,s)ds = 0$. Then any solution $x(t)$ of (2.9.24) satisfies $|x(t) - a(t)| \to 0$ as $t \to \infty$.

Proof. For any $T > 0$ we have from (2.9.24) that

$$\begin{aligned}
|x(t) - a(t)| &\le \int_0^t |C(t,s)||g(s,x(s))|ds \\
&= \int_0^T |C(t,s)||g(s,x(s))|ds \\
&\quad + \int_T^t |C(t,s)||g(s,x(s))|ds \\
&\le \sqrt{\int_0^T C^2(t,s)ds \int_0^T g^2(s,x(s))ds} \\
&\quad + \sqrt{\int_T^t C^2(t,s)ds \int_T^t g^2(s,x(s))ds} \\
&\le \sqrt{L\int_0^T C^2(t,s)ds} + \sqrt{M\int_T^t g^2(s,x(s))ds}.
\end{aligned}$$

Now, for a given $\epsilon > 0$ choose T so large that

$$M\int_T^\infty g^2(s,x(s))ds < \epsilon^2/4.$$

Having chosen T, take t so large that

$$L\int_0^T C^2(t,s)ds < \epsilon^2/4.$$

This completes the proof.

Notes The material in Section 2.9 is taken from Burton (2010a). It was first published in the *E*lectronic Journal of Qualitative Theory of Differential Equations from the University of Szeged, Hungary. Full details are given in the bibliography.

2.10 Singular small kernels

We consider a scalar integral equation of the form

$$x(t) = a(t) - \int_0^t D(t,s)[x(s) + G(s,x(s))]ds \tag{2.10.1}$$

in which D is a weakly singular kernel in the sense of Definition 1.2.1, while $G : [0,\infty) \times \Re \to \Re$ and $a : [0,\infty) \to \Re$ are continuous.

We have seen in Section 2.8 that (2.10.1) can be decomposed into

$$y(t) = a(t) - \int_0^t D(t,s)y(s)ds \tag{2.10.2}$$

and two variation of parameters formulae

$$y(t) = a(t) - \int_0^t R(t,s)a(s)ds$$

$$x(t) = y(t) - \int_0^t R(t,s)G(s,x(s))ds \tag{2.10.3}$$

where x solves (2.10.1), y solves (2.10.2), and $R(t,s)$ is the resolvent solving

$$R(t,s) = D(t,s) - \int_s^t D(t,u)R(u,s)du. \tag{2.10.4}$$

We will construct a Liapunov functional for a nonlinear form of (2.10.2), namely,

$$z(t) = a(t) - \int_0^t D(t,s)g(s,z(s))ds \tag{2.10.2a}$$

which will yield our first main result. It will be necessary to develop the properties of R in two steps owing to the fact that both terms in the integral in (2.10.4) are singular, making it difficult to interchange the order of integration in one crucial step. A similar problem is encountered in establishing (2.10.3). These are both solved by the Tonelli-Hobson test. As we saw in Section 1.2 of Chapter 1, Equations (1.2.18) - (1.2.20),

$$R(t,s) = D(t,s) - R_1(t,s) \tag{2.10.5}$$

where $R_1(t,s)$ is the continuous solution of the equation

$$R_1(t,s) = D^*(t,s) - \int_s^t D(t,u)R_1(u,s)du \tag{2.10.6}$$

when

$$D^*(t,s) = \int_s^t D(t,u)D(u,s)du \tag{2.10.7}$$

is continuous. As $R_1(t,s)$ is continuous the aforementioned difficulty of interchange of order of integration will vanish upon application of the Tonelli-Hobson test. Moreover, we showed in Proposition 1.2.2 that the integration in (2.10.7) can be very simple.

We will construct a Liapunov functional for (2.10.6) in Theorem 2.10.3 yielding certain integrability properties of R_1. Those properties in (2.10.5) will yield properties of R since D is given. With this information in hand and with properties of y known from an earlier Liapunov functional, we will be able to offer important properties of the solution of (2.10.1). Finally, we will use R itself in a Liapunov functional for (2.10.3). For each integral equation, its kernel will be used in the same way to construct a Liapunov functional for the integral equation.

A Liapunov Functional for (2.10.2*a*)

Before we start with our analysis, we need to mention two problems.

Existence For Equation (2.10.2a) we suppose that $g : [0,\infty) \times \Re \to \Re$ is continuous, that there is a $K > 0$ with

$$|g(t,x)| \le |x| \text{ and } |g(t,x) - g(t,y)| \le K|x-y|, \tag{2.10.8}$$

and that whenever $\phi : [0,\infty) \to \Re$ is continuous, then both $\int_0^t |D(t,s)|ds$ and $\int_0^t D(t,s)\phi(s)ds$ are continuous. Moreover, we ask that $D(t,s)$ be continuous for $0 \le s < t < \infty$. With these assumptions we are set up to give a contraction mapping argument with weighted norm. But for the weight we need to also suppose that for each $T > 0$ there is a $\gamma > 0$ and an $\eta < 1$ with

$$\int_0^t e^{-\gamma(t-s)}|D(t,s)|ds \leq \eta$$

for $0 \leq t \leq T$. This is enough to ensure the existence of a unique solution of (2.10.2), (2.10.2a), and (2.10.6) when (2.10.7) holds. However, (2.10.1) offers several additional difficulties. $G(t,x)$ is a perturbation which may represent uncertainties and it would be unsuitable to ask for a Lipschitz condition. Moreover, because of the singularity if we only ask continuity, then the classical existence proof of Tonelli, Theorem 1.3.3, would be troublesome, although the difficulties might be overcome. What seems to be best is to use the fixed point theorem of Krasnoselskii (See Smart (1980; p. 31).) in which we define a mapping from (2.10.1) by

$$(P\phi)(t) = \left(a(t) - \int_0^t D(t,s)\phi(s)ds\right) - \int_0^t D(t,s)G(s,\phi(s))ds$$

with the first term a contraction and the second a compact map.

Our work here is primarily an illustration of the use of a number of Liapunov functionals to solve a complex problem. It would be a distraction to develop that fixed point theory here. Instead, when we discuss (2.10.1) we will state that any solution existing on $[0,\infty)$ satisfies the conclusions stated in the theorem.

Interchange of order of integration When we decompose (2.10.1), when we integrate the derivative of any of our Liapunov functionals, and when we discuss the classical relations of C and R, we always need to interchange the order of integration. A problem arises in every case because of the singularity. We avoid those problems by using the Hobson-Tonelli test. For that, we need some of the assumptions in our **existence** discussion, as well as the condition that there is an $\epsilon_0 > 0$ so that if $0 \leq \epsilon \leq \epsilon_0$, then

$$\int_s^t |D(u+\epsilon,s)|du$$

exists. Notice that $\epsilon = 0$ is included. These problems were encountered in Section 2.5.

Theorem 2.10.1. *Let $xg(t,x) \geq 0$. Suppose there are positive numbers α, β and an even integer $p > 0$ so that*

$$\beta + (p-1)\alpha < p, \tag{2.10.9}$$

that for each $\epsilon > 0$ we have

$$\sup_{t\geq 0}\int_{\epsilon}^{\infty}|D(u+t,t)|du \leq \beta, \tag{2.10.10}$$

and that for $t \geq 0$ then

$$\int_{0}^{\infty}|D(t,s)|ds \leq \alpha. \tag{2.10.11}$$

Moreover, assume that there exists $\mu > 0$ with

$$\mu \in (0, p-\beta-(p-1)\alpha) \tag{2.10.12}$$

such that for all sufficiently small $\epsilon > 0$ then

$$\sup_{s\in[0,\infty)}\int_{s}^{\infty}|D(u+\epsilon,s)-D(u,s)|du < \mu. \tag{2.10.13}$$

If $a \in L^p[0,\infty)$ and if z solves (2.10.2a) on $[0,\infty)$ then $g(t,z(t)) \in L^p[0,\infty)$.

Proof. For $\epsilon > 0$ satisfying (2.10.13) and for $t \geq 0$ define

$$V(t,\epsilon) = \int_{0}^{t}\left[\int_{t-s+\epsilon}^{\infty}|D(u+s,s)|du\right]|g(s,z(s))|^{p}ds \tag{2.10.14}$$

so that $u \geq t-s+\epsilon \geq \epsilon$ since $0 \leq s \leq t$; that is, the integrand is continuous. In preparation for V' we derive two relations. First, since $z(t)$ is a solution of (2.10.2a), it is true that

$$pg^{p-1}(t,z(t))\left[a(t)-z(t)-\int_{0}^{t}D(t,s)g(s,z(s))ds\right] = 0.$$

Next, due to

$$-|D(t+\epsilon,s)| \leq -|D(t,s)| + |D(t+\epsilon,s)-D(t,s)|$$

we have

$$\begin{aligned}
V'(t,\epsilon) &= \int_{\epsilon}^{\infty}|D(u+t,t)|du|g(t,z)|^{p} - \int_{0}^{t}|D(t+\epsilon,s)||g(s,z(s))|^{p}ds \\
&\leq \beta|g(t,z)|^{p} - \int_{0}^{t}|D(t,s)||g(s,z(s))|^{p}ds \\
&+ \int_{0}^{t}|D(t,s)-D(t+\epsilon,s)||g(s,z(s))|^{p}ds \\
&+ pg^{p-1}(t,z(t))\left[a(t)-z(t)-\int_{0}^{t}D(t,s)g(s,z(s))ds\right].
\end{aligned}$$

Denote by H the last line in V'; that is,

$$\begin{aligned}H &= pg^{p-1}(t,z(t))\left[a(t) - z(t) - \int_0^t D(t,s)g(s,z(s))ds\right] \\ &= pg^{p-1}(t,z(t))a(t) - pg^{p-1}(t,z(t))z(t) \\ &- p\int_0^t D(t,s)g(s,z(s))g^{p-1}(t,z(t))ds\end{aligned}$$

and note that by (2.10.8) we have

$$-g^{p-1}(t,z(t))z(t) \leq -g^p(t,z(t)).$$

Next, note that for $p \geq 2$ we have

$$\frac{1}{\frac{p}{p-1}} + \frac{1}{p} = 1$$

for use in Young's inequality:

$$ab \leq \frac{a^p}{p} + \frac{b^q}{q},$$

where $a \geq 0$, $b \geq 0$, and $q = p/(p-1)$. For

$$\gamma \in \left(0, \frac{p-(p-1)\alpha - \beta - \mu}{p-1}\right)$$

and for M satisfying

$$M^{\frac{1}{p}}\gamma^{\frac{p-1}{p}} \geq 1,$$

we apply the inequality to

$$M^{\frac{1}{p}}|a(t)| \cdot \gamma^{\frac{p-1}{p}}|g(t,z(t))|^{p-1}$$

obtaining

$$\begin{aligned}|g(t,z(t))|^{p-1}|a(t)| &\leq M^{\frac{1}{p}}|a(t)| \cdot \gamma^{\frac{p-1}{p}}|g(t,z(t))|^{p-1} \\ &\leq M\frac{a^p(t)}{p} + \gamma\frac{g^p(t,z(t))}{\frac{p}{p-1}}.\end{aligned}$$

Then this, along with Young's inequality also applied to the integrand below, yields

$$\begin{aligned}
H &\leq p|g(t,z(t))|^{p-1}|a(t)| - pg^{p-1}(t,z(t))z(t) \\
&+ p\int_0^t |D(t,s)||g(s,z(s))||g(t,z(t))|^{p-1}ds \\
&\leq pM\frac{a^p(t)}{p} + p\gamma\frac{g^p(t,z(t))}{\frac{p}{p-1}} - pg^p(t,z(t)) \\
&+ p\int_0^t |D(t,s)|\left(\frac{g^p(t,z(t))}{\frac{p}{p-1}} + \frac{g^p(s,z(s))}{p}\right)ds \\
&= Ma^p(t) + \gamma(p-1)g^p(t,z(t)) - pg^p(t,z(t)) \\
&+ (p-1)\int_0^t |D(t,s)|dsg^p(t,z(t)) + \int_0^t |D(t,s)|g^p(s,z(s))ds.
\end{aligned}$$

Putting this back into V' yields

$$\begin{aligned}
V'(t,\epsilon) &\leq \beta g^p(t,z(t)) - \int_0^t |D(t,s)|g^p(s,z(s))ds \\
&+ \int_0^t |D(t+\epsilon,s) - D(t,s)|g^p(s,z(s))ds \\
&+ Ma^p(t) + \gamma(p-1)g^p(t,z(t)) - pg^p(t,z(t)) \\
&+ (p-1)\left[\int_0^t |D(t,s)|ds\right]g^p(t,z(t)) + \int_0^t |D(t,s)|g^p(s,z(s))ds \\
&\leq \left[\beta + \gamma(p-1) - p + (p-1)\int_0^t |D(t,s)|ds\right]g^p(t,z(t)) \\
&+ Ma^p(t) + \int_0^t |D(t+\epsilon,s) - D(t,s)|g^p(s,z(s))ds.
\end{aligned}$$

Taking (2.10.11) into consideration

$$\begin{aligned}
V'(t,\epsilon) &\leq [\beta + \gamma(p-1) - p + (p-1)\alpha]g^p(t,z(t)) \\
&+ Ma^p(t) + \int_0^t |D(t+\epsilon,s) - D(t,s)|g^p(s,z(s))ds. \quad (2.10.15)
\end{aligned}$$

If we integrate the last term from 0 to t and interchange the order of integration we obtain

$$\begin{aligned}
&\int_0^t \int_0^u |D(u+\epsilon, s) - D(u,s)| g^p(s, z(s)) ds du \\
&= \int_0^t \left[\int_s^t |D(u+\epsilon, s) - D(u,s)| du \right] g^p(s, z(s)) ds \\
&\leq \mu \int_0^t g^p(s, z(s)) ds.
\end{aligned}$$

Using (2.10.15) this yields

$$\begin{aligned}
&V(t,\epsilon) - V(0,\epsilon) \leq [\beta + (p-1)\alpha - p + \gamma(p-1)] \int_0^t g^p(s, z(s)) ds \\
&+ M \int_0^t a^p(s) ds + \int_0^t \int_0^u |D(u+\epsilon, s) - D(u,s)| g^p(s, z(s)) ds du \\
&\leq [\beta + (p-1)\alpha - p + \gamma(p-1)] \int_0^t g^p(s, z(s)) ds + M \int_0^t a^p(s) ds \\
&+ \mu \int_0^t g^p(s, z(s)) ds \\
&\leq [\beta + (p-1)\alpha - p + \gamma(p-1) + \mu] \int_0^t g^p(s, z(s)) ds \\
&+ M \int_0^t a^p(s) ds.
\end{aligned}$$

As

$$\begin{aligned}
\mu^* &:= \beta + (p-1)\alpha - p + \gamma(p-1) + \mu \\
&< \beta + (p-1)\alpha - p + \frac{p - (p-1)\alpha - \beta - \mu}{p-1}(p-1) + \mu \\
&= \beta + (p-1)\alpha - p + p - (p-1)\alpha - \beta - \mu + \mu = 0,
\end{aligned}$$

it follows that $\mu^* < 0$ and

$$0 \leq V(t,\epsilon) \leq V(0,\epsilon) + \mu^* \int_0^t g^p(s, z(s)) ds + M \int_0^t a^p(s) ds, \quad (2.10.16)$$

as required.

Context Everything we do here will center on variants of (2.10.10) and (2.10.11), with particular attention paid to the constants α and β when either of them (or both) is greater than or equal to 1. This section is entirely about small kernels, that is, kernels satisfying variants of conditions such as (A) or (B) given below; and it is restricted to such kernels because there are absolutely no sign restrictions on any of the functions in (2.10.1). Not only are such results interesting in their own right, but there is a peculiarity about Liapunov functionals and integral equations. We may add kernels and add Liapunov functionals. So that if we have an integral equation of interest and a Liapunov functional for it, then we can add one of our small kernels to cover uncertain perturbations and add our Liapunov functionals developed here to the aforementioned Liapunov functional and have a ready-made perturbation result. We have discussed equations with convex kernels and weak singularities in Section 2.5 using Liapunov functionals. Those kernels can be large, but there are very pronounced sign restrictions which are frequently realized in real-world problems. Early classical results for (2.10.2)

$$y(t) = a(t) - \int_0^t D(t,s)y(s)ds$$

(and nonlinear analogs) ask

$$\sup_{0\le t<\infty} \int_0^t |D(t,s)|ds = \alpha < 1. \tag{A}$$

When (A) holds there are three central results which we have encountered in Chapters 1 and 2.

(i) If $a \in L^\infty$, so is y.

(ii) Also, $\sup_{0\le t<\infty} \int_0^t |R(t,s)| \le \frac{\alpha}{1-\alpha}$.

(iii) If for each $T > 0$ we have $\lim_{t\to\infty} \int_0^T |D(t,s)|ds = 0$, then the same is true for R.

In Section 2.8 and in Burton-Haddock (2010) we find a parallel theory for integration of the first coordinate of D. If there is a $\beta < 1$ with

$$\sup_{0\le s\le t<\infty} \int_s^t |D(u,s)|du \le \beta < 1 \tag{B}$$

then there are L^p results and a pleasant parallel to (ii) in the form

$$\sup_{0\le s\le t<\infty} \int_s^t |R(u,s)|du \le \frac{\beta}{1-\beta}.$$

Both (ii) and this last result depend on D being continuous because of a needed interchange of order of integration. However, in case D has weak singularities of the type discussed here, the equations (2.10.5), (2.10.6), and (2.10.7) will allow us to obtain good substitutes.

With this context, we have four claims. First, write (2.10.9) as

$$\frac{\beta}{p} + \left(1 - \frac{1}{p}\right)\alpha < 1. \tag{2.10.9a}$$

Claim 1. Inequality (2.10.9) cannot hold for $\alpha \geq 2$.

Here are the details. Since p is an even integer with $p > 0$, we see that

$$\frac{1}{p} \leq \frac{1}{2}$$

and that

$$1 - \frac{1}{p} \geq 1 - \frac{1}{2} = \frac{1}{2};$$

hence

$$\frac{1}{2}\alpha \leq \left(1 - \frac{1}{p}\right)\alpha < 1,$$

from which we have $\alpha < 2$, as required.

Claim 2. From (2.10.9a) we have

$$\frac{\beta - \alpha}{p} < 1 - \alpha$$

and it follows that for any $\alpha \in (0, 1)$ and any $\beta > 0$ we may always find an even integer $p \geq 2$ such that (2.10.9) holds true.

Claim 3. If $\alpha \in (1, 2)$, then (2.10.9) holds for a positive even integer p only if $\beta < \alpha$. Furthermore, a necessary and sufficient condition for the existence of such an integer so that (2.10.9) holds is $\beta + \alpha < 2$.

To see this, we first note that (2.10.9), namely $\beta + (p - 1)\alpha < p$ is equivalent to $\beta - \alpha < p(1 - \alpha)$. Thus, $p > 0$ and $\alpha > 1$ implies $\beta < \alpha$.

If (2.10.9) holds for an even integer $p \geq 2$, then as $1 - \alpha < 0$, it follows that $p(1-\alpha) \leq 2(1-\alpha)$. Hence, $\beta - \alpha < 2(1-\alpha)$, which implies $\beta + \alpha < 2$. Conversely, if $\beta + \alpha < 2$, then $\beta - \alpha < 2(1 - \alpha)$. Consequently, (2.10.9) holds with $p = 2$.

Claim 4. If $\alpha = 1$ then a necessary and sufficient condition for the existence of some positive even integer such that (2.10.9) holds is $\beta < \alpha = 1$. In this case (2.10.9) holds for any positive even integer.

Finally, we have a corollary to Theorem 2.10.1.

Corollary. *Let $xg(t,x) \geq 0$ and assume that there exist α, β such that for each $\epsilon > 0$ we have*

$$\sup_{t\geq 0} \int_{\epsilon}^{\infty} |D(u+t,t)|du \leq \beta, \tag{2.10.10}$$

and for $t \geq 0$

$$\int_0^{\infty} |D(t,s)|ds \leq \alpha < 2. \tag{2.10.11}$$

Moreover, suppose that for some even integer p with

$$\beta + (p-1)\alpha < p \tag{2.10.9}$$

we have

$$\sup_{s\in[0,\infty)} \int_s^{\infty} |D(u+\epsilon, s) - D(u,s)|du < p - \beta - (p-1)\alpha. \tag{2.10.13}$$

If $a \in L^p[0,\infty)$ and if z solves (2a) on $[0,\infty)$ then $g(t,z(t)) \in L^p[0,\infty)$.

Note that if $\alpha \in (0,1)$ and if the integral in (2.10.13) is bounded then in view of the monotonicity of $p - \beta - (p-1)\alpha$ in p (or Claim 2) we may see that there always exists a (smallest) positive even integer p_0 such that (2.10.9) and (2.10.13) hold true for all $p \geq p_0$. It follows that if $\alpha \in (0,1)$, if $\beta < 1$ and p_0 is the smallest positive even integer such that (2.10.9) and (2.10.13) hold true, then $a \in L^p[0,\infty)$ implies that $g(t,z(t)) \in L^p[0,\infty)$ for any even integer $p \geq p_0$.

A Liapunov Functional for (2.10.2a) when $p = 1$

Important equations are missed when $p \geq 2$ and we also need to prepare for the resolvent equation where we will pick up the case of z bounded when $a(t)$ is bounded in the convolution case. Thus, we return to

$$z(t) = a(t) - \int_0^t D(t,s)g(s,z(s))ds, \quad t \geq 0, \tag{2.10.2a}$$

with the existence assumptions detailed above.

Theorem 2.10.2. *Let $z(t)$ be a solution of (2.10.2a) for $0 \le t < \infty$. Assume that there exists a continuous function $h : [0,\infty) \to [0,\infty)$ with*

$$|g(t,x)| \le h(t)|x|, \text{ for all } (t,x) \in [0,\infty) \times \Re, \tag{2.10.8a}$$

a function $\beta : [0,\infty) \to [0,\infty)$, and an $\epsilon > 0$ such that for all $t \ge 0$ we have the convergent integral satisfying

$$\int_{\epsilon}^{\infty} |D(u+t,t)|du \le \beta(t). \tag{2.10.10a}$$

Moreover, suppose that there is a positive constant T with

$$\sup_{t \ge T} h(t)[\beta(t) + \phi(t)] < 1 \tag{2.10.12a}$$

where the continuous function ϕ is defined by the convergent integral

$$\int_{s}^{\infty} |D(u+\epsilon,s) - D(u,s)|du =: \phi(s). \tag{2.10.13a}$$

Then $a \in L^1[0,\infty)$ implies $z \in L^1[0,\infty)$. In fact, the integral in (2.10.11) need not be bounded.

Proof. For the selected $\epsilon > 0$ we define

$$V(t,\epsilon) = \int_0^t \left[\int_{t-s+\epsilon}^{\infty} |D(u+s,s)|du \right] |g(s,z(s))|ds, \quad t \ge 0. \tag{2.10.14a}$$

Since $0 \le s \le t$ we have $u \ge t-s+\epsilon \ge \epsilon$, so V is well defined and the integrand is continuous.

The derivative of V yields

$$\begin{aligned} V'(t,\epsilon) &= \int_{\epsilon}^{\infty} |D(u+t,t)|du|g(t,z(t))| \\ &\quad - \int_0^t |D(t+\epsilon,s)||g(s,z(s))|ds \\ &\le \beta(t)|g(t,z(t))| - \int_0^t |D(t,s)||g(s,z(s))|ds \\ &\quad + \int_0^t |D(t+\epsilon,s) - D(t,s)||g(s,z(s))|ds \\ &\le \beta(t)h(t)|z(t)| - |a(t) - z(t)| \\ &\quad + \int_0^t |D(t+\epsilon,s) - D(t,s)||g(s,z(s))|ds. \end{aligned}$$

In the final lines of our computation below we will need to see that

$$\begin{aligned}\int_0^t \int_0^u &|D(u+\epsilon,s)-D(u,s)||g(s,z(s))|dsdu\\ &= \int_0^t \left[\int_s^t |D(u+\epsilon,s)-D(u,s)|du\right]|g(s,z(s))|ds\\ &\le \int_0^t \phi(s)|g(s,z(s))|ds.\end{aligned}$$

Integrating this estimate for V' yields

$$\begin{aligned}V(t,\epsilon)-V(0,\epsilon) &\le \int_0^t [\beta(s)h(s)|z(s)|+|a(s)|-|z(s)|]ds\\ &+\int_0^t\int_0^u |D(u+\epsilon,s)-D(u,s)||g(s,z(s))|dsdu\\ &= \int_0^t [\beta(s)h(s)|z(s)|+|a(s)|-|z(s)|]ds\\ &+\int_0^t\int_s^t |D(u+\epsilon,s)-D(u,s)|duh(s)|z(s)|ds\\ &= \int_0^t [\beta(s)h(s)|z(s)|+\phi(s)h(s)|z(s)|]ds\\ &+\int_0^t |a(s)|ds-\int_0^t |z(s)|ds.\end{aligned}$$

In view of (2.10.12a) for $\mu = \sup_{t\ge T}\{h(t)[\beta(t)+\phi(t)]\} < 1$ we have for $t \ge T$

$$\begin{aligned}V(t,\epsilon)-V(0,\epsilon) &\le \int_0^T h(s)[\beta(s)+\phi(s)]|z(s)|ds\\ &+\int_T^t h(s)[\beta(s)+\phi(s)]|z(s)|ds+\int_0^t |a(s)|ds-\int_0^t |z(s)|ds\\ &< \int_0^T h(s)[\beta(s)+\phi(s)]|z(s)|ds+\mu\int_T^t |z(s)|ds\\ &+\int_0^t |a(s)|ds-\int_T^t |z(s)|ds\\ &= \int_0^T h(s)[\beta(s)+\phi(s)]|z(s)|ds+\int_0^t |a(s)|ds-(1-\mu)\int_T^t |z(s)|ds.\end{aligned}$$

As the solution exists for all $t \ge 0$, the third-to-last integral is a finite number, while the last integral yields the result.

A Liapunov Functional for the Resolvent

Our focus here is on the resolvent equation

$$R_1(t,s) = D^*(t,s) - \int_s^t D(t,u)R_1(u,s)du \tag{2.10.6}$$

where $R_1(t,s)$ and

$$D^*(t,s) = \int_s^t D(t,u)D(u,s)du \tag{2.10.7}$$

are both continuous, while D satisfies the existence and interchange conditions stated above. We now present a result which is parallel to Theorem 2.10.2 with $h(t) = 1$, D^* replacing $a(t)$, and yielding the conclusion that there is a $\mu^* > 0$ with

$$\int_s^t |R_1(u,s)|du \le \mu^* \int_s^t |D^*(u,s)|du. \tag{2.10.17}$$

From (2.10.3) and (2.10.5) we see that

$$y(t) = a(t) - \int_0^t [D(t,s) - R_1(t,s)]a(s)ds. \tag{2.10.18}$$

If D is of convolution type, so are R and R_1 so that when $\mu^* \int_s^t |D^*(u,s)|du \le M$ for some finite value, then (2.10.17) becomes

$$\int_0^{t-s} |R_1(v)|dv = \int_s^t |R_1(u-s)|du \le M.$$

In other words, $R_1 \in L^1[0,\infty)$.

If D is also in $L^1[0,\infty)$, then $R \in L^1[0,\infty)$ from (2.10.5) and so from (2.10.3) we have that

$$a \in L^\infty \implies y \in L^\infty.$$

Thus, in the convolution case we get both $a \in L^1$ and $a \in L^\infty$ imply the same for y.

Continuing in the same vein, if $R \in L^1[0,\infty)$ and if $a(t) \to 0$ as $t \to \infty$, from (2.10.3) we get that $y(t) \to 0$ as $t \to \infty$.

The rewards continue as we look at the other equation in (2.10.3).

It would be a real *coup* to advance that to the non-convolution case, but we will see parallel L^p results.

Theorem 2.10.3. *Suppose there is a constant $\beta < 1$ such that for each sufficiently small $\epsilon > 0$ we have*

$$\sup_{t\geq 0}\int_{\epsilon}^{\infty}|D(u+t,t)|du \leq \beta \tag{2.10.10b}$$

and

$$\sup_{s\in[0,\infty)}\int_{s}^{\infty}|D(u+\epsilon,s)-D(u,s)|du =: \mu < 1-\beta. \tag{2.10.13b}$$

Then for

$$\mu^* = \frac{1}{1-\beta-\mu}$$

we have

$$\int_s^t |R_1(u,s)|du \leq \mu^* \int_s^t |D^*(u,s)|du. \tag{2.10.19}$$

Proof. For $\epsilon > 0$ so small that (2.10.10b) and (2.10.13b) hold, define

$$V(t,\epsilon) = \int_s^t \int_{t-u+\epsilon}^{\infty} |D(v+u,u)|dv|R_1(u,s)|du$$

so that the derivative of V along (2.10.6) is

$$\begin{aligned}
V'(t,\epsilon) &= \int_{\epsilon}^{\infty}|D(v+t,t)|dv|R_1(t,s)| - \int_s^t |D(t+\epsilon,u)||R_1(u,s)|du\\
&\leq \beta|R_1(t,s)| - \int_s^t |D(t,u)||R_1(u,s)|du\\
&\quad + \int_s^t |D(t+\epsilon,u)-D(t,u)||R_1(u,s)|du\\
&\leq \beta|R_1(t,s)| - |R_1(t,s)| + |D^*(t,s)|\\
&\quad + \int_s^t |D(t+\epsilon,u)-D(t,u)||R_1(u,s)|du.
\end{aligned}$$

Integration of the last term yields

$$\begin{aligned}
&\int_s^t\int_s^v |D(v+\epsilon,u)-D(v,u)||R_1(u,s)|dudv\\
&= \int_s^t\int_u^t |D(v+\epsilon,u)-D(v,u)|dv|R_1(u,s)|du\\
&\leq \int_s^t \mu|R_1(u,s)|du.
\end{aligned}$$

Thus, if we integrate V' from s to t we have

$$\begin{aligned} V(t,\epsilon) &\leq V(s,\epsilon) - (1-\beta)\int_s^t |R_1(v,s)|dv \\ &\quad + \mu\int_s^t |R_1(v,s)|dv + \int_s^t |D^*(u,s)|du \\ &=: -\lambda\int_s^t |R_1(v,s)|dv + \int_s^t |D^*(u,s)|du. \end{aligned}$$

Taking $\mu^* = 1/\lambda$ completes the proof.

The Nonlinear Equation

We now consider (2.10.1) in the form of

$$x(t) = y(t) - \int_0^t R(t,s)G(s,x(s))ds. \tag{2.10.3}$$

Theorem 2.10.4. *Let $x(t)$ solve (2.10.1) on $[0,\infty)$. Suppose that $R(t,s) = R(t-s)$, $R \in L^1[0,\infty)$, and that $y \in L^\infty$. Assume also that*

$$|G(t,x)| \leq \phi(t)|x| \tag{2.10.20}$$

where $\phi : [0,\infty) \to \Re$ is continuous. If $\phi(t) \to 0$ as $t \to \infty$, then $x \in L^\infty$. If $y(t) \to 0$ as $t \to \infty$ and if $\phi(t) \to 0$ as $t \to \infty$, then $x(t) \to 0$ as $t \to \infty$.

Proof. Let $y \in L^\infty$, $R \in L^1[0,\infty)$, and let $\phi(t) \to 0$. If x is not bounded, then there is a sequence $\{t_n\} \uparrow \infty$ such that $|x(t)| \leq |x(t_n)|$ if $0 \leq t \leq t_n$. Let $\|\cdot\|$ denote the supremum norm. Then for large n we have $|x(t_n)|$ unbounded and

$$|x(t_n)| \leq \|y\| + |x(t_n)|\int_0^{t_n} |R(t_n - s)|\phi(s)ds \leq \|y\| + (1/2)|x(t_n)|,$$

a contradiction.

Next, if $R \in L^1[0,\infty)$ and if $y(t) \to 0$ then we have x bounded and

$$|x(t)| \leq |y(t)| + \|x\|\int_0^t |R(t-s)|\phi(s)ds \to 0.$$

In preparation for our next result, note from Theorem 2.10.3 that

$$\mu^* = \frac{1}{1-\beta-\mu}$$

where β and μ are defined in (2.10.10b) and (2.10.13b).

Theorem 2.10.5. *Let $y(t)$ solve (2.10.2) and let (2.10.10b) and (2.10.13b) hold. If*

$$\int_0^t \left[\int_s^t [|D(u,s)| + \mu^*|D^*(u,s)|]du\right]|a(s)|ds$$

is bounded for $t \geq 0$, then $a(t) \in L^1[0,\infty)$ implies that $y \in L^1[0,\infty)$.

Proof. Note that $y(t) = a(t) - \int_0^t R(t,s)a(s)ds$ and $R(t,s) = D(t,s) - R_1(t,s)$. From this we have

$$\begin{aligned}
&\int_0^t |y(u)|du - \int_0^t |a(u)|du \\
&\leq \int_0^t \int_0^u |D(u,s) - R_1(u,s)||a(s)|dsdu \\
&= \int_0^t \int_s^t |D(u,s) - R_1(u,s)||a(s)|duds \\
&\text{(but by (2.10.19)} \int_s^t |R_1(u,s)|du \leq \mu^* \int_s^t |D^*(u,s)|du \text{ so)} \\
&\leq \int_0^t \int_s^t (|D(u,s)| + \mu^*|D^*(u,s)|)du|a(s)|ds,
\end{aligned}$$

as required

Notice that for Theorem 2.10.1 we were forced to take p larger than 1 to satisfy (2.10.9) and we obtained $g(t,z(t)) \in L^p$. Thus, suppose that in those results we have

$$g(t,z) = z + G(t,z)$$

and we have obtained $z \in L^2$. Can we force that back to $z \in L^1$? In Section 2.5 of Chapter 2 we studied a convex kernel with singularity and we obtained $g(t,x) \in L^2$ with no way to force it back into L^1. Moreover, using the properties of the Liapunov functional we also showed ways to get $x(t) - a(t)$ bounded.

There are many times when we want L^1 instead of L^2. The next two theorems show us how to meet such needs. We do it here for $x \in L^2$, but using Hölder's inequality it can be extended, as we will note parenthetically in one of the proofs. We also show how to use x bounded for the same conclusion.

Theorem 2.10.6. *Let x solve (2.10.1), y solve (2.10.2), $y \in L^1[0,\infty)$, $x \in L^2[0,\infty)$, and let (2.10.10b), (2.10.13b), and (2.10.20) hold. If, in addition,*

$$\sup_{t\geq 0}\int_0^t \left[\int_s^t (|D(u,s)| + \mu^*|D^*(u,s)|)du\right]^2 \phi^2(s)ds < \infty, \tag{*}$$

then $x \in L^1[0,\infty)$.

Proof. We have

$$\int_0^t |x(s)|ds - \int_0^t |y(s)|ds \leq \int_0^t \int_0^u |D(u,s) - R_1(u,s)|\phi(s)|x(s)|dsdu$$
$$= \int_0^t \int_s^t |D(u,s) - R_1(u,s)|du\phi(s)|x(s)|ds$$
$$\leq \int_0^t \int_s^t (|D(u,s)| + \mu^*|D(u,s)|)du\phi(s)|x(s)|ds$$

(we use x^2, but we could use x^p, Hölder's inequality, and change (*))

$$\leq \sqrt{\int_0^t \left[\int_s^t (|D(u,s)| + \mu^*|D^*(u,s)|)du\right]^2 \phi^2(s)ds \int_0^t x^2(s)ds},$$

as required.

The same computation yields the following result.

Corollary. *Let (2.10.10b), (2.10.13b), (2.10.20), and (*) hold. Then $x \in L^2[0,\infty)$ implies $y - x \in L^1[0,\infty)$.*

Theorem 2.10.7. *Let x solve (2.10.1), y solve (2.10.2), $y \in L^1[0,\infty)$, x be bounded, and let (2.10.10b), (2.10.13b), and (2.10.20) hold. If, in addition,*

$$\sup_{t\geq 0}\int_0^t \left[\int_s^t (|D(u,s)| + \mu^*|D^*(u,s)|)du\right] \phi(s)ds < \infty, \tag{**}$$

then $x \in L^1[0,\infty)$.

Proof. Follow the proof of Theorem 2.10.6 down to the relation

$$\int_0^t |x(s)|ds - \int_0^t |y(s)|ds \leq \int_0^t \int_s^t (|D(u,s)| + \mu^*|D^*(u,s)|)du\phi(s)|x(s)|ds.$$

As x is bounded, the conclusion is immediate.

Corollary 1. *Let (2.10.10b), (2.10.13b), (2.10.20), and (**) hold. Then x bounded implies $y - x \in L^1[0, \infty)$.*

Corollary 2. *If (2.10.10b), (2.10.13b), (2.10.20) hold and if*

$$\sup_{0 \le s \le t < \infty} \left[\int_s^t (|D(u,s)| + \mu^* |D^*(u,s)|) du \phi(s) \right] \le \gamma < 1,$$

then $y \in L^1[0, \infty)$ implies $x \in L^1[0, \infty)$.

In the last proof we see that upon integration of that last display we obtain

$$\int_0^t |x(s)| ds \le \int_0^t |y(s)| ds + \int_0^t \gamma |x(s)| ds,$$

yielding the result. Corollary 2, with Theorem 2.10.5, yields the next result.

Corollary 3. *Let the conditions of Corollary 2 and*

$$\sup_{t \ge 0} \int_0^t \left[\int_s^t (|D(u,s)| + \mu^* |D^*(u,s)|) du \right] |a(s)| ds < \infty$$

hold. Then $a \in L^1[0, \infty)$ implies $x \in L^1[0, \infty)$.

A Liapunov Functional Based on $R(t, s)$

Equation (2.10.3) with (2.10.5) is

$$x(t) = y(t) - \int_0^t [D(t,s) - R_1(t,s)] G(s, x(s)) ds.$$

If $R_1 = 0$ and if $y \in L^p$ we would define

$$V_1(t, \epsilon) = \int_0^t \int_{t-s+\epsilon}^\infty |D(u+s, s)| du |G(s, x(s))| ds.$$

If $D = 0$ we would define

$$V_2(t) = \int_0^t \int_{t-s}^\infty |R_1(u+s, s)| du |G(s, x(s))| ds.$$

One of the surprising aspects of Liapunov's direct method for investigators well acquainted with Liapunov theory for differential equations is that when we add kernels then we can add Liapunov functionals. That will be illustrated here.

In preparation for construction of a Liapunov functional based on the unknown function $R_1(t,s)$ we note two relations. First,

$$\int_0^\infty |R_1(v+t,t)|dv = \int_t^\infty |R_1(w,t)|dw$$

and if the conditions of Theorem 2.10.3 hold, then we consider (2.10.19)

$$\int_s^t |R_1(u,s)|du \le \mu^* \int_s^t |D^*(u,s)|du$$

and ask that there is a $\Lambda > 0$ with

$$\int_t^\infty |R_1(u,t)|du \le \mu^* \int_t^\infty |D^*(u,t)|du \le \Lambda. \tag{2.10.21}$$

Next, by a change of variable we see that

$$\int_{t-s}^\infty |R_1(u+s,s)|du \le \int_{t-t}^\infty |R_1(u+s,s)|du = \int_s^\infty |R_1(v,s)|dv \le \Lambda.$$

This means that when the conditions of Theorem 2.10.3 and (2.10.21) hold then

$$V(t,\epsilon) = \int_0^t \left[\int_{t-s+\epsilon}^\infty |D(u+s,s)|du + \int_{t-s}^\infty |R_1(u+s,s)|du\right] |G(s,x(s))|ds \tag{2.10.22}$$

is well-defined.

Theorem 2.10.8. *If the conditions of Theorem 2.10.3, (2.10.20), and (2.10.21) hold, if there is a $\gamma < 1$ with*

$$[\beta + \Lambda + \mu]\phi(t) \le \gamma, \quad 0 \le t < \infty, \tag{2.10.23}$$

then $y \in L^1[0,\infty)$ implies $x \in L^1[0,\infty)$.

Proof. With V defined in (2.10.22) we have

$$V'(t,\epsilon) = \left[\int_{\epsilon}^{\infty} |D(u+t,t)|du + \int_0^{\infty} |R_1(u+t,t)|du\right] |G(t,x(t))|$$
$$- \int_0^t [|D(t+\epsilon,s)| + |R_1(t,s)|]|G(s,x(s))|ds$$
$$\leq [\beta+\Lambda]|G(t,x)| - \int_0^t [|D(t,s)| + |R_1(t,s)|]|G(s,x(s))|ds$$
$$+ \int_0^t |D(t+\epsilon,s) - D(t,s)||G(s,x(s))|ds$$

where β is defined in (2.10.10b) and Λ is defined in (2.10.21)

$$\leq [\beta+\Lambda]|G(t,x)| + |y(t)| - |x(t)| + \int_0^t |D(t+\epsilon,s)$$
$$- D(t,s)||G(s,x(s))|ds$$
$$\leq \{[\beta+\Lambda]\phi(t) - 1\}|x(t)| + |y(t)|$$
$$+ \int_0^t |D(t+\epsilon,s) - D(t,s)|\phi(s)|x(s)|ds.$$

Integrate the last term to obtain

$$\int_0^t \int_0^u |D(u+\epsilon,s) - D(u,s)|\phi(s)|x(s)|dsdu$$
$$= \int_0^t \int_s^t |D(u+\epsilon,s) - D(u,s)|du\phi(s)|x(s)|ds$$
$$\leq \int_0^t \mu\phi(s)|x(s)|ds$$

where μ is defined in (2.10.13b).

Thus, integration of V' yields

$$0 \leq V(t,\epsilon) \leq V(0,\epsilon) + \int_0^t \{[\beta+\Lambda+\mu]\phi(s) - 1\}|x(s)|ds + \int_0^t |y(s)|ds$$
$$\leq -(1-\gamma)\int_0^t |x(s)|ds + \int_0^t |y(s)|ds,$$

as required.

Corollary. *Let the conditions of Theorem 2.10.3 and (2.10.20) hold with* $\phi(t) \leq 1$. *If, in addition,*

$$\sup_{t\geq 0} \int_t^\infty |D^*(u,t)|du < (1-\beta-\mu)^2,$$

then $y \in L^1[0,\infty)$ *implies that* $x \in L^1[0,\infty)$.

Proof. Since $\phi(t) \leq 1$, if we set

$$\psi(t) = \sup_{t\geq 0} \frac{|G(t,x)|}{|x|}$$

then $\psi(t) \leq 1$ for $t \geq 0$. Moreover, (2.10.21) holds with

$$\Lambda = \mu^* \sup_{t\geq 0} \int_t^\infty |D^*(u,t)|du.$$

Thus, taking into consideration $\mu^*(1-\beta-\mu) = 1$ (from the definition of μ^*) we have for $t \geq 0$

$$\begin{aligned}
[\beta + \Lambda + \mu]\psi(t) \leq \gamma &:= \left[\beta + \mu^* \sup_{s\geq 0} \int_s^\infty |D^*(u,s)|du + \mu\right] \sup_{t\geq 0} \psi(t) \\
&\leq \left[\beta + \mu^* \sup_{s\geq 0} \int_s^\infty |D^*(u,s)|du + \mu\right] \cdot 1 \\
&< \beta + \mu^*(1-\beta-\mu)^2 + \mu \\
&= \beta + 1 - \beta - \mu + \mu \\
&= 1;
\end{aligned}$$

that is, (2.10.23) is satisfied.

Integrations

While the reader may work through the presentation with approval at each step, there is the nagging problem of all those integrations which start with D^* in (2.10.7) and progress to (**) in Theorem 2.10.7. Fortunately, they turn out to be fairly simple, even for the deep problems which occur throughout so much of applied mathematics.

Let $g(t,x)$ be continuous on $[0,\infty)\times\Re$ and consider the scalar fractional differential equation of Caputo type

$$^cD^q x(t) = -g(t,x(t)), \quad 0<q<1, \quad x(0)\in\Re,$$

which is inverted as the ordinary integral equation

$$x(t) = x(0) - \frac{1}{\Gamma(q)}\int_0^t (t-s)^{q-1} g(s,x(s))ds$$

where Γ is the gamma function. A myriad of real-world problems take this form and the value $q=1/2$ is at the forefront. That is emphasized here because when $q=1/2$ that intimidating integral D^* in (2.10.7) is simply a constant. In Proposition 1.2.1 of Chapter 1 and Section 2, we showed how to compute D^* for this kernel. We will now show how to use that for the calculations in Corollary 3 to Theorem 2.10.7 with a very different kernel. From those details the reader will see that a whole class of problems is solved for smaller kernels when the one calculation is done.

Example. Review Proposition 1.2.1 and see that if

$$D(t,s) = \frac{1}{\Gamma(q)}(t-s)^{q-1}, \quad 0<q<1,$$

then for $0\le s<t$ we have

$$D^*(t,s) = \int_s^t D(t,u)D(u,s)du = \frac{1}{\Gamma(2q)}(t-s)^{2q-1}.$$

In particular, if $q=1/2$ then $D^*(t,s) = 1/\Gamma(2q) = 1$.

Now continue on to Theorems 2.10.6 and 2.10.7 and see that we are also needing D to be an L^1 kernel. Recall that through transformations we map our fractional differential equation into an equation having an L^1 kernel, so this is no real surprise. We now want to parlay the above example into D^* with an L^1 kernel. There is actually a simple way to do this. The point can be made in the following example and the reader will see that there is some generality in the method.

Example. The kernel

$$D(t,s) = \frac{1}{\Gamma(q)}\frac{(t-s)^{q-1}}{(t-s+1)^2}, \quad 0<q<1,$$

satisfies the conditions of Corollary 3 to Theorem 2.10.7.

Proof. Use the change of variable in the proof of Proposition 1.2.1 and note that $0 \leq v \leq 1$ so $1 - v$ or v is always as large as $1/2$. Thus for $t > s$ we have

$$\begin{aligned} D^*(t,s) &= \frac{1}{\Gamma^2(q)} \int_s^t \frac{(t-u)^{q-1}(u-s)^{q-1}}{(t-u+1)^2(u-s+1)^2} du \\ &= \frac{1}{\Gamma^2(q)} \int_0^1 \frac{(t-s)^{q-1}(1-v)^{q-1}(t-s)^{q-1}v^{q-1}(t-s)}{[(t-s)(1-v)+1]^2[(t-s)v+1]^2} dv \\ &\leq \frac{(t-s)^{2q-1}}{\Gamma^2(q)[(1/2)(t-s)+1]^2} \int_0^1 v^{q-1}(1-v)^{q-1} dv \\ &= \frac{(t-s)^{2q-1}}{[(1/2)(t-s)+1]^2\Gamma(2q)}. \end{aligned}$$

Since

$$\begin{aligned} \int_s^t |D^*(u,s)|du &\leq \frac{1}{\Gamma(2q)} \int_s^t \frac{(u-s)^{2q-1}}{[(1/2)(u-s)+1]^2} du \\ &= \frac{1}{\Gamma(2q)} \int_0^{t-s} \frac{w^{2q-1}}{[(1/2)w+1]^2} dw \end{aligned}$$

and $2q - 1 > -1$ so this integral converges at the lower limit. It also converges as the upper limit tends to ∞ since $2 - 2q + 1 > 1$. Thus, in Corollary 3 we have

$$\int_0^t \int_s^t |D^*(u,s)|du|a(s)|ds$$

and this is finite since $a \in L^1[0,\infty)$.

Next,

$$\int_s^t |D(u,s)|du = \frac{1}{\Gamma(q)} \int_s^t \frac{(u-s)^{q-1}}{(u-s+1)^2} du = \frac{1}{\Gamma(q)} \int_0^{t-s} \frac{w^{q-1}}{(w+1)^2} dw.$$

As $q - 1 > -1$, this converges at the lower limit. As the denominator is of order w^2, $w^{2-q+1} = w^{3-q}$ so the integral converges as $t - s \to \infty$. As $a \in L^1[0,\infty)$, the integral condition in Corollary 5.8 is satisfied.

Notes The material in this section is from Becker, Burton, and Purnaras (2012). It was first published by Elsevier in *N*onlinear Analysis: TMA. Full details are given in the bibliography.

2.11 Singular periodic kernels

We consider a scalar integral equation

$$x(t) = a(t) - \int_{-\infty}^{t} C(t,s)g(s,x(s))ds \tag{2.11.1}$$

for which there is a $T > 0$ so that

$$a(t+T) = a(t),\ g(t+T,x) = g(t,x),\ C(t+T,s+T) = C(t,s) \tag{2.11.2}$$

for all $t \in \Re$ and $s < t$ with a and g continuous. We denote by $(\mathcal{P}_T, \|\cdot\|)$ the Banach space of continuous T-periodic functions.

If g is Lipschitz and if C is small enough then a contraction mapping will yield a periodic solution. If C is convex then Liapunov arguments will produce *a priori* bounds. Under compactness conditions, Schaefer's fixed point theorem will yield a periodic solution. A collection of such results are found in Burton (2006b). A recent n-dimensional result is given in Zhang (2009).

In this section we ask that g satisfies

$$|g(t,x) - g(t,y)| \leq K|x-y| \tag{2.11.3}$$

for all $x, y \in \Re$ and some $K > 0$, while C satisfies a truncated convexity condition, but has a significant singularity at $t = s$. We derive a set of conditions measuring the magnitude of the singularity that will still permit proof of the existence of a periodic solution using a combination Krasnoselskii-Schaefer fixed point theorem which will be proved here.

A Fixed Point Theorem

The first task is to prove a fixed point theorem of Krasnoselskii-Schaefer type in which the mapping function has the form $Px = Bx + Ax$ with A being compact and $(I-B)^{-1}$ continuous on an appropriate subset M of a Banach space S. The theorem resembles that of Burton-Kirk (1998) without having a λ term in B. See Gao-Li-Zhang (2011), Krasnoselskii (1958), Liu-Li (2006) (2008), Park (2007), Schaefer (1955), Smart (1980) for work on Krasnoselskii and Schaefer theorems and their extended forms.

Since P is the sum of two operators, it is in general a non-self map; that is, P may not necessarily map a closed convex subset M of S into itself. To prove the existence of a fixed point of P, we apply topological degree theory or transversality method by constructing a homotopy U_λ on M with $U_1 = P$. It is assumed that $U_\lambda(\phi) = U(\lambda, \phi)$ is a continuous mapping of $[0,1] \times M$ into a compact subset of S. In many applications, U_0 is a constant map sending M to a point $p \in M/\partial M$. In this case, U_0 is an "essential" map. If $U_\lambda(\phi)$ is fixed point free on ∂M for all $\lambda \in (0,1)$, then $U_1(\phi)$ is essential having a fixed point property in M (Granas and Dugundji (2003; pp. 120-123)). This fact is often written in the form of Leray-Schauder Principle or its nonlinear alternatives which states that either

(A_1) U_1 has a fixed point in M or

(A_2) there exists $x \in \partial M$ and $\lambda \in (0,1)$ with $x = U_\lambda(x)$.

Theorem 2.11.1. *Let $(S, \|\cdot\|)$ be a Banach space, $A, B : S \to S$ such that A is continuous with A mapping bounded sets into compact sets, $(I-B)^{-1}$ exists and is continuous on $(I-B)S$ with $\lambda A(M) \subset (I-B)S$ for each closed convex subset $M \subset S$ and $\lambda \in [0,1]$. Then either*

(i) $x = Bx + \lambda Ax$ has a solution in S for $\lambda = 1$, or

(ii) the set of all such solutions, $0 < \lambda < 1$, is unbounded.

Proof. Since $\lambda A(M) \subset (I-B)S$, we have $0 \in (I-B)S$. If $x^* = (I-B)^{-1}(0)$, then x^* is the unique fixed point of B. For each positive integer n, define a closed and bounded set

$$M_n = \{x \in S : \|x\| \leq n\}.$$

We choose n sufficiently large so that $x^* \in M_n/\partial M_n$. Now $(I-B)^{-1}$ exists and is continuous on $(I-B)S$. Since A is continuous with A mapping M_n into a compact set, so is $(I-B)^{-1}(\lambda A)$ for each $\lambda \in [0,1]$. Define $U : [0,1] \times M_n \to S$ by

$$U(\lambda, \phi) = (I-B)^{-1}(\lambda A\phi).$$

Then $U_\lambda(\phi) = U(\lambda, \phi)$ is a continuous mapping of $[0,1] \times M_n$ into a compact subset of S. Indeed, set $\Gamma = \{\lambda A\phi : \lambda \in [0,1],\ \phi \in M_n\}$ and let $\{(\lambda_k, \phi_k)\}$ be a sequence in $[0,1] \times M_n$. We may assume that $\lambda_k \to \lambda_0 \in [0,1]$ as $k \to \infty$. Since AM_n is contained in a compact subset of S, there exists a convergent subsequence $\{A\phi_{k_j}\}$ of $\{A\phi_k\}$. Now $\{\lambda_{k_j} A\phi_{k_j}\}$ converges in S. This implies that Γ is pre-compact, and so is $(I-B)^{-1}\Gamma$. Observe that for all $\phi \in M_n$,

$$U_0(\phi) = (I-B)^{-1}(0) = x^*$$

is a constant map. Moreover, $x^* \in M_n/\partial M_n$. By the statement of nonlinear alternatives (A_1) and (A_2) above, either U_1 has a fixed point in M_n

or there exists $x_n \in \partial M_n$ such that $x_n = U_\lambda(x_n)$ for some $\lambda \in (0,1)$. This implies that either $x = Bx + Ax$ has a solution in M_n or there exists $x_n \in \partial M_n$ with $x_n = Bx_n + \lambda Ax_n$ for some $\lambda \in (0,1)$. In the later case, we have $\|x_n\| = n$. Thus, if (i) does not hold, then $\|x_n\| \to \infty$ as $n \to \infty$ and (ii) must hold. This completes the proof.

Remark It is known that if B is a contraction mapping with contraction constant $0 < \alpha < 1$, then $(I-B)^{-1}$ exists and is continuous on S. A proof is found in Burton (1996). Many generalized or nonlinear contractions satisfy this condition (see Boyd-Wong (1969), Burton (1996), Gao-Li-Zhang (2011), Liu-Li (2006)(2008), Meir-Keeler (1969), Park (2007).

Technical Conditions

We now introduce the conditions which will produce the *a priori* bound needed in the fixed point theorem, as well as the required compactness. The kernel, $C(t,s)$, can have a singularity at $t = s$, but we ask that there exists a fixed $\epsilon > 0$ so that

$$C(t,s) \geq 0,\ C_s(t,s) \geq 0,\ C_t(t,s) \leq 0,\ C_{st}(t,s) \leq 0 \tag{2.11.4}$$

provided that

$$-\infty < s \leq t - \epsilon,\ t < \infty. \tag{2.11.5}$$

Moreover, if $x \in \mathcal{P}_T$, then

$$\int_{-\infty}^{t-\epsilon} C(t,s)g(s,x(s))ds \text{ and } \int_{t-\epsilon}^{t} C(t,s)g(s,x(s))ds \text{ are continuous.} \tag{2.11.6}$$

The ϵ will play a central role. First, assume that there is an $\eta < 1$ with

$$K\int_{t-\epsilon}^{t} |C(t,s)|ds \leq \eta,\ t \in \Re. \tag{2.11.7}$$

Next, there are positive constants α and β with $2\alpha + \beta < 2$ so that both

$$\int_{s}^{s+\epsilon} [\epsilon C_s(u,u-\epsilon) + C(u,u-\epsilon) + |C(u,s)|]du < \alpha,\ s \in \Re \tag{2.11.8}$$

and

$$C(t,t-\epsilon)\epsilon + \int_{t-\epsilon}^{t} |C(t,s)|ds < \beta,\ t \in \Re. \tag{2.11.9}$$

Relations (2.11.7)–(2.11.9) specify the strength of the singularity. For a weak singularity such as $C(t,s) = [t-s]^{-p}$, $0 < p < 1$, then (2.11.4),

(2.11.5), (2.11.7)–(2.11.9) are satisfied for any $K > 0$ when it is allowed that ϵ can be taken sufficiently small. But (2.11.6) would fail. The following function satisfies (2.11.4)-(2.11.9) with $0 < \epsilon \leq 1$ and an appropriate constant $k > 0$

$$C(t,s) = \frac{k}{(t-s)(1 + |\ln(t-s) - \ln \epsilon|)^2}.$$

We now define for $0 \leq \lambda \leq 1$ a companion equation to (2.11.1)

$$x(t) = \lambda \left[a(t) - \int_{-\infty}^{t-\epsilon} C(t,s)g(s,x(s))ds \right] - \int_{t-\epsilon}^{t} C(t,s)g(s,x(s))ds. \quad (2.11.1_\lambda)$$

The mappings $A, B : \mathcal{P}_T \to \mathcal{P}_T$ mentioned in the theorem are defined by $\phi \in \mathcal{P}_T$ implies that

$$(A\phi)(t) := a(t) - \int_{-\infty}^{t-\epsilon} C(t,s)g(s,\phi(s))ds \quad (2.11.10)$$

and

$$(B\phi)(t) := -\int_{t-\epsilon}^{t} C(t,s)g(s,\phi(s))ds. \quad (2.11.11)$$

By (2.11.6), if $\phi \in \mathcal{P}_T$ then ϕ is continuous so these integrals are continuous functions. To see that $A\phi, B\phi \in \mathcal{P}_T$ we note that

$$\begin{aligned}(A\phi)(t+T) &= a(t+T) - \int_{-\infty}^{t+T-\epsilon} C(t+T,s)g(s,\phi(s))ds \\ &= a(t) - \int_{-\infty}^{t-\epsilon} C(t+T,s+T)g(s+T,\phi(s+T))ds = (A\phi)(t)\end{aligned}$$

while

$$\begin{aligned}(B\phi)(t+T) &= -\int_{t+T-\epsilon}^{t+T} C(t+T,s)g(s,\phi(s))ds \\ &= -\int_{t-\epsilon}^{t} C(t+T,s+T)g(s+T,\phi(s+T))ds = (B\phi)(t).\end{aligned}$$

Moreover, by (2.11.3) and (2.11.7), B is a contraction.

A Liapunov Functional

We begin with the assumption that there is an $L > 0$ with

$$xg(t, x) \geq 0 \text{ for } |x| \geq L \tag{2.11.12}$$

and that

$$\lim_{s \to -\infty} (t - s)C(t, s) = 0 \text{ for fixed } t. \tag{2.11.13}$$

Then define a Liapunov functional by

$$V(t, \epsilon) = \lambda \int_{-\infty}^{t-\epsilon} C_s(t, s) \left(\int_s^t g(v, x(v))dv \right)^2 ds. \tag{2.11.14}$$

Lemma 2.11.1. *If $x \in \mathcal{P}_T$ solves (2.11.1$_\lambda$) then $V'(t, \epsilon)$ satisfies*

$$\begin{aligned} V'(t, \epsilon) &\leq \lambda C_s(t, t - \epsilon) \left(\int_{t-\epsilon}^t g(v, x(v))dv \right)^2 \\ &+ 2g(t, x(t)) \left[\lambda C(t, t - \epsilon) \int_{t-\epsilon}^t g(v, x(v))dv - \int_{t-\epsilon}^t C(t, s)g(s, x(s))ds \right] \\ &+ 2g(t, x(t))[\lambda a(t) - x(t)]. \end{aligned} \tag{2.11.15}$$

Proof. Taking into account that $C_{st} \leq 0$ we have

$$\begin{aligned} V'(t, \epsilon) &\leq \lambda C_s(t, t - \epsilon) \left(\int_{t-\epsilon}^t g(v, x(v))dv \right)^2 \\ &\qquad + 2\lambda g(t, x(t)) \int_{-\infty}^{t-\epsilon} C_s(t, s) \int_s^t g(v, x(v))dvds \end{aligned}$$

If we integrate the last term by parts and use (2.11.13) in the lower limiting evaluation, keeping in mind that x is bounded, we obtain

$$\begin{aligned} V'(t, \epsilon) &\leq \lambda C_s(t, t - \epsilon) \left(\int_{t-\epsilon}^t g(v, x(v))dv \right)^2 \\ &+ 2\lambda g(t, x) \left[C(t, s) \int_s^t g(v, x(v))dv \Big|_{-\infty}^{t-\epsilon} + \int_{-\infty}^{t-\epsilon} C(t, s)g(s, x(s))ds \right] \\ &= \lambda C_s(t, t - \epsilon) \left(\int_{t-\epsilon}^t g(v, x(v))dv \right)^2 \\ &+ 2\lambda g(t, x(t)) \left[C(t, t - \epsilon) \int_{t-\epsilon}^t g(v, x(v))dv \right] \\ &+ 2g(t, x(t)) \left[\lambda \int_{-\infty}^{t-\epsilon} C(t, s)g(s, x(s))ds + \int_{t-\epsilon}^t C(t, s)g(s, x(s))ds \right] \\ &- 2g(t, x(t)) \int_{t-\epsilon}^t C(t, s)g(s, x(s))ds. \end{aligned}$$

Using ($2.11.1_\lambda$) in the next-to-last term yields (2.11.15).

We will integrate (2.11.15) to relate $g(t, x(t))$ to $a(t)$ and then use that relation in a lower bound on the Liapunov functional to obtain the *a priori* bound. We now obtain that lower bound.

Lemma 2.11.2. *For any $q > 0$, if $x \in \mathcal{P}_T$ solves (1_λ), then*

$$\begin{aligned}(x(t) - \lambda a(t))^2 &\le 2(1+q^{-1}) \int_{-\infty}^{t-\epsilon} C_s(t,s)ds\, V(t,\epsilon) \\ &+ 2(1+q^{-1})\epsilon C^2(t, t-\epsilon) \int_{t-\epsilon}^{t} g^2(s, x(s))ds \\ &+ (1+q)\left(\int_{t-\epsilon}^{t} |C(t,s)|ds\right)^2 \left(K\|x\| + \sup_{0\le u\le T} |g(u,0)|\right)^2.\end{aligned} \tag{2.11.16}$$

Proof.
Let $q > 0$ be fixed and define $H = (1+\lambda q)\left(\int_{t-\epsilon}^{t} C(t,s)g(s,x(s))ds\right)^2$ so that from ($2.11.1_\lambda$) we obtain

$$\begin{aligned}&(x(t) - \lambda a(t))^2 \\ &= \left(\lambda \int_{-\infty}^{t-\epsilon} C(t,s)g(s,x(s))ds + \int_{t-\epsilon}^{t} C(t,s)g(s,x(s))ds\right)^2 \\ &\le \lambda(1+q^{-1})\left(\int_{-\infty}^{t-\epsilon} C(t,s)g(s,x(s))ds\right)^2 + H \\ &= \lambda(1+q^{-1})\left(-C(t,s)\int_{s}^{t} g(u,x(u))du\Big|_{-\infty}^{t-\epsilon}\right. \\ &\quad \left.+ \int_{-\infty}^{t-\epsilon} C_s(t,s)\int_{s}^{t} g(u,x(u))duds\right)^2 + H \\ &\text{(using (2.11.13) and } x \in \mathcal{P}_T) \\ &= \lambda(1+q^{-1})\left(-C(t,t-\epsilon)\int_{t-\epsilon}^{t} g(u,x(u))du\right. \\ &\quad \left.+ \int_{-\infty}^{t-\epsilon} C_s(t,s)\int_{s}^{t} g(u,x(u))duds\right)^2 + H \\ &\le 2\lambda(1+q^{-1})C^2(t,t-\epsilon)\left(\int_{t-\epsilon}^{t} g(u,x(u))du\right)^2 \\ &\quad + 2\lambda(1+q^{-1})\left(\int_{-\infty}^{t-\epsilon} C_s(t,s)\int_{s}^{t} g(u,x(u))duds\right)^2 + H\end{aligned}$$

$$
\begin{aligned}
&\leq 2\lambda(1+q^{-1})C^2(t,t-\epsilon)\epsilon \int_{t-\epsilon}^{t} g^2(u,x(u))du + H \\
&\quad + 2\lambda(1+q^{-1}) \int_{-\infty}^{t-\epsilon} C_s(t,s)ds \int_{-\infty}^{t-\epsilon} C_s(t,s)\left(\int_s^t g(u,x(u))du\right)^2 ds \\
&\leq 2(1+q^{-1})C^2(t,t-\epsilon)\epsilon \int_{t-\epsilon}^{t} g^2(u,x(u))du \\
&\quad + 2(1+q^{-1}) \int_{-\infty}^{t-\epsilon} C_s(t,s)ds\, V(t,\epsilon) \\
&\quad + (1+q)\left(\int_{t-\epsilon}^{t} |C(t,s)|ds\right)^2 \left(K\|x\| + \sup_{0\leq u\leq T} |g(u,0)\|\right)^2,
\end{aligned}
$$

as required.

Lemma 2.11.3. *If*

$$|g(t,x)| \leq |x| \text{ for } |x| \geq L \tag{2.11.17}$$

where L is defined in (2.11.12), then for any $\gamma > 0$ there is an $M > 0$ such that for any solution of (2.11.1$_\lambda$)) in $\mathcal{P}_T$ we have

$$
\begin{aligned}
V'(t,\epsilon) &\leq Ma^2(t) + [\gamma+\beta-2]g^2(t,x(t)) + M \\
&\quad + \int_{t-\epsilon}^{t} [|C(t,s)| + \epsilon C_s(t,t-\epsilon) + C(t,t-\epsilon)]g^2(s,x(s))ds.
\end{aligned}
\tag{2.11.18}
$$

Proof. By the Cauchy inequality, for any $\gamma > 0$, there is an $M > 0$ such that

$$2g(t,x)a(t) \leq \gamma g^2(t,x) + Ma^2(t).$$

By (2.11.17), we may choose M so large that

$$-2g(t,x)x \leq -2g^2(t,x) + M$$

for all $t \geq 0$ and $x \in \Re$. Now from (2.11.15) we have

$$\begin{aligned}
V'(t,\epsilon) &\leq \gamma g^2(t,x(t)) + Ma^2(t) \\
&\quad - 2g^2(t,x(t)) + M + C_s(t,t-\epsilon)\epsilon \int_{t-\epsilon}^{t} g^2(v,x(v))dv \\
&\quad + C(t,t-\epsilon) \int_{t-\epsilon}^{t} [g^2(t,x(t)) + g^2(v,x(v))]dv \\
&\quad + \int_{t-\epsilon}^{t} |C(t,s)|[g^2(t,x(t)) + g^2(s,x(s))]ds \\
&= Ma^2(t) + g^2(t,x(t))\left[\gamma - 2 + \epsilon C(t,t-\epsilon) + \int_{t-\epsilon}^{t} |C(t,s)|ds\right] \\
&\quad + M + \int_{t-\epsilon}^{t} [\epsilon C_s(t,t-\epsilon) + C(t,t-\epsilon) + |C(t,s)|]g^2(s,x(s))ds \\
&\quad \text{by (2.11.9)} \\
&\leq Ma^2(t) + g^2(t,x(t))[\gamma + \beta - 2] + M \\
&\quad + \int_{t-\epsilon}^{t} [\epsilon C_s(t,t-\epsilon) + C(t,t-\epsilon) + |C(t,s)|]g^2(s,x(s))ds,
\end{aligned}$$

as required.

Lemma 2.11.4. *If (2.11.17) holds, if $\epsilon \leq T$, and if γ is small enough then there is a $\mu > 0$ so that if x solves (1_λ) and $x \in \mathcal{P}_T$ then*

$$\int_0^T g^2(s,x(s))ds \leq (M/\mu)\int_0^T a^2(s)ds + TM/\mu. \tag{2.11.19}$$

Proof. We are going to integrate (2.11.18) from 0 to T and note that $0 = V(T,\epsilon) - V(0,\epsilon)$. First, we estimate the integral of the last term in (2.11.18) as follows. We have

$$\begin{aligned}
&\int_0^T \int_{t-\epsilon}^{t} [|C(t,s)| + \epsilon C_s(t,t-\epsilon) + C(t,t-\epsilon)]g^2(s,x(s))dsdt \\
&\leq \int_{-\epsilon}^{T} \int_{s}^{s+\epsilon} [|C(t,s)| + \epsilon C_s(t,t-\epsilon) + C(t,t-\epsilon)]dtg^2(s,x(s))ds \\
&\text{by (2.11.8)} \\
&\leq \alpha \int_{-\epsilon}^{T} g^2(s,x(s))ds \leq 2\alpha \int_0^T g^2(s,x(s))ds.
\end{aligned}$$

With this information we now integrate (2.11.18) and obtain

$$\begin{aligned} 0 = V(T,\epsilon) - V(0,\epsilon) &\leq M\int_0^T a^2(s)ds + TM \\ &+ \int_0^T [\gamma - 2 + \beta + 2\alpha] g^2(s,x(s))ds \\ &\leq M\int_0^T a^2(s)ds - \mu\int_0^T g^2(s,x(s))ds + TM \end{aligned}$$

since $\beta + 2\alpha < 2$ and γ can be made as small as we please.

Lemma 2.11.5. *Let the conditions of Lemma 2.11.4 hold and suppose there is a $Q > 0$ with*

$$\int_{-\infty}^{t-\epsilon} C_s(t,s))(t+T-s)^2 ds \leq Q. \tag{2.11.20}$$

Then there is a $Q^* > 0$ with $V(t,\epsilon) \leq Q^*$.

Proof. We have

$$\begin{aligned} V(t,\epsilon) &= \lambda\int_{-\infty}^{t-\epsilon} C_s(t,s)\left(\int_s^t g(u,x(u))du\right)^2 ds \\ &\leq \int_{-\infty}^{t-\epsilon} C_s(t,s)(t-s)\int_s^t g^2(u,x(u))duds \\ &\leq \int_{-\infty}^{t-\epsilon} C_s(t,s)(t-s)\left[\int_s^{t+T}(M/\mu)a^2(u)du + (t-s+T)TM/\mu\right]ds \\ &\leq \int_{-\infty}^{t-\epsilon} C_s(t,s)(t+T-s)^2 ds[(M/\mu)\|a^2\| + TM/\mu] \end{aligned}$$

from which the result follows.

Lemma 2.11.6. *Let the conditions of Lemma 2.11.5 hold. Then there exists a constant $J > 0$ such that $\|x\| < J$ whenever x is T-periodic solution of $(2.11.1_\lambda)$ for $0 < \lambda \leq 1$.*

Proof. By (2.11.9) and (2.11.13), we have

$$\int_{-\infty}^{t-\epsilon} C_s(t,s)ds = C(t,t-\epsilon) \leq \beta/\epsilon.$$

If $x \in \mathcal{P}_T$ solves (1_λ), then (2.11.19) holds, and by Lemma 2.11.5, $V(t,\epsilon) \leq Q^*$. Now taking into account that (2.11.7) holds with $\eta < 1$, we obtain from (2.11.16) that

$$\begin{aligned}(x(t) - \lambda a(t))^2 \leq\ & 2(1+q^{-1})(\beta/\epsilon)Q^* \\ & + 2(1+q^{-1})(\beta^2/\epsilon)TM(\|a^2\|+1)/\mu \\ & + (1+q)(\eta\|x\| + \beta g^*)^2\end{aligned}$$

where $g^* = \|g(t,0\|$. Since $\eta < 1$, we may choose $q > 0$ small enough so that $(1+q)\eta^2 < 1$, and hence, there exists $J > 0$ such that $\|x\| < J$. The proof is complete.

Continuity and Compactness

We select part of (2.11.10) and define the mapping $U : \mathcal{P}_T \to \mathcal{P}_T$ by $\phi \in \mathcal{P}_T$ implies that

$$(U\phi)(t) = \int_{-\infty}^{t-\epsilon} C(t,s)g(s,\phi(s))ds. \tag{2.11.21}$$

Then U is well defined on P_T by (2.11.6). By a change of variable we have

$$(U\phi)(t) = \int_{-\infty}^{t} C(t,s-\epsilon)g(s-\epsilon,\phi(s-\epsilon))ds$$

with a fully convex kernel.

Lemma 2.11.7. *Suppose that $\int_{-\infty}^{t-\epsilon}[|C(t,s)| + |C_t(t,s)|]ds$ is bounded for all $t \in \Re$. Then U is continuous on P_T and for each $J > 0$, $\Gamma = \{U(\phi) : \phi \in \mathcal{P}_T|, \|\phi\| \leq J\}$ is uniformly bounded and equicontinuous.*

Proof. First, there is a J^* such that $\phi \in \Gamma$ implies that $|g(t,\phi(t))| \leq J^*$ and there is a C^* with

$$\int_{-\infty}^{t-\epsilon} [|C(t,s)| + |C_t(t,s)|]ds \leq C^*, \quad t \in \Re. \tag{2.11.22}$$

It is clear that $U\phi \in P_T$ by (2.11.6) and the argument following (2.11.10). We now show that U is continuous on P_T. If $\tilde{\phi}$, $\phi \in P_T$, then

$$\begin{aligned} |U(\phi)(t) - U(\tilde{\phi})(t)| &= \left| \int_{-\infty}^{t-\epsilon} C(t,s)g(s,\phi(s))ds \right. \\ &\quad \left. - \int_{-\infty}^{t-\epsilon} C(t,s)g(s,\tilde{\phi}(s))ds \right| \\ &= \left| \int_{-\infty}^{t-s} C(t,s)\left[g(s,\phi(s)) - g(s,\tilde{\phi}(s))\right] ds \right|. \end{aligned} \tag{2.11.23}$$

Since g is uniformly continuous on $[0,T] \times \{x \in R : |x| \leq \|\tilde{\phi}\| + 1\}$, for any $\epsilon > 0$, there exists $0 < \delta < 1$ such that $\|\phi - \tilde{\phi}\| < \delta$ implies $|g(s,\phi(s)) - g(s,\tilde{\phi}(s))| < \varepsilon$ for all $s \in [0,T]$. It follows from (2.11.23) that $\|U(\phi) - U(\tilde{\phi})\| \leq \epsilon C^*$. Thus, F is continuous on P_T.

Next, for an arbitrary $\phi \in \Gamma$ we have

$$\frac{d}{dt}(U\phi)(t) = C(t,t-\epsilon)g(t-\epsilon,\phi(t-\epsilon)) + \int_{-\infty}^{t-\epsilon} C_t(t,s)g(s,x(s))ds.$$

and this derivative is bounded by

$$C(t,t-\epsilon)J^* + J^* \int_{-\infty}^{t-\epsilon} |C_t(t,s)|ds \leq J^* \sup_{0\leq t\leq T} \|C(t,t-\epsilon)\| + J^*C^*.$$

This implies that Γ is equicontinuous. The uniform boundedness of Γ follows from the inequality

$$|U(\phi)(t)| \leq \int_{-\infty}^{t-\epsilon} |C(t,s)||g(s,\phi(s))|ds \leq J^*C^*.$$

Periodic Solutions

We will show the existence of T-periodic solutions of (2.11.1) by applying Theorem 2.11.1. By (2.11.10) and (2.11.11), we see that $x \in P_T$ is a solution of ($2.11.1_\lambda$) if and only if it is a fixed point of $B + \lambda A$.

Theorem 2.11.2. *If (2.11.2)-(2.11.9), (2.11.12)-(2.11.13), (2.11.17), (2.11.20), and (2.11.22) hold with $\epsilon \leq T$, then (2.11.1) has a T-periodic solution.*

Proof. Let the mappings A and B be defined in (2.11.10) and (2.11.11) with $S = P_T$. Then B is a contraction mapping with contraction constant η, and hence, $(I - B)^{-1}$ exists and is continuous on $(I - B)S = S$. By Lemma 2.11.7 and the Ascoli-Arzela Theorem, we see that A is continuous and maps bounded sets into compact sets. It is also clear that $\lambda A(M) \subset (I - B)S$ for each closed convex subset $M \subset S$ and $\lambda \in [0, 1]$. Now by Lemma 2.11.6, the set of solutions to $x = Bx + \lambda Ax$ is bounded. Therefore, the alternative (i) of Theorem 2.11.1 must hold; that is, $B + A$ has a fixed point in P_T which is a T-periodic solution of (2.11.1).

Remark Observe that the continuity of $C(t, s)$ with respect to s for $t - \epsilon < s < t$ is not required for fixed t. One may readily verify that the function $C(t, s)$ defined by $C(t, s) = k(t - s)^{-p}$ for $t - s \geq \epsilon$ and $C(t, s) = (t - s)^{-q}$ for $0 < t - s < \epsilon$ with $p > 2, 0 < q < 1, 0 < \epsilon \leq 1, k > 0$ satisfy all conditions of Theorem 2.11.2 for an appropriately chosen constant k.

Notes The material in Section 2.11 is from Burton and Zhang (2011). It was first published by the Informath Publishing Group in the journal *Nonlinear Dynamics and Systems Theory*. Full details are given in the bibliography.

Chapter 3

Integrodifferential equations

3.1 Introduction

In Chapter 0 we quoted the statement of Miller (1971a; p. 337) that if an integral equation of the form

$$x(t) = f(t) + \int_0^t g(t, s, x(s))ds$$

can be written in the differentiated form

$$x'(t) = f'(t) + g(t, t, x(t)) + \int_0^t g_t(t, s, x(s))ds, \quad x(0) = f(0),$$

then a Liapunov functional can be united with it by means of the chain rule. Thus, the next section begins with a study of doing exactly that and constructing Liapunov functionals. But there are many other ways in which we arrive at an integrodifferential equation from a given integral equation and some of these are pursued in subsequent Sections 3 and 4. In Section 5 we develop Floquet theory for integrodifferential equations, noting that periodic properties of the equation often allow us to treat general kernels as if they are of convolution type.

In Section 6 we develop a Liapunov functional for an integrodifferential equation with convex kernel and periodic forcing function. From the derivative of that Liapunov functional, and not from the Liapunov functional itself, we show that there is an *a priori* bound on all possible periodic solutions of the equation with parameter. We then apply Schaefer's fixed point theorem to show that there is a periodic solution. This is an example of the work which we mentioned in our sketch of a parallel problem for integral equations near the end of Section 2.4 and in Lemma 2.11.4.

This gives us a selection of integrodifferential equations which might have originated as integral equations. With this background in hand and with the understanding that integrodifferential equations might come from integral equations in ways that we do not anticipate, we proceed to the next two sections. In Section 3.7 we develop general Liapunov theory for integrodifferential equations, introducing substantial stability theory, all for continuous kernels. We discuss nonlinearities in Section 3.8 and that is followed in Section 3.9 with theory for weakly singular kernels of convex type. A reference is then given for small weakly singular kernels.

Looking ahead to Chapter 4 we introduce the idea of starting with an integral equation, differentiating it, and constructing a new equation from the combination $x' + kx$ which gives many new properties which greatly enhance study of the subject of integral equations.

3.2 The differentiated equation and resolvent

This section is intended to flow out from Sections 1 and 2 of Chapter 2. Thus, the reader may wish to go back for a very brief review. Here, we will assume that C has at least one partial derivative and often we will differentiate a as well. Frequently we will ask that a' be bounded or L^p and we will conclude that the solution of (2.1.1) is at least bounded. The applied mathematician will correctly object that uncertainties and even stochastic elements make such behavior of a difficult or impossible to detect. But there is a practical way around this objection. For a fixed $C(t,s)$ there are many simple ways to establish

$$\sup_{t\geq 0}\int_0^t |R(t,s)|ds < \infty. \tag{3.2.0}$$

The tactic then is as follows. Fix $C(t,s)$ and obtain (3.2.0) independent of the forcing function. Now our real problem of interest is $x(t) = b(t) - \int_0^t C(t,s)x(s)ds$, where $b(t)$ may be a large and badly behaved function. Select a nice function, $a(t)$, which is "close" to $b(t)$ and satisfies one of our subsequent results; that is, the solution of $y(t) = a(t) - \int_0^t C(t,s)y(s)ds$ is at least bounded. We suppose that there is a $K > 0$ with $|a(t) - b(t)| \leq K$ for all $t \geq 0$. Then using the same $R(t,s)$ which depends only on $C(t,s)$ we have

$$x(t) = b(t) - \int_0^t R(t,s)b(s)ds$$

and

$$y(t) = a(t) - \int_0^t R(t,s)a(s)ds.$$

Now,

$$\begin{aligned}|x(t) - y(t)| &\leq |a(t) - b(t)| + \int_0^t |R(t,s)||a(s) - b(s)|ds \\ &\leq K\left[1 + \sup_{t\geq 0}\int_0^t |R(t,s)|ds\right].\end{aligned}$$

Thus, our results here may seem to demand much from $a(t)$, but when (3.2.0) holds they apply to a much larger class of functions. We are concerned with the scalar equations

$$x(t) = a(t) - \int_0^t C(t,s)x(s)ds, \tag{3.2.1}$$

$$R(t,s) = C(t,s) - \int_s^t C(t,u)R(u,s)du, \tag{3.2.2}$$

and

$$x(t) = a(t) - \int_0^t R(t,s)a(s)ds \tag{3.2.3}$$

where $a : [0,\infty) \to \Re$ and C is a continuous scalar function defined for $0 \leq s \leq t < \infty$. We have often asked that

$$\sup_{t\geq 0}\int_0^t |C(t,s)|ds \leq \alpha < 1$$

but if there is an additive constant or an additive function of t that condition is almost certain to fail. That kernel can often be cleansed by differentiation. We begin with differentiation with respect to t, denoted by C_1 or C_t, and in the next section use differentiation with respect to s. In the first case we will also need to differentiate a, but as we are often seeking properties of R we can treat a as a test function and bestow upon it any properties which are convenient. In other cases, the exact properties of a are crucial. Thus, we differentiate (3.2.1) and obtain

$$x'(t) = a'(t) - C(t,t)x(t) - \int_0^t C_1(t,s)x(s)ds. \tag{3.2.4}$$

When (3.2.4) is related to the solution of (3.2.1) then $x(0) = a(0)$. Moreover, unless otherwise stated solutions of (3.2.4) are always on $[0,\infty)$ with the initial function being only $x(0)$.

We will use the classical resolvent equation for (3.2.4) because we know it is $Z(t,s)$ and the equation itself will announce that Z_s exists which we will need for integration by parts. The resolvent equation for (3.2.4) as seen in Chapter 1, Section 1.2.1, is (1.2.1.8) which becomes

$$Z_s(t,s) = Z(t,s)C(s,s) + \int_s^t Z(t,u)C_1(u,s)du,\ Z(t,t) = 1, \quad (3.2.5)$$

and the variation of parameters formulae are

$$\begin{aligned} x(t) &= Z(t,0)x(0) + \int_0^t Z(t,s)a'(s)ds \\ &= Z(t,0)[x(0) - a(0)] + a(t) - \int_0^t Z_s(t,s)a(s)ds. \end{aligned} \quad (3.2.6)$$

In conjunction with this, keep in mind that the solution of (3.2.1) is also given by (3.2.3).

Suppose that

$$\int_0^t C(s,s)ds \to \infty$$

as $t \to \infty$ and use the variation of parameters formula to write (3.2.4) as

$$\begin{aligned} x(t) &= x(0)e^{\int_0^t -C(s,s)ds} + \int_0^t e^{\int_u^t -C(s,s)ds}a'(u)du \\ &\quad - \int_0^t e^{\int_u^t -C(s,s)ds} \int_0^u C_1(u,s)x(s)dsdu. \end{aligned} \quad (3.2.7)$$

Remark 3.2.1. *Thus, our equation is again an integral equation and it will require the integral of the second coordinate of $C_1(t,s)$ to be small; C_1 is cleansed of any additive constants or additive functions of t which might have conflicted with (2.1.6). But, perhaps more to the point, any such constants are now transferred to the exponential which can help the subsequent contraction condition.*

In order to make (3.2.6) and (2.1.3) more symmetric we begin with a proposition showing $Z(t,0)$ bounded.

Theorem 3.2.1. *Suppose that $\int_0^t C(s,s)ds$ is bounded below and that there is an $\alpha < 1$ with*

$$\sup_{t \geq 0} \int_0^t e^{\int_u^t -C(s,s)ds} \int_0^u |C_1(u,s)|dsdu \leq \alpha. \quad (3.2.8)$$

Then $Z(t,0)$ in (3.2.6) is bounded. Moreover:

(a) Every solution of (3.2.1) is bounded for every $a(t)$ with $a'(t)$ bounded and continuous if and only if

$$\sup_{t\geq 0}\int_0^t |Z(t,s)|ds < \infty.$$

(b) Every solution of (3.2.1) is bounded for every bounded and continuous $a(t)$ if and only if

$$\sup_{t\geq 0}\int_0^t |Z_s(t,s)|ds < \infty.$$

Proof. In (3.2.7) to deal with $Z(t,0)$ we let $a(t) = 0$ and use (3.2.7) to define a mapping $Q : \mathcal{BC} \to \mathcal{BC}$. It is a contraction by (3.2.8) with fixed point $x(t) = Z(t,0)x(0)$ which is bounded for each $x(0)$. Parts (a) and (b) now follow exactly as in the proof of Theorem 2.1.2 using (3.2.6).

We will now use this to obtain a remarkable property of the resolvent.

Theorem 3.2.2. *Suppose that (3.2.8) holds, that $\int_0^t C(s,s)ds$ is bounded below, and that*

$$\int_0^t e^{\int_u^t -C(s,s)ds}du \text{ is bounded for } t \geq 0. \tag{3.2.9}$$

Then the unique solution of (3.2.1) is bounded for each continuous function $a(t)$ with $a'(t)$ bounded and continuous; thus, from (3.2.6) we see that $Z_s(t,s)$ and $R(t,s)$ generate an approximate identity on the space of functions $\phi : [0,\infty) \to \Re$ for which $\phi'(t)$ is bounded.

Proof. Use (3.2.7) to define a mapping $Q : \mathcal{BC} \to \mathcal{BC}$ to prove that the solution of (3.2.4) (and, hence, of (3.2.1)) is bounded for every bounded and continuous $a'(t)$. But remember that $x(t) = a(t) - \int_0^t R(t,s)a(s)ds$ which is bounded for every a with $a' \in \mathcal{BC}$; thus, refer to (2.1.4) and see that $R(t,s)$ generates an approximate identity on the space of functions a with $a' \in \mathcal{BC}$. In the second line of (3.2.6) we use the fact that $Z(t,0)$ is bounded and that x is bounded for every a with $a \in \mathcal{BC}$ to see that $a(t) - \int_0^t Z_s(t,s)a(s)ds$ is bounded for all such a. This shows that Z_s also generates an approximate identity on that space.

Example 3.2.1. *Let $a(t) = \ln(t+1)$ so that $a'(t) = 1/(t+1)$ and obtain from (3.2.6) with $x(0) = a(0)$ that*

$$x(t) = a(t) - \int_0^t Z_s(t,s)a(s)ds = a(t) - \int_0^t R(t,s)a(s)ds$$

is bounded. Thus, $\int_0^t R(t,s)\ln(s+1)ds$ is a fair approximation to $\ln(t+1)$; and, it will get better.

Remark 3.2.2. *This theorem tells us that x is bounded when a' is bounded and that is very powerful result. But to implement the opening statement of this section we must have (3.2.0) and, hence, we must know that x is bounded for every bounded a. There is a companion to Theorem 3.2.2 which does exactly that and we consider it in the next section. It turns out that by integrating $\int_0^t C(t,s)x(s)ds$ by parts we obtain an integrodifferential equation parallel to (3.2.4), but containing $a(t)$ instead of $a'(t)$ and the counterpart of Theorem 3.2.2 will yield $\int_0^t x(s)ds$ bounded for each bounded $a(t)$; this, in turn, can be parlayed into $x(t)$ bounded for every bounded $a(t)$.*

Remark 3.2.3. *If the same resolvent $R(t,s)$ generates an approximate identity on vector spaces V_1 and V_2, then it generates an approximate identity on the space V_3 in the sense that if $\phi \in V_3$ and if we find $\phi_1 \in V_1$ and $\phi_2 \in V_2$ with $\phi = \phi_1 + \phi_2$, then $P\phi \in \mathcal{BC}$. For example, $\phi(t) = t^{1/3}$ is neither bounded, as in Theorem 3.1.3, nor is ϕ' bounded and continuous, as in Theorem 3.2.2. However, we can write*

$$\phi_1(t) = \begin{cases} t^{1/3} - (1/3)t, & \text{for } 0 \le t \le 1 \\ (2/3) & \text{for } t \ge 1, \end{cases}$$

and

$$\phi_2(t) = \begin{cases} (1/3)t & \text{for } 0 \le t \le 1 \\ t^{1/3} - (2/3) & \text{for } t \ge 1. \end{cases}$$

Theorem 3.2.3. *Let the conditions of Theorem 3.2.2 hold, $a'(t) \to 0$ as $t \to \infty$, and suppose there is a constant $\lambda > 0$ with $-C(t,t) \le -\lambda$. Suppose also that there is a continuous function $\Phi : [0,\infty) \to [0,\infty)$ with $\Phi \in L^1[0,\infty)$ and $\Phi(u-s) \ge |C_1(u,s)|$. Then the solution $x(t)$ of (3.2.1) tends to zero as $t \to \infty$ and so does $Z(t,0)$. Finally, under these additional conditions the conclusion of Theorem 3.2.2 changes to asymptotic identity.*

Proof. In our mapping we add to the mapping set $\mathcal{BC}$ the condition that $\phi(t) \to 0$ as $t \to \infty$. Then notice that

$$\int_0^u |C_1(u,s)\phi(s)|ds \le \int_0^u |\Phi(u-s)|\phi(s)|ds,$$

is the convolution of an L^1–function with a function tending to zero so it tends to zero by the Convolution Lemma. Then

$$\int_0^t e^{\int_u^t -C(s,s)ds} \int_0^u |C_1(u,s)\phi(s)|ds \le \int_0^t e^{-\lambda(t-u)} \int_0^u |C_1(u,s)\phi(s)|ds$$

which tends to zero for the same reason. Finally,

$$\int_0^t e^{\int_u^t -C(s,s)ds}|a'(u)|du \leq \int_0^t e^{-\lambda(t-u)}|a'(u)|du$$

which tends to zero. This will then show that the modified $\mathcal{BC}$ set will be mapped into itself in (3.2.7).

Theorem 3.2.4. *Let (3.2.8) hold and suppose that*

$$\int_0^t e^{\int_u^t -C(s,s)ds}du$$

is bounded for $t \geq 0$. In addition, let $a'(t) \to 0$ as $t \to \infty$, suppose that for each $T > 0$

$$\int_0^T e^{\int_u^t -C(s,s)ds}\int_0^u |C_1(u,s)|dsdu \to 0$$

as $t \to \infty$, and that

$$\int_0^T e^{\int_u^t -C(s,s)ds}du \to 0$$

as $t \to \infty$. Then every solution of (3.2.1) tends to zero as $t \to \infty$ and so does $\int_0^t Z(t,s)\phi(s)ds$ for every continuous ϕ which tends to 0 as $t \to \infty$.

Proof. The mapping is defined from (3.2.7) and, as in the proof of Theorem 3.2.3, we add to the mapping set M the condition that $\phi \in M$ implies that $\phi(t) \to 0$. Now from (3.2.7) we write for $\phi \in M$ the equation

$$(P\phi)(t) = x(0)e^{\int_0^t -C(s,s)ds} + \int_0^t e^{\int_u^t -C(s,s)ds}a'(u)du$$
$$- \int_0^t e^{\int_u^t -C(s,s)ds}\int_0^u C_1(u,s)\phi(s)dsdu.$$

We will show that the last term tends to zero as $t \to \infty$.

Let $\phi \in M$ be fixed, $\epsilon > 0$ be given, and find $T > 0$ with $|\phi(t)| < \epsilon/\alpha$ for $t \geq T$. Thus,

$$\int_T^t e^{\int_u^t -C(s,s)ds} \int_0^u |C_1(u,s)\phi(s)|dsdu \leq \epsilon$$

for $t \geq T$. For that $\phi \in M$, there is a $J > 0$ with $|\phi(t)| \leq J$. For $t > T$ we have

$$\begin{aligned}\left|\int_0^t e^{\int_u^t -C(s,s)ds} \int_0^u C_1(u,s)\phi(s)dsdu\right| \\ &\leq \int_0^T e^{\int_u^t -C(s,s)ds} \int_0^u |C_1(u,s)|dsduJ \\ &\quad + \int_T^t e^{\int_u^t -C(s,s)ds} \int_0^u |C_1(u,s)\phi(s)|dsdu.\end{aligned}$$

The first term on the right is bounded by ϵ for large t and so is the second.

We examine the second term of $P\phi$ and for $|a'(t)| \leq J$ write

$$\left|\int_0^t e^{\int_u^t -C(s,s)ds} a'(u)du\right| \leq \int_0^T e^{\int_u^t -C(s,s)ds} duJ + \int_T^t e^{\int_u^t -C(s,s)ds}|a'(u)|du.$$

The first term on the right tends to zero and the second term can be made small by taking T large since $|a'(t)| \to 0$ and $\int_0^t e^{\int_u^t -C(s,s)ds}du$ is bounded. This means that $P : M \to M$ so the fixed point tends to zero. As

$$x(t) = Z(t,0)x(0) + \int_0^t Z(t,s)a'(s)ds,$$

we argue that for every continuous $a'(t)$ which tends to zero then $x(t) \to 0$, while $Z(t,0) \to 0$ because that corresponds to the case $a'(t) = 0$. Thus, $\int_0^t Z(t,s)\phi(s)ds \to 0$ for every continuous ϕ which tends to zero. This completes the proof.

Example 3.2.2. *Let $g : (-\infty, \infty) \to (0, \infty)$ be bounded and locally Lipschitz. Consider the integral equation*

$$x(t) = t + \int_0^t g(x(s))ds - \int_0^t C(t,s)x(s)ds$$

where C satisfies (3.2.8) and (3.2.9). Standard existence theory (given in Chapter 1) will yield a unique solution on $[0, \infty)$ so it is possible to define a unique continuous function

$$a(t) = t + \int_0^t g(x(s))ds$$

with $a'(t)$ being bounded and $a(t) \geq t$. The conditions of Theorem 3.2.2 are satisfied and we have then a list of properties.

(a) The solution $x(t)$ is bounded. This means that $a(t)$ and $\int_0^t C(t,s)x(s)ds$ differ by at most a bounded function. Recall that $a(t) \geq t$.

(b) The variation of parameters formula for the solution is

$$x(t) = a(t) - \int_0^t R(t,s)a(s)ds$$

where R is the resolvent from (3.2.2). That integral differs from $a(t)$ by at most a bounded function and, again, $a(t) \geq t$.

(c) From the second equation in (3.2.6) we have

$$x(t) = t + \int_0^t g(x(s))ds - \int_0^t Z_s(t,s)\left[s + \int_0^s g(x(u))du\right]ds$$

and that quantity is bounded since Z_s generates an approximate identity on functions with bounded derivative. Again, $a(t) \geq t$ and $a(t) - \int_0^t Z_s(t,s)a(s)ds$ is bounded.

(d) From the first equation in (3.2.6) we have

$$x(t) = Z(t,0)x(0) + \int_0^t Z(t,s)[1 + g(x(s))]ds$$

where the last term is bounded. That integral differs from $1 + g(x(t))$ by at most a bounded function, while $Z(t,0)$ is bounded.

Remark 3.2.4. *Here is the typical objective. If we can show that $x \in L^2$ for every $a' \in L^2$ then $R(t,s)$ (or $Z_s(t,s)$) generates an L^2 approximate identity on the vector space W of functions $a(t)$ for which $a'(t)$ is continuous and is in L^2. Thus, for example, $a(t) = \ln(t+1)$ qualifies and the solution of (3.2.1) is L^2, but it does not follow $a(t)$ in any sense at all. The classical idea of investigators is that for well-behaved kernels the solution of (3.2.1) follows $a(t)$. Our kernel can be arbitrarily well-behaved and the solution simply does not follow $a(t)$. There is an entirely different principle at work in all such problems. The principle is that, however complicated $R(t,s)$ may be, the integral strips away the complication and $\int_0^t R(t,s)a(s)ds$ may approximate $a(t)$.*

In the result below, notice that the solution $x(t)$ of (3.2.1) is in L^1, but $a(t)$ is merely bounded. Thus, $x(t)$ is not following $a(t)$ and the kernel can be very well-behaved in almost any sense.

Theorem 3.2.5. *Let $\int_{t-s}^{\infty}|C_1(u+s,s)|du$ be continuous for $0 \le s \le t$. If there is an $\alpha > 0$ such that*

$$-C(t,t) + \int_0^{\infty} |C_1(u+t,t)|du \le -\alpha$$

then $a' \in L^1$ implies that the solution $x(t)$ of (2.2.1) is in L^1 and is bounded. Hence, $x \in L^p$ for $p \in [1,\infty]$ and $R(t,s)$ generates an L^p approximate identity on the vector space W of functions ϕ with $\phi' \in L^1$. If $\alpha = 0$ then $x(t)$ is bounded.

Proof. Define

$$V(t) = |x(t)| + \int_0^t \int_{t-s}^{\infty} |C_1(u+s,s)|du|x(s)|ds.$$

Along solutions of (3.2.4) we have

$$\begin{aligned}
V'(t) &\le |a'(t)| - C(t,t)|x(t)| + \int_0^t |C_1(t,s)x(s)|ds \\
&\quad + \int_0^{\infty} |C_1(u+t,t)|du|x(t)| - \int_0^t |C_1(t,s)x(s)|ds \\
&= \left[-C(t,t) + \int_0^{\infty} |C_1(u+t,t)|du\right]|x(t)| + |a'(t)| \\
&\le -\alpha|x(t)| + |a'(t)|.
\end{aligned}$$

An integration and use of V yields

$$|x(t)| \le V(t) \le V(0) + \int_0^{\infty} |a'(s)|ds - \alpha \int_0^t |x(s)|ds.$$

The result follows from this.

NOTE The case $\alpha = 0$ can allow large $C(t,s)$ in line with that allowed in Theorem 3.2.7.

Theorem 3.2.6. *Let $\int_{t-s}^{\infty}|C_1(u+s,s)|du$ be continuous for $0 \le s \le t$. If there exists $\alpha > 0$ such that*

$$-2C(t,t) + \int_0^{\infty} |C_1(u+t,t)|du + \int_0^t |C_1(t,s)|ds \le -\alpha$$

and if $a' \in L^2$, then the solution $x(t)$ of (3.2.1) is bounded and satisfies $x \in L^p$ for $p \in [2,\infty]$. Thus, $R(t,s)$ generates an L^p approximate identity on the vector space W of functions ϕ with $\phi' \in L^2$.

Proof. Define

$$V(t) = x^2(t) + \int_0^t \int_{t-s}^{\infty} |C_1(u+s,s)|du x^2(s)ds.$$

For any $\epsilon > 0$ there is an $M > 0$ with $2a'(t)x \leq Ma'(t)^2 + \epsilon x^2$. Thus, along any solution of (3.2.4) we have

$$\begin{aligned}
V'(t) &= 2a'(t)x(t) - 2C(t,t)x^2(t) - 2x(t)\int_0^t C_1(t,s)x(s)ds \\
&\quad + \int_0^{\infty} |C_1(u+t,t)|du x^2(t) - \int_0^t |C_1(t,s)|x^2(s)ds \\
&\leq 2a'(t)x(t) - 2C(t,t)x^2(t) + \int_0^t |C_t(t,s)|(x^2(t) + x^2(s))ds \\
&\quad + \int_0^{\infty} |C_1(u+t,t)|du x^2(t) - \int_0^t |C_1(t,s)|x^2(s)ds \\
&\leq Ma'(t)^2 + \epsilon x^2(t) - 2C(t,t)x^2(t) \\
&\quad + \int_0^t |C_t(t,s)|ds x^2(t) + \int_0^{\infty} |C_1(u+t,t)|du x^2(t) \\
&\leq -(\alpha/2)x^2(t) + Ma'(t)^2
\end{aligned}$$

for $\epsilon = \alpha/2$. It readily follows that $x \in L^2$ and is bounded so the conclusion follows.

We now present a result which asks more about the derivatives of C, but no upper bound on the magnitude.

Theorem 3.2.7. *Suppose that $C(t,t) \geq \alpha > 0$ and for $H(t,s) = C_t(t,s)$ we suppose that*

$$H(t,s) \geq 0,\ H_s(t,s) \geq 0,\ H_{st}(t,s) \leq 0,\ H_t(t,0) \leq 0. \tag{3.2.10}$$

If, in addition, $a' \in L^2$, then any solution $x(t)$ of (3.2.4) on $[0,\infty)$ is also in L^2. Thus, $R(t,s)$ generates an L^2 approximate identity on the space of functions ϕ with $\phi' \in L^2$.

Proof. We have

$$x'(t) = a'(t) - C(t,t)x(t) - \int_0^t H(t,s)x(s)ds$$

and we define

$$V(t) = x^2(t) + \int_0^t H_s(t,s)\left(\int_s^t x(u)du\right)^2 ds + H(t,0)\left(\int_0^t x(s)ds\right)^2$$

so that

$$\begin{aligned} V'(t) &= \int_0^t H_{st}(t,s)\left(\int_s^t x(u)du\right)^2 ds + 2x(t)\int_0^t H_s(t,s)\int_s^t x(u)duds \\ &\quad + H_t(t,0)\left(\int_0^t x(s)ds\right)^2 + 2x(t)H(t,0)\int_0^t x(s)ds \\ &\quad + 2x(t)a'(t) - 2C(t,t)x^2(t) - 2x(t)\int_0^t H(t,s)x(s)ds. \end{aligned}$$

Integrate the second term on the right-hand-side and obtain

$$\begin{aligned} &2x(t)\left[H(t,s)\int_s^t x(u)du\Big|_0^t + \int_0^t H(t,s)x(s)ds\right] \\ &= 2x(t)\left[-H(t,0)\int_0^t x(u)du + \int_0^t H(t,s)x(s)ds\right]. \end{aligned}$$

This yields

$$V'(t) \leq 2x(t)a'(t) - 2C(t,t)x^2(t) \leq Da'(t)^2 - Ex^2(t)$$

for appropriate positive numbers D and E. Noting that $x^2 \leq V(t)$ and integrating we obtain $x \in L^\infty$, $x \in L^2$. Thus, as $a' \in L^2$ we obtain $a(t) - \int_0^t R(t,s)a(s)ds \in L^2$.

Exercise 3.2.1. Combine Theorems 3.2.6 and 3.2.7 by considering

$$x' = a'(t) - C(t,t)x - D(t,t)x - \int_0^t C_t(t,s)x(s) - \int_0^t D_t(t,s)x(s)ds$$

and

$$\begin{aligned} V(t) &= x^2 + \int_0^t \int_{t-s}^\infty |D_t(u+s,s)|dux^2(s)ds \\ &+ \int_0^t H_s(t,s)\left(\int_s^t x(u)du\right)^2 ds + H(t,0)\left(\int_0^t x(s)ds\right)^2. \end{aligned}$$

Obtain a result yielding $a' \in L^2$ implies $x \in L^2$.

We will see that for $a' \in L^p$ then $x \in L^p$ so that (3.2.3) will assure us that the integral so faithfully duplicates $a(t)$ that the error in that duplication is an L^p function. If we were to think of averages, as time goes on the duplication becomes so precise on average that we can hardly tell the difference between the integral in (3.2.3) and $a(t)$. From that point of view, the large function, $a(t) = (t+1)^\beta$, has such a small effect on the solution of (3.2.1) it is almost as if it were absent. The same is true for $a(t) = \sin(t+1)^\beta$ when $0 < \beta < 1$.

Theorem 3.2.8. *Let $H(t,s) := C_t(t,s)$, and suppose there is an $\alpha > 0$ with $C(t,t) \geq \alpha$, and let (3.2.10) hold.*

(i) If a' is bounded and if there is an $M > 0$ with

$$\int_0^t H_s(t,s)(t-s)\int_s^t |a'(u)|^2 du ds + H(t,0)t\int_0^t |a'(u)|^2 du \leq M, \quad (3.2.11)$$

then any solution of (3.2.1) or (3.2.4) on $[0,\infty)$ is bounded. Thus, the resolvent for (3.2.4) satisfies $\sup_{t\geq 0}\int_0^t |Z(t,s)|ds < \infty$ and $Z(t,0)$ is bounded.

(ii) If, in addition to the conditions of (i), we have

$$|C(t,t)| + \int_0^t |C_t(t,s)|ds$$

bounded, then $Z(t,0) \to 0$ as $t \to \infty$.

Proof. Define V as in the last proof and obtain the derivative of V along that same equation as

$$V'(t) \leq 2a'(t)x - 2C(t,t)x^2 \leq D|a'(t)|^2 - Ex^2(t)$$

for positive constants D and E, exactly as in the last proof. Note that x is bounded if V is bounded.

Now, assume $a'(t)$ bounded and let (3.2.11) hold; we will bound V and, hence, x. From V' and the boundedness of $|a'|$ we see that there is a $\mu > 0$ such that if $V'(t) > 0$ then $|x(t)| < \mu$. Suppose, by way of contradiction, that V is not bounded. Then there is a sequence $\{t_n\} \uparrow \infty$ with $V'(t_n) \geq 0$ and $V(t_n) \geq V(s)$ for $0 \leq s \leq t_n$; thus, $|x(t_n)| \leq \mu$. If $0 \leq s \leq t_n$ then

$$0 \leq V(t_n) - V(s) \leq -\int_s^{t_n} Ex^2(u)du + D\int_s^{t_n} |a'(u)|^2 du.$$

Using these values in the formula for V, taking $|x(t_n)| \leq \mu$, and applying the Schwarz inequality yields at $t = t_n$ the inequality

$$V(t) \leq \mu^2 + \int_0^t H_s(t,s)(t-s) \int_s^t (D/E)|a'(u)|^2 du ds$$
$$+ H(t,0)t(D/E) \int_0^t |a'(u)|^2 du = \mu^2 + (D/E)M.$$

Thus, $V(t)$ and $x(t)$ are bounded.

Now the variation of parameters formula for (3.2.4) is

$$x(t) = Z(t,0)x(0) + \int_0^t Z(t,s)a'(s)ds.$$

If $a'(t) \equiv 0$, then $V'(t) \leq -Ex^2(t)$ so $x^2 \in L^1[0,\infty)$ and $V(t)$ is bounded so $x(t)$ is bounded. This means that $Z(t,0)$ is bounded and, hence, $\int_0^t Z(t,s)a'(s)ds$ is bounded for every bounded and continuous $a'(t)$. By Perron's theorem, $\sup_{t\geq 0} \int_0^t |Z(t,s)|ds < \infty$. If $|C(t,t)| + \int_0^t |C_t(t,s)|ds$ is bounded, then $x'(t)$ is bounded so $Z(t,0) \to 0$.

It is a very useful result. The solution is bounded and (3.2.11) will yield a computable bound in spite of $a(t)$ being unbounded.

Example 3.2.3. *Let $C(t,s) = 2 - e^{-(t-s)}$ and $a(t) = (t+1)^{1/2}$. Then $C(t,t) = 1 =: \alpha > 0$, $C_t(t,s) = e^{-(t-s)} =: H(t,s)$ so (3.2.10) holds. Also, (3.2.11) is*

$$\int_0^t e^{-(t-s)}(t-s) \int_s^t \frac{1}{4(u+1)} du ds + e^{-t} t \int_0^t \frac{1}{4(u+1)} du$$

which is bounded. By part (i), $x(t)$ is bounded. Hence, $\int_0^t R(t,s)a(s)ds$ closely follows $a(t)$, but $a(t)$ diverges far from $x(t)$. Consider (i) and note that $a(t) = 3t$ qualifies and the solution is bounded. Moreover, if $b(t)$ is any continuous function so that $|a(t) - b(t)|$ is bounded and if (3.2.0) holds then the solution of $y(t) = b(t) - \int_0^t C(t,s)y(s)ds$ is bounded. Notice that $x(t) = 3t - \int_0^t R(t,s)3sds$ is bounded. That resolvent has extremely strong properties enabling the integral to closely approximate $3t$.

In preparation for the next result, we note that Young's inequality states that if p and q are numbers with $p > 1$, $q > 1$, and $(1/p) + (1/q) = 1$, then

$$|ab| \leq \frac{|a|^p}{p} + \frac{|b|^q}{q}.$$

For our repeated application below we will have n a positive integer, $p = \frac{2n}{2n-1}$, and $q = 2n$.

Theorem 3.2.9. *Suppose there is a positive integer n with $a'(t) \in L^{2n}[0,\infty)$, a constant $\alpha > 0$, and a constant $N > 0$ with*

$$\frac{2n-1}{2nN^{\frac{2n}{2n-1}}} - C(t,t) + \frac{2n-1}{2n}\int_0^t |C_t(t,s)|ds + \frac{1}{2n}\int_0^\infty |C_t(u+t,t)|du \le -\alpha.$$

Then the unique solution x of (3.2.1) is bounded and $x \in L^{2n}[0,\infty)$. From (3.2.3) we then have $x(t) = a(t) - \int_0^t R(t,s)a(s)ds \in L^{2n}$.

Proof. For a fixed solution of (3.2.4) on $[0,\infty)$ we define the function

$$V(t) = \frac{x^{2n}(t)}{2n} + \frac{1}{2n}\int_0^t \int_{t-s}^\infty |C_1(u+s,s)|du x^{2n}(s)ds.$$

Compute the derivative along a solution of (3.2.4) on $[0,\infty)$ by the chain rule as

$$\begin{aligned}
V'(t) &= -C(t,t)x^{2n}(t) - \int_0^t C_t(t,s)x(s)x^{2n-1}(t)ds + x^{2n-1}(t)a'(t)\\
&\quad + \frac{1}{2n}\int_0^\infty |C_1(u+t,t)|du x^{2n}(t) - \frac{1}{2n}\int_0^t |C_1(t,s)|x^{2n}(s)ds\\
&\quad \text{(Use Hölder's inequality on the 2nd \& 3rd terms.)}\\
&\le \frac{(2n-1)x^{2n}(t)}{2nN^k} + \frac{(Na'(t))^{2n}}{2n} - C(t,t)x^{2n}(t)\\
&\quad + \int_0^t |C_t(t,s)|\left[\frac{(2n-1)x^{2n}(t)}{2n} + \frac{x^{2n}(s)}{2n}\right]ds\\
&\quad + \frac{1}{2n}\int_0^\infty |C_1(u+t,t)|du x^{2n}(t) - \frac{1}{2n}\int_0^t |C_1(t,s)|x^{2n}(s)ds\\
&= \frac{(Na'(t))^{2n}}{2n} + x^{2n}(t)\left[\frac{(2n-1)}{2nN^p} - C(t,t)\right.\\
&\quad \left. + \frac{2n-1}{2n}\int_0^t |C_t(t,s)|ds + \frac{1}{2n}\int_0^\infty |C_1(u+t,t)|du\right]\\
&\le -\alpha x^{2n}(t) + \frac{N^{2n}}{2n}|a'(t)|^{2n}
\end{aligned}$$

for large N.

It follows that

$$\frac{x^{2n}(t)}{2n} \leq V(t) \leq V(0) - \alpha \int_0^t x^{2n}(s)ds + k^* \int_0^\infty (a'(s))^{2n} ds$$

for some $k^* > 0$. This is true for every solution of (3.2.4) on $[0, \infty)$ and, hence, for (3.2.1).

As we continue to study the behavior of the resolvent, it increasingly seems to be a question of stability.

Definition 3.2.1. *A function R mapping $[0, \infty) \times [0, \infty)$ into the reals is said to be L^N-stable with respect to a vector space W of specified continuous functions ϕ mapping $[0, \infty)$ into the reals if for each $\phi \in W$ there is an integer n with*

$$\phi(t) - \int_0^t R(t, s)\phi(s)ds \in L^n[0, \infty).$$

In our last theorem, the vector space W consisted of those functions ϕ such that $\phi' \in L^p$ for some $p \in (0, \infty)$.

Remark 3.2.4. *Of course, this is a classical stability concept. For suppose that ϕ_1 and ϕ_2 are functions with $\phi_1 - \phi_2 \in W$. Then for*

$$x_{\phi_1}(t) = \phi_1(t) - \int_0^t C(t, s)x(s)ds$$

and

$$x_{\phi_2}(t) = \phi_2(t) - \int_0^t C(t, s)x(s)ds$$

we have

$$x_{\phi_1} - x_{\phi_2} = \phi_1 - \phi_2 - \int_0^t R(t, s)[\phi_1(s) - \phi_2(s)]ds \in L^n[0, \infty).$$

We are saying that if $\phi_1 - \phi_2 \in W$ then they are "close" and the solutions generated are "close." In our examples we find that the functions $\sin(t+1)^\beta$, $(t + 1)^\beta$, and $(t + 1)$ are "close." Notice that if our examples are based on Theorem 3.2.9 with $\phi_1' \in L^p$ and $\phi_2' \in L^q$ where $p < q$, since $x^{2n}/2n \leq V(t)$ we have $x(t) = \phi_1(t) - \int_0^t R(t, s)\phi_1(s)ds \in L^p$ and also $x \in L^q$.

We now introduce a Razumikhin technique which begins with a Liapunov function and then deals with functionals by keeping track of past behaviors. Compare this with Theorem 2.6.2.1.

Theorem 3.2.10. *Suppose that there is a $K > 0$ with*

$$\sup_{t\geq 0}\{[|a'(t)| - C(t,t)K + K\int_0^t |C_1(t,s)|ds]\} < 0.$$

Then every solution of (3.2.4) on $[0, \infty)$ (and hence (3.2.1)) is bounded.
Now suppose there is an $\alpha > 0$ with

$$-C(t,t) + \int_0^t |C_1(t,s)|ds \leq -\alpha.$$

(i) Then every solution of (3.2.4) and (3.2.1) on $[0, \infty)$ is bounded for every $a(t)$ with bounded $a'(t)$; moreover, the resolvent for (3.2.4), $Z(t, s)$, satisfies $Z(t, 0)$ is bounded and there is a $J > 0$ with $\int_0^t |Z(t,s)|ds < J$.
(ii) If, in addition, $C(t,t) \leq M$ for some $M > 0$ and if $a'(t)$ is continuous (not necessarily bounded), then (1.2.14) holds ($x(t) = a(t) - \int_0^t Z_s(t,s)a(s)ds$) and there is an $L > 0$ with $\int_0^t |Z_s(t,s)|ds \leq L$ for $t \geq 0$.

Proof. Let $x(t)$ be a solution of (3.2.4) on $[0, \infty)$ and suppose there is a $K^* \geq K$ and $t^* > 0$ with $|x(s)| < K^*$ for $0 \leq s < t^*$, but $|x(t^*)| = K^*$. Clearly, $C(t,t) \geq 0$. Then for $V(t) = |x(t)|$ and $0 \leq t < t^*$ we have

$$V'(t) \leq |a'(t)| - C(t,t)|x(t)| + \int_0^t |C_1(t,s)||x(s)|ds$$

and at $t = t^*$ we have

$$\begin{aligned} V'(t^*) &\leq |a'(t^*)| - C(t^*,t^*)K^* + K^* \int_0^{t^*} |C_1(t^*,s)|ds \\ &\leq |a'(t^*)| - C(t^*,t^*)K + K \int_0^{t^*} |C_1(t^*,s)|ds < 0, \end{aligned}$$

a contradiction to $V'(t^*) \geq 0$.

To prove (i), for every bounded continuous $a'(t)$ we find K so that $\alpha K > \sup_{t\geq 0} |a'(t)|$ and the first condition is satisfied and so every solution of (3.2.1) and (3.2.4) on $[0, \infty)$ is bounded for every bounded and continuous $a'(t)$. In particular, if $a'(t) = 0$ then every solution of (3.2.4) is bounded so $Z(t, 0)$ is bounded, as seen in (3.2.6). But for any continuous $a'(t)$ any solution of (3.2.4) on $[0, \infty)$ is given by (3.2.6) and, since $x(t)$ and $Z(t, 0)$ are bounded, so is the integral in (3.2.6). By Perron's theorem $\sup_{t\geq 0} \int_0^t |Z(t,s)|ds < \infty$.

To prove (ii), under these conditions (1.2.14) does hold and we consider the Grossman-Miller resolvent, (1.2.13), and recall that $H = Z$ by Theorem 1.2.6.1. An integration of that resolvent yields

$$\begin{aligned}\int_0^t |Z_s(t,s)|ds &\le M\int_0^t |Z(t,s)|ds + \int_0^t\int_s^t |Z(t,u)C_t(u,s)|duds \\ &\le MJ + \int_0^t\int_0^u |C_t(u,s)|ds|Z(t,u)|du \\ &\le MJ + M\int_0^t |Z(t,u)|du \le 2MJ.\end{aligned}$$

Refer back to Theorem 3.2.5. We are now going to let a' be bounded and get $x(t)$ bounded. We started with

$$x(t) = a(t) - \int_0^t C(t,s)x(s)ds$$

and

$$x(t) = a(t) - \int_0^t R(t,s)a(s)ds.$$

If we prove that $x \in L^p$ for every $a \in L^q$ then

$$(P\phi)(t) = \phi(t) - \int_0^t R(t,s)\phi(s)ds$$

maps L^q into L^p and $R(t,s)$ generates an L^p approximate identity on the vector space L^q. The resolvent $R(t,s)$ is determined from $C(t,s)$ alone.

If $x \in L^p$ for each a with $a^{(q)} \in L^d$, then $R(t,s)$ generates an L^p approximate identity on the vector space W of functions ϕ such that $\phi^{(q)} \in L^d$. Thus, if $\phi^{(q)} \in L^d$, then ϕ can have arbitrarily rapid growth as $q \to \infty$.

We have seen resolvents generate L^p approximate identities on spaces of functions which grow as fast as $t^{1/2}$. Can we continue and obtain L^p approximate identities on spaces of functions with arbitrarily rapid growth? Our ability to prove such behavior is limited only by our ability to prove that $x \in L^p$ when $a^{(q)} \in L^d$. That is simply a technical problem. Here, we contrive such a problem, allowing a with $a'' \in L^1$. At the same time we show how the Levin functional can be used in a variety of ways.

The following result concerns (3.2.1) in which

$$C(t,t) = \alpha > 0,\ C_1(t,t) = \beta > 0,$$

$$C_{11}(t,s) =: f(t-s) < 0,\ \beta + \int_0^\infty f(u)du > 0, \tag{3.2.12}$$

and

$$\int_{t-s}^\infty f(u)du =: F(t-s),\ F(t-s) < 0,$$

$$F_s(t-s) < 0,\ F_{st}(t-s) > 0,\ F(t) < 0,\ F_t(t) > 0. \tag{3.2.13}$$

A function satisfying these conditions is

$$C(t,s) = 2 + 3(t-s) - e^{-(t-s)}.$$

Moreover, taking $a(t) = t$ will show the desired behavior and will avoid much of the work below.

Theorem 3.2.11. *If (3.2.12) and (3.2.13) hold, then the solution x of (3.2.1) is in L^2 whenever $a'' \in L^1$. Thus, $R(t,s)$ generates an L^2 approximate identity on the vector space of functions $\phi : [0,\infty) \to R$ with $\phi''(t) \in L^1[0,\infty)$.*

Proof. We have

$$x'(t) = a'(t) - C(t,t)x(t) - \int_0^t C_1(t,s)x(s)ds$$

and

$$\begin{aligned} &x''(t) \\ &= a''(t) - \alpha x'(t) - C_1(t,t)x(t) - \int_0^t C_{11}(t,s)x(s)ds \\ &= a''(t) - \alpha x'(t) - \beta x(t) - \int_0^t f(t-s)x(s)ds. \end{aligned}$$

Write

$$x'(t) = y(t) - \alpha x(t) + \int_0^t \int_{t-s}^\infty f(u)du x(s)ds$$

so

$$\begin{aligned} &x''(t) \\ &= y' - \alpha x'(t) + \int_0^\infty f(u)du x(t) - \int_0^t f(t-s)x(s)ds \\ &= a''(t) - \alpha x'(t) - \beta x(t) - \int_0^t f(t-s)x(s)ds \end{aligned}$$

or

$$y' = a''(t) - \beta x(t) - \int_0^\infty f(u)du x(t).$$

This yields the system

$$x'(t) = y(t) - \alpha x(t) + \int_0^t F(t-s)x(s)ds$$
$$y'(t) = a''(t) - \left(\beta + \int_0^\infty f(u)du\right)x(t).$$

A suitable Liapunov functional is

$$V(t) = \frac{x^2(t)}{2} + \frac{y^2(t)}{2\left(\beta + \int_0^\infty f(u)du\right)}$$
$$- (1/2)\int_0^t F_s(t-s)\left(\int_s^t x(u)du\right)^2 ds - (1/2)F(t)\left(\int_0^t x(u)du\right)^2$$

so that

$$V'(t) = xy - \alpha x^2 + x\int_0^t F(t-s)x(s)ds + \frac{a''(t)y}{\beta + \int_0^\infty f(u)du}$$
$$- xy - (1/2)\int_0^t F_{st}(t-s)\left(\int_s^t x(u)du\right)^2 ds$$
$$- x\int_0^t F_s(t-s)\int_s^t x(u)duds$$
$$- (1/2)F_t(t)\left(\int_0^t x(u)du\right)^2 - xF(t)\int_0^t x(u)du.$$

Integrating the third-to-last term by parts yields

$$-x\left[F(t-s)\int_s^t x(u)du\Big|_0^t + \int_0^t F(t-s)x(s)ds\right]$$
$$= -x\left[-F(t)\int_0^t x(u)du + \int_0^t F(t-s)x(s)ds\right].$$

Hence,

$$\begin{aligned}V'(t) &= -\alpha x^2(t) + \frac{a''(t)y(t)}{\beta + \int_0^\infty f(u)du} \\ &- (1/2)\int_0^t F_{st}(t-s)\Big(\int_s^t x(u)du\Big)^2 ds \\ &- (1/2)F_t(t)\Big(\int_0^t x(u)du\Big)^2 \le -\alpha x^2(t) + \frac{a''(t)y(t)}{\beta + \int_0^\infty f(u)du} \\ &\le -\alpha x^2(t) + K|a''(t)|[V(t)+1]\end{aligned}$$

for an appropriate constant K.

We integrate that differential inequality and obtain

$$V(t) \le V(0)e^{\int_0^t K|a''(s)|ds} - \int_0^t \alpha x^2(s)ds + e^{\int_0^t K|a''(s)|ds}.$$

This yields $x \in L^2$ whenever $a'' \in L^1$.

Notes The material for Section 3.2 comes from Burton (2006a) (2007a,b,c). Part of the material was first published in the *Electronic Journal of Qualitative Theory of Differential Equations* from The University of Szeged, Hungary. Part was first published by Elsevier in *Nonlinear Analysis: TMA* and *Mathematics and Computer Modelling.* Part was first published by the University of Baia Mara (Romania) in the *Carpathian Journal of Mathematics.* It was also presented at a confernce in Baia Mara. Full details are given in the bibliography.

3.3 Integration by parts

The differentiated form (3.2.4) has been studied for more than one hundred years. It enabled us to prove Theorem 3.2.2 which yields every solution of (3.2.1) bounded for every a with a' bounded. But in order to utilize the stability scheme described in the opening of Section 3.2 we must know that (3.2.0) holds and that requires that every solution of (3.2.1) be bounded for every bounded a. There is a little known companion of (3.2.4) with $a(t)$ instead of $a'(t)$ and we will be able to prove a counterpart of Theorem 3.2.2 for it, thereby providing the machinery necessary for that stability argument. In fact, every theorem proved in the last section can be proved for our central equation here simply by changing a' to a, $C_t(t,s)$ to $-C_s(t,s)$, and the conclusion from properties of x to the same properties of $\int_0^t x(s)ds$.

Thus, our work here will be somewhat abbreviated, but enough detail will be given so that the interested reader may make the indicated changes.

Once more we begin with the scalar equations

$$x(t) = a(t) - \int_0^t C(t,s)x(s)ds, \tag{3.3.1}$$

$$R(t,s) = C(t,s) - \int_s^t C(t,u)R(u,s)du, \tag{3.3.2}$$

and

$$x(t) = a(t) - \int_0^t R(t,s)a(s)ds \tag{3.3.3}$$

with $a(t)$, $C(t,s)$, and $C_s(t,s)$ continuous.

If $C(t,s)$ contains an additive function of t, perhaps a constant, then

$$\int_0^t |C(t,s)|ds \leq \alpha < 1$$

must often fail. But such problems can be effectively removed if $C_s(t,s)$ is continuous. In that case we write (3.3.1) as

$$\begin{aligned} x(t) &= a(t) - \int_0^t C(t,s)x(s)ds \\ &= a(t) - C(t,s)\int_0^s x(u)du\Big|_0^t + \int_0^t C_2(t,s)\int_0^s x(u)duds \\ &= a(t) - C(t,t)\int_0^t x(u)du + \int_0^t C_2(t,s)\int_0^s x(u)duds. \end{aligned}$$

If we let $y(t) = \int_0^t x(u)du$ our equation becomes

$$y'(t) = a(t) - C(t,t)y(t) + \int_0^t C_2(t,s)y(s)ds. \tag{3.3.4}$$

There is good independent reason for studying $\int_0^t x(u)du$, as is discussed by Feller (1941) concerning the renewal equation. The resolvent equation for (3.3.4) is

$$Z_s(t,s) = Z(t,s)C(s,s) - \int_s^t Z(t,u)C_2(u,s)du, \; Z(t,t) = 1 \tag{3.3.5}$$

with resolvent $Z(t,s)$ and with y satisfying the variation of parameters formula

$$y(t) = Z(t,0)y(0) + \int_0^t Z(t,s)a(s)ds \tag{3.3.6}$$

and (remembering that $y(0) = 0$ since $y(t) = \int_0^t x(u)du$) we have, upon integration by parts,

$$y(t) = \int_0^t a(s)ds - \int_0^t Z_s(t,s)\int_0^s a(u)duds. \tag{3.3.7}$$

In this problem we will see $y(t)$ bounded even when $\int_0^t a(s)ds$ is unbounded, meaning that $Z_s(t,s)$ generates an approximate identity on a space of unbounded functions. In the next result we will get an asymptotic identity. But the crucial point is to note from (3.3.4) that if $y(t)$ is bounded and if $C(t,t)$ and $\int_0^t |C_2(t,s)|ds$ are bounded, then $y'(t) = x(t)$ is bounded.

Theorem 3.3.1. *Let $Z(t,s)$ be the solution of (3.3.5). Every solution $y(t) = \int_0^t x(u)du$ of (3.3.4) is bounded for every bounded continuous $a(t)$ if and only if*

$$\sup_{t\geq 0}\int_0^t |Z(t,s)|ds < \infty. \tag{3.3.8}$$

Moreover, if (3.3.8) holds then $Z_s(t,s)$ generates an approximate identity on the vector space of continuous functions $\phi : [0,\infty) \to \Re$ for which $\phi'(t)$ is bounded. Finally, if in addition, $y(t) \to 0$ for every $a(t)$ which tends to zero, then $Z_s(t,s)$ generates an asymptotic identity on the vector space of continuous functions $\phi : [0,\infty) \to \Re$ for which $\phi'(t) \to 0$.

Proof. The proof of the first part is like that of Theorem 3.2.1 using (3.3.6) with $y(0) = 0$. The next part, $Z_s(t,s)$ generates an approximate identity, follows from (3.3.7) when we recall that $y(t)$ is bounded for bounded $a(t)$. The last conclusion follows in the same way.

Next, recall that $y(0) = 0$ and write

$$y(t) = \int_0^t e^{\int_u^t -C(s,s)ds} a(u)du + \int_0^t e^{\int_u^t -C(s,s)ds}\int_0^u C_2(u,s)y(s)dsdu. \tag{3.3.9}$$

Theorem 3.3.2. *Suppose that $a(t)$ is bounded, that $\int_0^t e^{\int_u^t -C(s,s)ds}du$ is bounded, and that there exists $\alpha < 1$ with*

$$\sup_{t\geq 0} \int_0^t e^{\int_u^t -C(s,s)ds} \int_0^u |C_2(u,s)|dsdu \leq \alpha. \tag{3.3.10}$$

Then for $x(t)$ the solution of (3.3.1) we have $\int_0^t x(s)ds$ bounded. Thus, $Z_s(t,s)$ of (3.3.5) generates an approximate identity on the space of functions ϕ such that ϕ' is bounded. If $|C(t,t)| + \int_0^t |C_2(t,s)|ds$ is bounded, then $x(t)$ is also.

Proof. Use (3.3.9) and the supremum norm to define a mapping $Q : \mathcal{BC} \to \mathcal{BC}$ by $\phi \in \mathcal{BC}$ implies that

$$(Q\phi)(t) = \int_0^t e^{\int_u^t -C(s,s)ds}a(u)du + \int_0^t e^{\int_u^t -C(s,s)ds} \int_0^u C_2(u,s)\phi(s)dsdu.$$

If $\phi \in \mathcal{BC}$, so is $Q\phi$ by assumption and (3.3.10). Also, Q is a contraction by (3.3.10). Hence, $y(t) = \int_0^t x(s)ds$ is bounded. The final conclusion follows from (3.3.4).

Theorem 3.3.3. *Let the conditions of Theorem 3.3.2 hold. Suppose that for each $T > 0$*

$$\int_0^T e^{\int_u^t -C(s,s)ds} \int_0^u |C_2(u,s)|dsdu \to 0$$

and

$$\int_0^T e^{\int_u^t -C(s,s)ds}du \to 0$$

as $t \to \infty$. If, in addition, $a(t) \to 0$ as $t \to \infty$ then the solution $x(t)$ of (3.3.1) satisfies $y(t) = \int_0^t x(u)du \to 0$ as $t \to \infty$. Also, for the $Z(t,s)$ of (3.3.5) we have $\int_0^t Z(t,s)\phi(s)ds \to 0$ as $t \to \infty$ for every continuous function ϕ which tends to zero as $t \to \infty$.

Proof. Use the mapping from the proof of Theorem 3.3.2, but replace $\mathcal{BC}$ by the complete metric space of continuous $\phi : [0,\infty) \to \Re$ such that $\phi(t) \to 0$ as $t \to \infty$. Use the assumptions and the Convolution Lemma to conclude that the convolution tends to zero. (See also the proof of Theorem 3.2.4.) This will show that $(Q\phi)(t) \to 0$ when $\phi(t) \to 0$. The mapping is a contraction as before with unique solution $y(t) = \int_0^t x(u)du \to 0$ as $t \to \infty$. The last conclusion is immediate.

Theorem 3.3.4. *Let $\int_{t-s}^{\infty} |C_2(u+s,s)|du$ be continuous. If there is a positive number α such that*

$$\int_0^{\infty} |C_2(u+s,s)|du - C(t,t) \leq -\alpha$$

and if $a \in L^1$, then $y \in L^1$ and bounded for any solution of (3.3.4). Hence

$$y(t) = \int_0^t a(u)du - \int_0^t Z_s(t,s) \int_0^s a(u)duds \in L^1.$$

Proof. Let

$$V(t) = |y(t)| + \int_0^t \int_{t-s}^{\infty} |C_2(u+s,s)|du|y(s)|ds$$

so that

$$\begin{aligned} V'(t) &\leq |a(t)| - C(t,t)|y(t)| + \int_0^t |C_s(t,s)||y(s))|ds \\ &+ \int_0^{\infty} |C_2(u+t,t)|du|y(t)| - \int_0^t |C_2(t,s)y(s)|ds \\ &\leq |a(t)| + \left[-C(t,t) + \int_0^{\infty} |C_2(u+t,t)|du\right]|y(t)| \\ &\leq |a(t)| - \alpha|y(t)|. \end{aligned}$$

Hence, $a \in L^1$ implies $y \in L^1$. Also, V bounded yields y bounded.

Keep in mind that these results are not to be confused with those in Section 3.2. Here, $y = \int_0^t x(s)ds$ and additional work is needed to obtain properties of x.

Theorem 3.3.5. *Let $\int_{t-s}^{\infty} |C_2(u+s,s)|du$ be continuous. If there exists $\alpha > 0$ such that*

$$\int_0^{\infty} |C_2(u+s,s)|du + \int_0^t |C_2(t,s)|ds - 2C(t,t) \leq -\alpha$$

and if $a \in L^2$ then any solution $y(t)$ of (3.3.4) on $[0,\infty)$ is also in L^2 and bounded. Notice that for $a(t) = 1/(t+1)$, then $\int_0^t a(u)du = \ln(t+1)$ so the terms in the variation of parameter formula (3.3.7) tend to ∞ and, yet, the difference in the terms is in L^2.

Proof. Let

$$V(t) = y^2(t) + \int_0^t \int_{t-s}^\infty |C_2(u+s,s)| du y^2(s) ds$$

so that for $\alpha/2 > 0$ there is a positive number M with $2y(t)a(t) \le (\alpha/2)y^2(t) + Ma^2(t)$ and

$$\begin{aligned}
V'(t) &= 2y(t)a(t) - 2C(t,t)y^2(t) + 2y(t)\int_0^t C_s(t,s)y(s)ds \\
&+ \int_0^\infty |C_2(u+t,t)| du y^2(t) - \int_0^t |C_2(t,s)| y^2(s) ds \\
&\le Ma^2(t) + (\alpha/2)y^2(t) - 2C(t,t)y^2(t) \\
&+ \int_0^t |C_s(t,s)|(y^2(t) + y^2(s)) ds \\
&+ \int_0^\infty |C_2(u+t,t)| du y^2(t) - \int_0^t |C_2(t,s)| y^2(s) ds \\
&= Ma^2(t) + \Big[(\alpha/2) - 2C(t,t) \\
&+ \int_0^\infty |C_2(u+t,t)| du + \int_0^t |C_s(t,s)| ds\Big] y^2(t) \\
&\le -(\alpha/2)y^2(t) + Ma^2(t).
\end{aligned}$$

This yields $y \in L^2$. As V is bounded, so is y.

The following result is a companion to Theorems 3.2.7 and 3.2.8 and is proved in exactly the same way. Notice that we obtain (3.2.0) so that the stability scheme of Section 3.2 can be realized.

Theorem 3.3.6. *Let $C(t,t) \ge \alpha > 0$, $H(t,s) = -C_s(t,s)$, and let*

$$H(t,s) \ge 0,\ H_t(t,s) \le 0,\ H_s(t,s) \ge 0,\ H_{st}(t,s) \le 0; \tag{3.3.11}$$

if $a \in L^2[0,\infty)$, then any solution y of (3.3.4) is bounded; also, $Z(t,0) \in L^2[0,\infty)$ and bounded so by (3.3.6) for (3.3.4), $\int_0^t Z(t,s)a(s)ds \in L^2[0,\infty)$ and bounded. Define $\lambda(t)$ by

$$\lambda(t) := \int_0^t H_s(t,s)(t-s)\int_s^t a^2(u)duds + H(t,0)t\int_0^t a^2(u)du. \tag{3.3.12}$$

If there is an M with $\lambda(t) < M$ and if $a(t)$ is bounded so is every solution of (3.3.4). If $C(t,t)$ and $\int_0^t C_s(t,s)ds$ are bounded, so is the solution of (3.3.1); in particular, then, $\sup_{t\ge 0}\int_0^t |R(t,s)|ds < \infty$.

Proof. Define V as in the proof of Theorem 3.2.8 with x replaced by y and get

$$V'(t) \leq -Ey^2 + Da^2(t)$$

for positive constants D and E. The constant M in (3.2.11) is defined with a' replaced by a. The theorem is repeated with y bounded. That means that $\int_0^t x(s)ds$ is bounded. The bound on x, the solution of (3.3.1), follows as stated in the theorem. Finally, in that last case x is bounded for every bounded and continuous $a(t)$ so by Perron's theorem (3.2.0) holds.

Notes The material for Section 3.3 comes from Burton (2006a) (2007a,c) which was detailed at the end of Section 3.2.

3.4 Contrasting effects of forcing functions

We begin once more with

$$x(t) = a(t) - \int_0^t C(t,s)x(s)ds, \tag{3.4.1}$$

$$R(t,s) = C(t,s) - \int_s^t C(t,u)R(u,s)du, \tag{3.4.2}$$

and

$$x(t) = a(t) - \int_0^t R(t,s)a(s)ds \tag{3.4.3}$$

with $a'(t)$, $C(t,s)$, and $C_t(t,s)$ continuous.

This continues with the notation from Theorem 2.1.6 and Lemma 2.1.1 and we also differentiate (3.4.1) to obtain

$$x' = a'(t) - C(t,t)x(t) - \int_0^t C_t(t,s)x(s)ds. \tag{3.4.4}$$

From (3.4.4) and the variation of parameters formula we have a new integral equation with $x(0) = a(0)$ in the form

$$x(t) = x(0)e^{-\int_0^t C(s,s)ds} + \int_0^t e^{-\int_u^t C(s,s)ds}\left[a'(u) - \int_0^u C_1(u,s)x(s)ds\right]du. \tag{3.4.5}$$

We assume that there is a $T > 0$ with

$$a(t+T) = a(t) \text{ and } C(t+T, s+T) = C(t,s). \tag{3.4.6}$$

Notice that (3.4.6) will bestow many properties on (3.4.5). For example, $C(t,t)$ is periodic since $C(t+T,t+T) = C(t,t)$. Thus,

$$\int_{u+T}^{t+T} C(s,s)ds = \int_u^t C(s,s)ds,$$

as will be needed to show periodicity later. We suppose that there is a number $c^* > 0$ with

$$C(t,t) \geq c^* \tag{3.4.7}$$

and an $\alpha < 1$ with

$$\int_0^t e^{-\int_u^t C(s,s)ds} \int_0^u |C_1(u,s)|dsdu \leq \alpha \tag{3.4.8}$$

which will make the mapping defined from (3.4.5) be a contraction on any space with the supremum norm. Notice again that (3.4.8) is almost like inequality (3.2.8) in Theorem 3.2.1. We are assuming that when we integrate $C_1(t,s)$ we get less than $C(t,s)$; something is lost in the differentiation, possibly a constant.

In order to prove that (3.4.5) has an asymptotically periodic solution in the same way we proved Theorem 2.1.6, (3.4.5) must be decomposed into the mappings A and B as in the proof of Theorem 2.1.6. Recall that in the notation we are using p is periodic and q tends to zero. See Lemma 2.1.1. Thus, we begin by writing $a'(t) = p^*(t) + q^*(t) \in Y$ and define a mapping from (3.4.5) by $\phi = p + q \in Y$ implies that

$$\begin{aligned}(P\phi)(t) &= a(0)e^{-\int_0^t C(s,s)ds} \\ &+ \int_0^t e^{-\int_u^t C(s,s)ds}\left[p^*(u) + q^*(u) - \int_0^u C_1(u,s)[p(s)+q(s)]ds\right]du.\end{aligned} \tag{3.4.9}$$

The decomposition will be done in the proof of Theorem 3.4.1.

Suppose that $\int_{-\infty}^0 |C_1(t,s)|ds$ is continuous, that

$$\int_{-\infty}^0 |C_1(t,s)|ds \to 0 \text{ as } t \to \infty, \tag{3.4.10}$$

and for $q \in Q$ then

$$\int_0^t C_1(t,s)q(s)ds \to 0 \text{ as } t \to \infty. \tag{3.4.11}$$

It may help to understand these by noting that if C were of convolution type then (3.4.10) would say that $C_1 \in L^1[0,\infty)$, while (3.4.11) would then be a form of the Convolution Lemma.

Lemma 3.4.1. *If (3.4.7) holds then*

$$\int_{-\infty}^{0} e^{-\int_u^t C(s,s)ds} du \to 0 \text{ as } t \to \infty \tag{3.4.12}$$

and for $q \in Q$ then

$$\int_{0}^{t} e^{-\int_u^t C(s,s)ds} q(u)du \to 0 \text{ as } t \to \infty. \tag{3.4.13}$$

Proof. We have $c^* = \min C(t,t)$ and $C(t,t) \in \mathcal{P}_T$. Thus,

$$\begin{aligned} \int_{-\infty}^{0} e^{-\int_u^t C(s,s)ds} du &\le (1/c^*) \int_{-\infty}^{0} C(u,u)e^{-\int_u^t C(s,s)ds} du \\ &= (1/c^*)e^{-\int_u^t C(s,s)ds}\Big|_{-\infty}^{0} \\ &= (1/c^*)e^{\int_0^t -C(s,s)ds} \end{aligned}$$

which tends to zero as $t \to \infty$.

Next,

$$\int_{0}^{t} e^{-\int_u^t C(s,s)ds} |q(u)|du \le \int_{0}^{t} e^{-c^*(t-u)} |q(u)|du$$

which tends to zero by the Convolution Lemma.

Theorem 3.4.1. *In (3.4.1) let a' and $C_1(t,s)$ be continuous. Let (3.4.7)-(3.4.8), and (3.4.10)-(3.4.11) hold. Suppose, in addition, that*

$$\int_{-\infty}^{t} |C_1(t,s)|ds \tag{3.4.14}$$

is bounded and continuous, while $C(t+T, s+T) = C(t,s)$. If $a' \in Y$ so is x, the unique solution of (3.4.1).

Proof. Using (3.4.5) we define a mapping $P : Y \to Y$ by $\phi = p + q \in Y$ implies that

$$(P\phi)(t) = a(0)e^{-\int_0^t C(s,s)ds} + \int_0^t e^{-\int_u^t C(s,s)ds}\left[a'(u) - \int_0^u C_1(u,s)\phi(s)ds\right] du.$$

By (3.4.8) it is clearly a contraction, but we must show that $P: Y \to Y$. Write $a' = p^* + q^*$ and then

$$\begin{aligned}(P\phi)(t) = &\int_{-\infty}^{t} e^{-\int_u^t C(s,s)ds}\left[p^*(u) - \int_{-\infty}^{u} C_1(u,s)p(s)ds\right]du \\ &- \int_{0}^{t} e^{-\int_u^t C(s,s)ds}\int_{0}^{u} C_1(u,s)q(s)dsdu + a(0)e^{-\int_0^t C(s,s)ds} \\ &- \int_{-\infty}^{0} e^{-\int_u^t C(s,s)ds}\left[p^*(u) - \int_{-\infty}^{u} C_1(u,s)p(s)ds\right]du \\ &+ \int_{0}^{t} e^{-\int_u^t C(s,s)ds}\int_{-\infty}^{0} C_1(u,s)p(s)dsdu \\ &+ \int_{0}^{t} e^{-\int_u^t C(s,s)ds}q^*(u)du.\end{aligned}$$

The first term on the right-hand-side is clearly in $\mathcal{P}_T$. In the second term, $\int_0^u C_1(u,s)q(s)ds \in Q$ by (3.4.11). Hence the second term is in Q by (3.4.13). The third term is in Q by (3.4.7). The fourth term is in Q by (3.4.12), (3.4.14), and the fact that $p^* \in \mathcal{P}_T$ and, hence, is bounded. The next to last term is in Q because of (3.4.10) followed by (3.4.11). The last term is in Q by (3.4.13). This completes the proof.

Remark 3.4.1. *Notice in the last result that a significant instability can occur at $\beta = 1$. Under conditions on $C(t,s)$ of Theorem 3.2.9 the integral of that resolvent has been faithfully following $\sin(t+1)^\beta$ so that the difference is an L^p function. Suddenly, that relationship breaks completely and the integral with the resolvent seems to "struggle along trying to catch up with $\sin(t+1)$" but always is out of step, lagging by a nontrivial periodic function plus a function tending to zero.*

Corollary 1. *If the conditions of Theorem 3.2.9 hold and if $0 < \beta < 1$ then $\sin(t+1)^\beta - \int_0^t R(t,s)\sin(s+1)^\beta ds \in L^s$ for some $s < \infty$. But at $\beta = 1$, under conditions on $C(t,s)$ of Theorem 3.4.1 then $s = \infty$ and that difference approaches a periodic function.*

We can now state the promised result, a corollary of Theorems 3.4.1 and 3.2.9.

Corollary 2. *Let the conditions on $C(t,s)$ of Theorems 3.4.1 and 3.2.9 hold. For fixed $\beta \in (0,1)$ there is a $p \in \mathcal{P}_T$, $q \in Q$, and $u \in L^s[0,\infty)$ for some $s > 0$ so that the solution of*

$$x(t) = \sin t + (t+1)^\beta - \int_0^t C(t,s)x(s)ds$$

may be written as

$$x(t) = p(t) + q(t) + u(t).$$

Proof. The solution is

$$x(t) = \sin t + (t+1)^{\beta} - \int_0^t R(t,s)[\sin s + (s+1)^{\beta}]ds.$$

But

$$(t+1)^{\beta} - \int_0^t R(t,s)(s+1)^{\beta}ds =: u(t) \in L^s[0,\infty),$$

while

$$\sin t - \int_0^t R(t,s)\sin s ds$$

is the solution described in Theorem 3.4.1 and it has the required form of $p+q$.

Here is an example which should awaken the most complacent among us. It is a mathematical David and Goliath.

Example 3.4.1. *Let the conditions on C of Theorem 3.4.1 hold and let $a(t) = t + \sin t$. Then $a' \in \mathcal{P}_T \subset Y$. The solution of (3.4.1) is asymptotically periodic. The large function t is simply absorbed, while $\sin t$ makes its presence felt forever.*

Example 3.4.2. *For a transparent example linking Theorem 3.2.9 and Theorem 3.4.1, let $k > 0$,*

$$C(t,s) = k + \sin^2 s + D(t-s),\ D(t) > 0,\ D'(t) \le 0.$$

We then have

$$\int_0^t -D'(s)ds = D(0) - D(t) < D(0)$$

and $C(t,t) = k + \sin^2 t + D(0)$ and so we readily verify the inequality in Theorem 3.2.9 holds for large N and n. To satisfy (3.4.8) we have

$$\int_0^t e^{-\int_u^t [k+\sin^2 s + D(0)]ds} \int_0^u -D'(u-s)dsdu < \frac{D(0)}{k+D(0)}$$

and conditions of Theorem 3.4.1 are satisfied.

Notice in Theorem 3.4.1 that we work with $C_1(t,s)$ and that conditions (3.4.10), (3.4.11), and (3.4.14) all concern $C_1(t,s)$. We now consider the transformation of Section 3.3 to obtain a completely parallel result by working with $C_2(t,s)$ and to avoid differentiating $a(t)$. This time one might think of $C(t,s)=k+\sin^2 t+D(t-s)$ and note that $C_s(t,s)=-D'(t-s)$; the kernel has been cleansed of $k+\sin^2 t$ so that conditions of Theorem 3.2.9 can possibly be satisfied.

Review the work in Section 3.3 where we obtain (3.3.4) and designate it here as

$$y'(t)=a(t)-C(t,t)y(t)+\int_0^t C_s(t,s)y(s)ds. \tag{3.4.15}$$

By the variation of parameters formula we have

$$y(t)=\int_0^t e^{-\int_u^t C(s,s)ds}\left[a(u)+\int_0^u C_s(u,s)y(s)ds\right]du \tag{3.4.16}$$

and we will need $\alpha<1$ with

$$\int_0^t e^{-\int_u^t C(s,s)ds}\int_0^u |C_s(u,s)|dsdu\le\alpha. \tag{3.4.17}$$

Parallel to (3.4.10) and (3.4.11) we ask that

$$\int_{-\infty}^0 |C_s(t,s)|ds\to 0 \text{ as } t\to\infty \tag{3.4.18}$$

and

$$\int_0^t C_s(t,s)q(s)ds\to 0 \text{ as } t\to\infty \text{ for } q\in Q. \tag{3.4.19}$$

Conditions (3.4.7), (3.4.12), (3.4.13), and Lemma 3.4.1 will be the same for both (3.4.4) and (3.4.15), while (3.4.14) will be replaced by the condition that

$$\int_{-\infty}^t |C_s(t,s)|ds \tag{3.4.20}$$

is bounded and continuous.

Theorem 3.4.2. *In (3.4.1) let $a(t)$ and $C_s(t,s)$ be continuous. Let (3.4.7), (3.4.17)–(3.4.20) hold. Let $C(t+T,s+T)=C(t,s)$ for some $T>0$. If $a\in Y$ so is the unique solution of (3.4.1) and of (3.4.16).*

Proof. The proof that $y \in Y$ is completely parallel to that of Theorem 3.4.1. Then consider (3.4.15) with $y \in Y$ so that $y = p + q$. We have

$$\int_0^t C_s(t,s)[p(s)+q(s)]ds = \int_{-\infty}^t C_s(t,s)p(s)ds - \int_{-\infty}^0 C_s(t,s)p(s)ds + \int_0^t C_s(t,s)q(s)ds \in Y.$$

It follows that $y' \in Y$.

Notes The material for Section 3.4 comes from Burton (2007b). That reference was described at the end of Section 3.2.

3.5 Floquet Theory

Consider once more (1.2.7), (1.2.8), and (1.2.10) which we designate here as

$$x'(t) = A(t)x(t) + \int_0^t B(t,u)x(u)du + f(t), \tag{3.5.1}$$

Becker's resolvent

$$\frac{\partial}{\partial t}Z(t,s) = A(t)Z(t,s) + \int_s^t B(t,u)Z(u,s)du, \quad Z(s,s) = I, \tag{3.5.2}$$

and variation of parameters formula

$$x(t) = Z(t,0)x(0) + \int_0^t Z(t,s)f(s)ds. \tag{3.5.3}$$

We examine the resolvent equation and notice that in $Z(t,s)$ it is the case that s is the initial time. Thus, when we ask that

$$\sup_{t\geq 0}\int_0^t |Z(t,s)|ds < \infty$$

then we are integrating with respect to the initial time and that is, at best, disturbing. If we think of the convolution case with $A(t)$ constant, we note that $Z(t,s) = Z(t-s)$ so that

$$\int_0^t |Z(t,s)|ds = \int_0^t |Z(t-s)|ds = \int_0^t |Z(s)|ds :$$

we are integrating with respect to the real time, not the initial time! For a case close at hand, look back at our Theorem 3.2.5 with Liapunov functional

$$V(t) = |x(t)| + \int_0^t \int_{t-s}^{\infty} |B(u+s,s)|du|x(s)|ds.$$

We can take $a \equiv 0$, $x(0) = 1$, and get

$$V'(t) \leq -\alpha|x(t)| = -\alpha|Z(t)|$$

so that instantly we have $\int_0^\infty |Z(t)|dt < \infty$. In other words, if our equation were only of convolution type and had A constant then we could dispense with that bounded function $a(t)$ which has caused so much trouble and we would never need to introduce Φ as we did in Theorem 2.1.12 and will do again later for an integrodifferential equation. Equation (1.2.10) would immediately give us x bounded for a' bounded.

Floquet theory shows us that whatever can be proved for a constant coefficient system can be proved for a periodic system. Indeed, there is a Liapunov transformation mapping $x' = A(t)x$ into a constant coefficient system, $y' = Ry$. Specifically, for the matrix system

$$x' = A(t)x, \quad A(t+T) = A(t), \quad \exists T > 0,$$

the principal matrix solution is $Z(t) = P(t)e^{Rt}$ where R is a constant matrix and P is periodic and nonsingular; hence, P and P^{-1} are bounded. The solution of $y' = A(t)y + F(t)$ with bounded F is

$$\begin{aligned} y(t) &= Z(t)y(0) + \int_0^t Z(t)Z^{-1}(s)F(s)ds \\ &= Z(t)y(0) + \int_0^t P(t)e^{R(t-s)}P^{-1}(s)F(s)ds \end{aligned}$$

with the integral bounded by $K\int_0^t |e^{Rs}|ds$ for some K. Again, we are able to get by simply by integrating the real time t instead of the initial time s.

Becker, Burton, and Krisztin (1988) set out to do the same for a Volterra equation

$$x' = A(t)x + \int_0^t B(t,s)x(s)ds \tag{3.5.4}$$

under the assumption that A and B are continuous and there is a $T > 0$ with

$$A(t+T) = A(t), \quad B(t+T, s+T) = B(t,s). \tag{3.5.5}$$

This was shown in Burton (2005b; p. 105) to imply that

$$Z(t+T, s+T) = Z(t, s). \tag{3.5.6}$$

The goal is a Floquet theory for Volterra equations in the sense that we can integrate the real time instead of the initial time. That is, we want to show that

$$\int_s^\infty |Z(t,s)|dt$$

bounded implies that

$$\sup_{t\geq 0} \int_0^t |Z(t,s)|ds < \infty.$$

In particular, if

$$\int_s^\infty |Z(t,s)|dt$$

is bounded for $0 \leq s \leq T$, then certainly

$$\int_0^T \int_s^\infty |Z(t,s)|dtds < \infty$$

and it turns out that this, and mild side conditions on B, does imply that $\sup_{t\geq 0} \int_0^t |Z(t,s)|ds < \infty$. This means that in our struggles in Section 3.2 we can set $a'(t) = 0$, resulting in simple arguments.

We see this struggle continue throughout the book. There are solutions to this difficulty in place of the periodicity assumptions. A review of Section 2.6 will produce some examples. Frequently, an integration of the resolvent equation, followed by change of the order of integration, can bring about exactly the situation we want. New results in this direction would be most welcome and we would expect such efforts to be very rewarding.

In preparation for the main result we first prove a special form of Sobolev's inequality.

If $g : [a, b] \to R^n$ has a continuous derivative, then

$$\int_a^t \left(|g(u)| + (b-a)|g'(u)|\right) du \ge (b-a) \max_{a \le u \le b} |g(u)| . \tag{3.5.7}$$

A simple proof proceeds as follows. Let $u_0, u_1 \in [a, b]$ and $m, M \in R$ be defined by

$$m = \min_{a \le u \le b} |g(u)| = |g(u_0)| , \quad M = \max_{a \le u \le b} |g(u)| = |g(u_1)| .$$

Then

$$\begin{aligned}
&\int_a^b \left(|g(u)| + (b-a)|g'(u)|\right) du \\
&\qquad \ge (b-a)m + (b-a)\left| \int_{u_0}^{u_1} g'(u)\, du \right| \\
&\qquad \ge (b-a)m + (b-a)|g(u_1) - g(u_0)| \\
&\qquad \ge (b-a)m + (b-a)\left(|g(u_1)| - |g(u_0)|\right) = (b-a)M .
\end{aligned}$$

One may verify that (3.5.7) holds for $n \times n$ matrices using the induced matrix norm.

Theorem 3.5.1. *Let A and B be continuous, $Z(t, s)$ satisfy (3.5.2), and let (3.5.5) hold.*

(i) *If there is a $J > 0$ such that*

$$\int_s^t |B(u, s)|\, du \le J \quad \text{for} \quad 0 \le s \le t < \infty \tag{3.5.8}$$

and

$$\int_0^T \int_s^\infty |Z(t, s)|\, dt\, ds < \infty , \tag{3.5.9}$$

then

$$\sup_{t \ge 0} \int_0^t |Z(t, s)|\, ds = M \quad \text{for some} \quad M > 0 . \tag{3.5.10}$$

(ii) *If there is a $K > 0$ such that*

$$\int_0^t |B(t, s)|\, ds \le K \quad \text{for} \quad t \ge 0 , \tag{3.5.11}$$

then (3.5.10) implies (3.5.9).

Proof. If we integrate (3.5.2) from s to t, $s \le t$, we obtain

$$\begin{aligned}\int_s^t &|\partial Z(u,s)/\partial u|\, du \\ &\le \|A\| \int_s^t |Z(u,s)|\, du + \int_s^t \int_s^u |B(u,v)|\, |Z(v,s)|\, dv\, du \\ &\le (\|A\| + J) \int_s^t |Z(u,s)|\, du\,,\end{aligned}$$

where $\|A\| = \max_{0 \le t \le T} |A(t)|$, as may be seen by interchanging the order of integration.

Let $t > T$ be fixed. There exists an integer $k \ge 1$ and an $\eta \in [0, T)$ with $t = kT + \eta$. Then,

$$\begin{aligned}\int_0^t |Z(t,s)|\, ds &= \int_0^{kT} |Z(kT+\eta, s)|\, ds + \int_{kT}^{kT+\eta} |Z(kT+\eta, s)|\, ds \\ &= \int_0^{kT} |Z(kT+\eta, s)|\, ds + \int_0^{\eta} |Z(\eta, u)|\, du \\ &\qquad \text{(using } Z(t+T, s+T) = Z(t,s) \\ &\qquad \text{and a variable change)} \\ &= \int_0^T \sum_{i=1}^k |Z(iT+\eta, s)|\, ds + \int_0^{\eta} |Z(\eta, u)|\, du \\ &\qquad \text{(by induction)} \\ &\le \alpha + \int_0^T \sum_{i=1}^k \max_{iT \le u \le (i+1)T} |Z(u,s)|\, ds\,,\end{aligned}$$

where $\alpha = \sup_{0 \le u \le T} \int_0^u |Z(u,s)|\, ds$. Applying (3.5.7), we have

$$\begin{aligned}\int_0^t &|Z(t,s)|\, ds \\ &\le \alpha + \int_0^T \sum_{i=1}^k \int_{iT}^{(i+1)T} \left[\frac{1}{T} |Z(u,s)| + \left|\frac{\partial Z(u,s)}{\partial u}\right|\right] du\, ds \\ &\le \alpha + \int_0^T \int_s^{\infty} \left[\frac{1}{T} |Z(u,s)| + \left|\frac{\partial Z(u,s)}{\partial u}\right|\right] du\, ds \\ &\le \alpha + \left(\|A\| + J + \frac{1}{T}\right) \int_0^T \int_s^{\infty} |Z(u,s)|\, du\, ds\,.\end{aligned}$$

Since t is arbitrary, (3.5.9) implies (3.5.10).

Now assume that (3.5.10) and (3.5.11) hold. In order to prove (3.5.9), by Fubini's theorem and the continuity of $Z(t,s)$, it suffices to show that

$$\int_T^\infty \int_0^T |Z(t,s)|\, ds\, dt < \infty\,.$$

Let

$$r(t) = \int_0^T |Z(t,s)|\, ds \quad \text{for} \quad t \geq T\,.$$

Then for $t_2 \geq t_1 \geq T$ we have

$$\begin{aligned} |r(t_2) - r(t_1)| &= \left| \int_0^T \big(|Z(t_2,s)| - |Z(t_1,s)|\big)\, ds \right| \\ &\leq \int_0^T |Z(t_2,s) - Z(t_1,s)|\, ds \\ &= \int_0^T \left| \int_{t_1}^{t_2} (\partial Z(t,s)/\partial t)\, dt \right| ds \\ &\leq \int_0^T \int_{t_1}^{t_2} |\partial Z(t,s)/\partial t|\, dt\, ds\,. \end{aligned}$$

Changing the order of integration yields

$$\begin{aligned} |r(t_2) - r(t_1)| &\leq \int_{t_1}^{t_2} \int_0^T |\partial Z(t,s)/\partial t|\, ds\, dt \\ &= \int_{t_1}^{t_2} \int_0^T \left| A(t)Z(t,s) + \int_s^t B(t,v)Z(v,s)\, dv \right| ds\, dt \\ &\leq \int_{t_1}^{t_2} \|A\| \int_0^T |Z(t,s)|\, ds\, dt + \int_{t_1}^{t_2} \int_0^t \int_s^t |B(t,v)|\, |Z(v,s)|\, dv\, ds\, dt \\ &\leq \int_{t_1}^{t_2} \|A\| \int_0^t |Z(t,s)|\, ds\, dt + \int_{t_1}^{t_2} \int_0^t \int_0^v |Z(v,s)|\, ds |B(t,v)|\, dv\, dt \\ &\leq \int_{t_1}^{t_2} (\|A\|M + KM)\, dt \leq (\|A\|M + KM)(t_2 - t_1)\,. \end{aligned}$$

This shows that $r(t)$ is Lipschitz continuous with Lipschitz constant $L = (\|A\| + K)M$.

Now (3.5.10) and $Z(t+T, s+T) = Z(t,s)$ imply that

$$\sum_{i=1}^{\infty} r(iT + \eta) \leq M$$

for all $\eta \in [0, T)$.

Let $k > 0$ be an integer. It follows from the Lipschitz condition on r that

$$r(iT + \eta + u) \le r(iT + \eta) + L(T/k)$$

for $i = 1, 2, \ldots, \eta \in [0, T)$, $u \in [0, T/k]$. Thus,

$$\begin{aligned}\int_T^{kT} r(t)dt &= \sum_{i=1}^{k-1}\sum_{j=0}^{k-1}\int_{iT+j(T/k)}^{iT+(j+1)(T/k)} r(t)\,dt\\ &\le \sum_{i=1}^{k-1}\sum_{j=0}^{k-1}(T/k)\{r(iT + j[T/k]) + L(T/k)\}\\ &\le \left\{(T/k)\sum_{j=0}^{k-1}\sum_{i=1}^{k-1} r(iT + j[T/k])\right\} + (k-1)k(T/k)L(T/k)\\ &\le TM + LT^2 < \infty\,.\end{aligned}$$

Since k is arbitrary, $r \in L^1[T, \infty)$ and the proof is complete.

Corollary. *Let A and B be continuous, $Z(t,s)$ satisfy (3.5.2), and let (3.5.5) and (3.5.8) hold. If there is an $E > 0$ such that*

$$\int_s^\infty |Z(t,s)|\,dt \le E \quad \text{for all} \quad s \in [0, T]\,, \tag{3.5.12}$$

then

$$\sup_{t\ge 0}\int_0^t |Z(t,s)|\,ds = M \quad \text{for some} \quad M > 0 \tag{3.5.10}$$

is satisfied.

Proof. If we integrate (3.5.12) from 0 to T, the value is bounded by ET.

Theorem 3.5.2. *Suppose that (3.5.5) and (3.5.8) hold with*

$$\int_0^\infty |Z(t,0)|\,dt < \infty\,.$$

Then $Z(t,0) \to 0$ as $t \to \infty$.

Proof. We showed in the proof of Theorem 3.5.1 that

$$\int_0^\infty |\partial Z(u,0)/\partial u|\, du \le (\|A\| + J) \int_0^\infty |Z(u,0)|\, du\,.$$

A similar result holds for each jth column of $Z(t,0)$, say $z(t,0,e_j)$. If the theorem is false, there is a j, an $\epsilon > 0$, and a sequence $\{t_n\} \to \infty$ with $|z(t_n,0,e_j)| \ge \epsilon$. Also,

$$z(t,0,e_j) = e_j + \int_0^t z'(u,0,e_j)\, du$$

so that $t_n \le t \le t_n + 1$ implies that

$$|z(t,0,e_j) - z(t_n,0,e_j)| \le \int_{t_n}^t |z'(u,0,e_j)|\, du < \epsilon/2$$

for large n. Hence $|z(t,0,e_j)| \ge \epsilon/2$ for $t_n \le t \le t_n + 1$, contradicting $z(t,0,e_j) \in L^1$. This completes the proof.

Example 3.5.1. *Consider (3.5.1) and suppose that $f(t) \equiv 0$. If there is an $\alpha > 0$ with*

$$A(t) + \int_0^\infty |B(u+t,t)|du \le -\alpha$$

then for $0 \le s \le t < \infty$ we have

$$\int_s^t |Z(u,s)|du \le 1/\alpha$$

and the conditions of the corollary are satisfied.

Proof. Becker's resolvent may be written as

$$z'(t) = A(t)z(t) + \int_s^t B(t,u)z(u)du, \quad z(s) = 1.$$

Thus, $z(t) = Z(t,s)$. Define the Liapunov functional

$$V(t) = |z(t)| + \int_s^t \int_{t-v}^\infty |B(u+v,v)|du|z(v)|dv$$

so that

$$\begin{aligned}V'(t) &\leq A(t)|z(t)| + \int_s^t |B(t,u)||z(u)|du \\ &+ \int_0^\infty |B(u+t,t)|du|z(t)| - \int_s^t |B(t,v)||z(v)|dv \\ &= \left[A(t) + \int_0^\infty |B(u+t,t)du\right]|z(t)| \\ &\leq -\alpha|z(t)|.\end{aligned}$$

An integration yields the result.

Here is one of the main applications of Theorem 3.5.1. It is known that there are examples of (3.5.1) which do have periodic solutions; indeed, when $B = 0$ they are very common. But for a general B they are rare. The reason for that is that the right-hand-side of (3.5.1) is generally not periodic even when f is periodic. But (3.5.1) can have solutions which are asymptotically periodic in the sense described below.

Under the conditions of Theorem 3.5.1, the following is shown in Burton (1985; p. 102) or (2005b; p. 105). Suppose that

$$\lim_{n\to\infty}\int_{-nT}^t |B(t,s)|\,ds = \int_{-\infty}^t |B(t,s)|\,ds \tag{3.5.13}$$

is bounded and continuous in t, that

$$\int_0^t |Z(t,s)|\,ds \leq M \quad \text{for} \quad t \geq 0\,, \tag{3.5.14}$$

that $Z(t,0) \to 0$ as $t \to \infty$, and that $f(t+T) = f(t)$. Then there exists a sequence of positive integers $\{n_j\}$ such that the function x defined in (3.5.3) satisfies

$$x(t+n_jT, 0, x_0) \to \int_{-\infty}^t Z(t,s)f(s)\,ds := y(t)\,, \quad j\to\infty\,, \tag{3.5.15}$$

where $y(t)$ is a $T-$periodic solution of

$$\frac{dy(t)}{dt} = A(t)y + \int_{-\infty}^t B(t,s)y(s)\,ds + f(t) \tag{3.5.16}$$

on $(-\infty,\infty)$.

Example 3.5.1 is closely related to work in Section 2.6 concerning Liapunov functionals for the resolvent equations themselves instead of on the differential equation.

Notes The material for Section 3.5 comes from Becker, Burton, Krisztin (1988). It was first published by the London Mathematical Society in the *Journal of the London Mathematical Society.* Full details are given in the bibliography.

3.6 A periodic solution

Consider the equation

$$x' = a(t) - \int_{-\infty}^{t} D(t,s)g(x(s))ds \tag{3.6.1}$$

with $g : \Re \to \Re$, g, a, and D continuous,

$$D(t,s) \geq 0, D_s(t,s) \geq 0, D_{st}(t,s) \leq 0, \tag{3.6.2}$$

and

$$a(t+T) = a(t), D(t+T, s+T) = D(t,s). \tag{3.6.3}$$

Levin introduced a Liapunov functional whose derivative was essentially $\int_{-\infty}^{t} D_{st}(t,s)(\int_s^t g(x(v))dv)^2 ds$ and it was then a great struggle to show that $x(t) \to 0$ even when $a(t)$ is very small. Our work here is motivated by the idea that this integral can be used very effectively whenever

$$-\int_{-\infty}^{t} \left[D_s^2(t,s)/D_{st}(t,s)\right] ds \leq L \text{ for some } L > 0. \tag{3.6.4}$$

Let

$$g(x)/x^2 \to 0 \text{ as } |x| \to \infty, xg(x) > 0 \text{ if } |x| > U \text{ for some } U > 0, \tag{3.6.5}$$

$$\int_{-\infty}^{t} \left[D(t,s) + D_s(t,s)(t-s)^2 + |D_{st}(t,s)|(t-s)^2\right] ds \text{ be continuous} \tag{3.6.6}$$

$$\lim_{s \to -\infty} (t-s)D(t,s) = 0 \text{ for fixed } t, \tag{3.6.7}$$

$$\int_0^T a(s)ds = 0 \tag{3.6.8}$$

and recall that by (3.6.3) and (3.6.6) we have

$$\text{there is a } B > 0 \text{ with } \int_{-\infty}^{t} D_s(t,s)ds \leq B.$$

Finally, suppose there is a $Q > 0$ such that

$$\int_{-\infty}^{t_1} |D(t_1, s) - D(t_2, s)|ds \le Q|t_1 - t_2|, \quad 0 \le t_1 \le t_2 \le T. \tag{3.6.9}$$

Theorem 3.6.1. *Let (3.6.2) – (3.6.9) hold. Then (3.6.1) has a T-periodic solution.*

Proof. To construct a homotopy for (3.6.1), let $0 \le \lambda \le 1$ and write

$$x' + x = \lambda x - \lambda \int_{-\infty}^{t} D(t, s)g(x(s))ds + \lambda a(t). \tag{3.6.1$_\lambda$}$$

If x is a T-periodic solution of (3.6.1$_\lambda$), then multiply (3.6.1$_\lambda$) by e^t and integrate from $-\infty$ to t obtaining

$$x(t) = \lambda \int_{-\infty}^{t} e^{-(t-v)} \left[x(v) - \int_{-\infty}^{v} D(v, s)g(x(s))ds + a(v)\right] dv. \tag{3.6.10}$$

Lemma 1. *There is a $K > 0$ such that any T-periodic solution x of (3.6.10) satisfies $\|x\| \le K$.*

Proof. Notice that if x is a T-periodic solution of (3.6.1$_\lambda$), then $x'(t)$ and $a(t)$ both have mean value zero, as does the remaining term

$$-(1 - \lambda)x(t) - \lambda \int_{-\infty}^{t} D(t, s)g(x(s))ds.$$

Since $D(t, s) \ge 0$ and $xg(x) \ge 0$ if $|x| \ge U$, that remaining term can not have mean value zero for $0 < \lambda < 1$ if $|x(t)| > U$ for all t. Hence,

$$\text{there is a } t_1 \in [0, T] \text{ with } |x(t_1)| \le U. \tag{3.6.11}$$

Define

$$V(t, x(\cdot)) = 2 \int_0^x g(s)ds + \lambda \int_{-\infty}^{t} D_s(t, s)\left(\int_s^t g(x(v))dv\right)^2 ds \tag{3.6.12}$$

so that along the T-periodic solution $x(t)$ of (3.6.1$_\lambda$) we have

$$\begin{aligned} V'(t, x(\cdot)) &= \lambda \int_{-\infty}^{t} D_{st}(t, s)\left(\int_s^t g(x(v))dv\right)^2 ds \\ &\quad + 2g(x(t))\left[(\lambda - 1)x(t) - \lambda \int_{-\infty}^{t} D(t, s)g(x(s))ds + \lambda a(t)\right] \\ &\quad + 2\lambda g(x(t)) \int_{-\infty}^{t} D_s(t, s) \int_s^t g(x(v))dv\, ds. \end{aligned}$$

If we integrate the last term by parts we get

$$2\lambda g(x) \int_{-\infty}^{t} D(t,s)g(x(s))ds$$

so that

$$\begin{aligned} V'(t,x(\cdot)) = \lambda \int_{-\infty}^{t} D_{st}(t,s)\bigg(\int_{s}^{t} g(x(v))dv\bigg)^2 ds \\ + 2g(x)\big[(\lambda-1)x + \lambda a(t)\big]. \end{aligned} \tag{3.6.13}$$

Now for $\lambda > 0$ we have

$$\begin{aligned} &(\lambda a(t) + (\lambda-1)x(t) - x'(t))^2 = \bigg(\lambda \int_{-\infty}^{t} D(t,s)g(x(s))ds\bigg)^2 \\ &= \bigg(\lambda \int_{-\infty}^{t} D_s(t,s) \int_{s}^{t} g(x(v))dv\, ds\bigg)^2 \\ &= \bigg[\lambda \int_{-\infty}^{t} [D_s(t,s) \frac{(-D_{st}(t,s))^{1/2}}{(-D_{st}(t,s))^{1/2}} \int_{s}^{t} g(x(v))dv\, ds\bigg]^2 \\ &\le -\lambda^2 \int_{-\infty}^{t} \bigg[\frac{D_s^2(t,s)}{D_{st}(t,s)}\bigg] ds \int_{-\infty}^{t} -D_{st}(t,s)\bigg(\int_{s}^{t} g(x(v))dv\bigg)^2 ds \\ &\le -L\lambda^2 \int_{-\infty}^{t} D_{st}(t,s)\bigg(\int_{s}^{t} g(x(v))dv\bigg)^2 ds. \end{aligned}$$

Using this in (3.6.13) yields

$$V'(t,x(\cdot)) \le -\frac{1}{\lambda L}\bigg((\lambda-1)x(t) - x'(t) + \lambda a(t)\bigg)^2 + 2\lambda a(t)g(x) + Q \tag{3.6.14}$$

where

$$Q = \max[2g(x)(\lambda-1)x] \text{ for } |x| \le U. \tag{3.6.15}$$

Now $1-\lambda = d \ge 0$ so we write (3.6.15) as

$$V'(t,x(\cdot)) \le Q + 2\lambda\|a\|\|g(x)\| - \frac{e^{-2dt}}{\lambda L}\bigg(|(xe^{dt})'| - |\lambda a(t)e^{dt}|\bigg)^2.$$

Since x is T-periodic, so is V, and an integration yields

$$0 = V(T, x(\cdot)) - V(0, x(\cdot)) \leq 2\lambda\|a\|\|g(x)\|T + QT$$
$$- \frac{e^{-2dT}}{\lambda LT}\left(\int_0^T |(xe^{dt})'|dt - \int_0^T |\lambda a(t)e^{dt}|dt\right)^2.$$

Thus,

$$\int_0^T |(xe^{dt})'|dt \leq \int_0^T |\lambda a(t)e^{dt}|dt$$
$$+ [\lambda LTe^{2dT}(QT + 2\lambda\|a\|\|g(x)\|T)]^{1/2}$$

and so by (3.6.11) since $|x(t_1)| = U$ we have $t_2 \in [0, T]$ with

$$\|x\| = |x(t_2)| \leq Ue^{dT} + \lambda T\|a\|e^{dT}$$
$$+ [\lambda LT^2e^{2dT}(Q + 2\lambda\|a\|\|g(x)\|)]^{1/2}.$$

If $\|g(x)\| = |g(x(t_3))|$ for some $t_3 \in [0, T]$ then $|x(t_2)| \geq |x(t_3)|$ and so

$$|x(t_3)| \leq |x(t_2)| \leq Ue^{dT} + \lambda\|a\|Te^{dT}$$
$$+ \left[2\lambda T^2Le^{2dT}(Q + 2\lambda\|a\||g(x(t_3))|)\right]^{1/2}.$$

Now this x is simply a representative T-periodic solution and we want an a priori bound on the whole class. If $|x(t_3)|$ is bounded over this class, but $\|x\| \to \infty$, then we divide the above inequality by $|x(t_2)|$ and get a contradiction for large $|x(t_2)|$. If $|x(t_3)| \to \infty$, then we divide the above inequality by $|x(t_3)|$ and get a contradiction to $g(x)/x^2 \to 0$ as $|x| \to \infty$. Hence, there is a $K > 0$ with $\|x\| \leq K$ and Lemma 1 is true.

Let $(P, \|\cdot\|)$ be the Banach space of continuous T-periodic functions and for $\varphi \in P$ define

$$H(\lambda, \varphi)(t) = \lambda \int_{-\infty}^t e^{-(t-v)}\left[\varphi(v) - \int_{-\infty}^v D(v, s)g(\varphi(s))ds + a(v)\right]dv. \tag{3.6.16}$$

Now $H : P \to P$ since a change of variable shows that $H(\lambda, \varphi)$ is T-periodic and, since φ is continuous, $H(\lambda, \varphi)$ is differentiable (and, thus, continuous).

Also, H maps bounded sets into compact sets. To see this, if $\|\varphi\| \leq J$, then there is a bound in terms of J on H and on the derivative of $H(\lambda, \varphi)$. The result now follows by Ascoli's theorem.

Lemma 2. *$H(\lambda, \varphi)$ is continuous in φ.*

Proof. Let $\varphi_1, \varphi_2 \in P$ so that $\|\varphi_i\| < J$ for some $J > 0$. Then by the uniform continuity of $g(t, x)$ and by (3.6.3) and (3.6.6) we can make

$$\begin{aligned}
&|H(\lambda, \varphi_1)(t) - H(\lambda, \varphi_2)(t)| \\
&\quad = \lambda \left| \int_{-\infty}^{t} D(t, s)\big[g(s, \varphi_1(s)) - g(s, \varphi_2(s))\big]\, ds \right|
\end{aligned}$$

as small as we please.

By Schaefer's theorem (Theorem 1.3.1.2), H has a fixed point for $\lambda = 1$.

Notes The material for Section 3.6 was taken from Burton (1993). It was first published by Research Square Publications in *Differential Equations and Dynamical Systems.* Full details are given in the bibliography.

3.7 General integrodifferential equations

3.7.1 Introduction

In this section we do not start with an integral equation. Instead, we study a scalar equation

$$x'(t) = A(t)x(t) + \int_0^t B(t, s)x(s)ds + a(t) \tag{3.7.1}$$

where A, B, and a are continuous. There is then Becker's resolvent equation

$$Z_t(t, s) = A(t)Z(t, s) + \int_s^t B(t, u)Z(u, s)du, \; Z(s, s) = 1, \tag{3.7.2}$$

and variation of parameters formula

$$x(t) = Z(t, 0)x(0) + \int_0^t Z(t, s)a(s)ds. \tag{3.7.3}$$

We must integrate the resolvent, Z, with respect to the initial time and it will be desirable to show that

$$\sup_{t \geq 0} \int_0^t |Z(t, s)|ds < \infty. \tag{3.7.4}$$

Our first series of results is aimed at presenting an example of a nonconvolution counterpart of the Convolution Lemma.

SCHEME

In this section each result is closely connected to the previous one so that an overall scheme is needed.

If A were constant and if B were of convolution type then we could show that Z is of convolution type. Thus, if in addition, $a(t) \to 0$ as $t \to \infty$ and if $\int_0^\infty |Z(u)|du < \infty$ then the Convolution Lemma would say that in (3.7.3) we have $\int_0^t Z(t-s)a(s)ds \to 0$ as $t \to \infty$. Our first main result here is to extend that to the case where Z is not of convolution type so that (3.7.4) (together with side conditions) can be substituted for $Z \in L^1$.

As a companion to that result we then give necessary and sufficient conditions for (3.7.4) to hold.

Those necessary and sufficient conditions require that we show that each solution of (3.7.1) is bounded for each bounded and continuous function $a(t)$. And that is hard to prove. We then give several techniques for showing that boundedness. Here, $a(t)$ is an arbitrary bounded and continuous function.

Next, to introduce the last subsection we review Section 3.5 with $A(t+T) = A(t)$ and $B(t+T, s+T) = B(t,s)$ for some $T > 0$ so that those necessary and sufficient conditions need only concern $\int_s^t |Z(u,s)|du$ bounded for $0 \le s \le T$; so $a(t)$ is not involved.

Finally, we derive substitute results in the nonperiodic case requiring only that $\int_s^t Z^2(u,s)du$ is bounded. Notice that in both of these cases we are integrating Z with respect to t, another simplification over the integration in (3.7.4) with respect to s.

3.7.2 Uniform Asymptotic Stability

To specify a solution of (3.7.1) we require a $t_0 \ge 0$ and a continuous initial function $\phi : [0, t_0] \to R$. There is then a unique solution $x(t, t_0, \phi)$ satisfying (3.7.1) for $t > t_0$ and with $x(t, t_0, \phi) = \phi(t)$ on $[0, t_0]$. We use the notation $|\phi|_{[0,t_0]} := \sup_{0 \le t \le t_0} |\phi(t)|$.

Our main definitions concern the case $a(t) = 0$ so that (3.7.1) becomes

$$x'(t) = A(t)x(t) + \int_0^t B(t,s)x(s)ds. \tag{3.7.5}$$

Definition 3.7.2.1. *The zero solution of (3.7.5) is uniformly stable (US) if for any $\epsilon > 0$ there exists a $\delta = \delta(\epsilon) > 0$ such that $[t_0 \ge 0,\ \phi \in C([0, t_0]),\ |\phi|_{[0,t_0]} < \delta,\ t \ge t_0]$ imply that $|x(t, t_0, \phi)| < \epsilon$.*

Definition 3.7.2.2. *The zero solution of (3.7.5) is uniformly asymptotically stable (UAS) if it is US and there exists a* $\delta_0 > 0$ *with the property that for each* $\epsilon > 0$ *there exists* $T = T(\epsilon)$ *such that* $[t_0 \geq 0, \phi \in C([0,t_0]), |\phi|_{[0,t_0]} < \delta_0, t \geq t_0 + T]$ *imply that* $|x(t, t_0, \phi)| < \epsilon$.

Becker (1979) has proved some very important results on UAS. Consider (3.7.5) and the resolvent $Z(t,s)$ defined in (3.7.2). First, define a scalar function

$$d(t,s) := \int_0^s \left| \int_s^t Z(t,\xi)B(\xi,u)d\xi \right| du, \quad 0 \leq s \leq t < \infty.$$

Definition 3.7.2.3. *A continuous function* $h(t,s)$ *defined for* $0 \leq s \leq t < \infty$ *is said to promote uniform asymptotic stability if* $h(t,s)$ *is bounded and if for each* $\eta > 0$, *there exists a number* $T(\eta) > 0$ *such that* $|h(t+s,s)| \leq \eta$ *for all* $s \geq 0$ *and* $t \geq T(\eta)$.

A main result of Becker (1979; p. 62) is:

Theorem 3.7.2.4. *The zero solution of (3.7.5) is UAS if and only if both* $Z(t,s)$ *and* $d(t,s)$ *promote UAS.*

The proposition is important in so many contexts. First, we will quote some results which will show that the zero solution is UAS. This will allow us to say that $Z(t,s)$ promotes UAS. That, in turn, will allow us to obtain a result akin to the Convolution Lemma; in that work we use $Z(t,s)$ as the analog of an L^1-function.

There is a major result by Zhang (1997) which shows that (3.7.4) is central to all of our work here. Under very mild assumptions, (3.7.4) characterizes UAS. In the next section we will show how to characterize (3.7.4). Our task then will be to give a variety of conditions under which (3.7.4) holds. The following are Zhang's conditions:

$$(H_1) \quad \sup_{t \geq 0}\left(|A(t)| + \int_0^t |B(t,s)|ds \right) < \infty.$$

(H_2) for any $\sigma > 0$, there exists an $S = S(\sigma) > 0$ such that

$$\int_0^{t-S} |B(t,u)|du < \sigma \text{ for all } t \geq S.$$

(H_3) $A(t)$ and $B(t, t+s)$ are bounded and uniformly continuous in $(t,s) \in \{(t,s) \in [0,\infty) \times K \mid -t \leq s \leq 0\}$ for any compact set $K \subset (-\infty, 0]$.

Theorem 3.7.2.5. *(Zhang (1997)) Let* (H_1), (H_2), (H_3) *hold. The zero solution of (3.7.5) is UAS if and only if (3.7.4) holds.*

The chain of reasoning in the next result is as follows. Condition (3.7.4), together with the (H_i), implies UAS. Now UAS implies that $Z(t,s)$ promotes UAS. Thus, (3.7.4) and the property that Z promotes UAS enable us to prove the following convolution-type result.

Theorem 3.7.2.6. *Suppose the resolvent,* $Z(t,s)$, *satisfies (3.7.4) and promotes UAS. If* $\phi : [0,\infty) \to R$ *is continuous and satisfies* $\phi(t) \to 0$ *as* $t \to \infty$, *then* $\int_0^t Z(t,s)\phi(s)ds \to 0$ *as* $t \to \infty$.

Proof. By (3.7.4),

$$M := \sup_{t\geq 0} \int_0^t |Z(t,s)|ds < \infty.$$

For a given $\epsilon > 0$ find $L > 0$ with $|\phi(t)| < \epsilon/(2M)$ if $t \geq L$. Then for $t \geq L$, we have

$$\begin{aligned}\int_L^t |Z(t,s)\phi(s)|ds &\leq (\epsilon/(2M)) \int_L^t |Z(t,s)|ds \\ &\leq (\epsilon/(2M)) \int_0^t |Z(t,s)|ds \leq \epsilon/2.\end{aligned}$$

Let $J = \max_{0\leq t\leq L} |\phi(t)|$. Then

$$\int_0^L |Z(t,s)\phi(s)|ds \leq J \int_0^L |Z(t,s)|ds = J \int_0^L |Z(s+\tau,s)|ds$$

where $\tau := t - s \geq 0$. By Definition 3.7.2.3, we can find $P > 0$ such that $|Z(s+\tau,s)| \leq \epsilon/(2LJ)$ for all $s \geq 0$ and $\tau \geq P$. Now let $T := P + L$. If $t \geq T$, then $\tau \geq P$ because for this integral $0 \leq s \leq L$. Consequently,

$$\int_0^L |Z(t,s)\phi(s)|ds \leq J \int_0^L \frac{\epsilon}{2LJ} ds = \epsilon/2.$$

It follows that

$$\left|\int_0^t Z(t,s)\phi(s)ds\right| \leq \int_0^L |Z(t,s)\phi(s)|ds + \int_L^t |Z(t,s)\phi(s)|ds < \epsilon$$

for $t \geq T$.

This completes the proof.

3.7.3 The Resolvent

In this subsection we offer seven sets of conditions ensuring (3.7.4) and, hence, forming the basis for UAS. This is done by using Liapunov functionals on Volterra equations with bounded continuous forcing functions. In the next section we use one of those Liapunov functionals to obtain a parallel result in the periodic case without that forcing function. It is so much easier without the forcing function. There then emerges the problem of trying to promote all of the following Liapunov functionals to that same framework in which the forcing function is avoided in the periodic case.

It turns out that there is a very nice characterization of (3.7.4). We denote by $(\mathcal{BC}, \|\cdot\|)$ the Banach space of bounded continuous functions $\phi : [0,\infty) \to \mathbb{R}$ with the supremum norm.

Theorem 3.7.3.1. *Suppose that $Z(t,0)$ is bounded for $t \geq 0$. Condition (3.7.4) holds if and only if every solution of (3.7.1) on $[0,\infty)$ is bounded for every $a \in \mathcal{BC}$. Moreover, if every solution of (3.7.1) on $[0,\infty)$ is bounded for every $a \in \mathcal{BC}$, then $Z(t,0)$ is bounded and (3.7.4) holds.*

Proof. If $x(t)$ is bounded for all such functions a then so is $\int_0^t Z(t,s)a(s)ds$. By Perron's theorem (1930) (or Burton(2005b; p. 116)) (3.7.4) follows. On the other hand, if (3.7.4) holds and if $Z(t,0)$ is bounded, then the boundedness of x follows for any bounded function a. For the last sentence, take $a(t) \equiv 0$. This completes the proof.

Notice! In this subsection our solutions are always on $[0,\infty)$ so that the solution is $x(t) := x(t,0,x(0))$. There is no initial function and the solution is differentiable for $t > 0$.

We will now examine four essentially different kinds of proofs showing that every solution of (3.7.1) is bounded for every bounded continuous function $a(t)$. These will range from the pedestrian to the intricate, but they will all illustrate nonconvolution properties. The first is a Razumikhin argument which is very simple and very demanding. But the focus is on the fact that we are integrating B with respect to s. The same result would hold if $B(t,s)$ were replaced by $B(t,s)\lambda(t)$ where $|\lambda(t)| \leq 1$ and there would be no change in the hypotheses or proof.

Theorem 3.7.3.2. *Suppose there is an $\alpha > 0$ such that*

$$\sup_{t\geq 0}\left[A(t) + \int_0^t |B(t,s)|ds\right] \leq -\alpha.$$

Then every solution of (3.7.1) on $[0,\infty)$ is bounded for every $a \in \mathcal{BC}$.

The proof is almost identical to that of Theorem 3.2.10.

Our next result is based on a contraction mapping and is only one of a class. Here we have $a \in L^\infty$, we use an L^∞ norm, our solution is in L^∞, as are $Z(t,0)$ and $\int_0^t Z(t,s)a(s)ds$. In a similar way, if $a \in L^p$ then we can use a norm with an L^p weight and get our solution and $\int_0^t Z(t,s)a(s)ds$ in L^p. The last result of this section is an example of that process.

Theorem 3.7.3.2 used pointwise conditions, but contractions use averaging conditions. Theorem 3.7.3.2 focused on integration of the second coordinate of $B(t,s)$, but Theorem 3.7.3.4 will focus on integration of both coordinates of B; each will be treated differently. Very often when we use a first order Liapunov functional it will require integration of only the first coordinate of B, but a second order Liapunov functional will require integration of both coordinates. All of these properties are hidden in the convolution case.

Theorem 3.7.3.3. *Suppose there is a $J > 0$ such that $\int_0^t A(u)du \leq J$, $\int_0^t e^{\int_u^t A(s)ds}du \leq J$, and suppose there is an $\alpha < 1$ with*

$$\sup_{t\geq 0}\int_0^t e^{\int_u^t A(s)ds}\int_0^u |B(u,s)|dsdu < \alpha.$$

Then every solution of (3.7.1) on $[0,\infty)$ is in $\mathcal{BC}$ when $a \in \mathcal{BC}$.

Proof. Use the variation of parameters formula for ordinary differential equations on (3.7.1) and from that define a mapping $P : \mathcal{BC} \to \mathcal{BC}$ by $\phi \in \mathcal{BC}$ implies that

$$(P\phi)(t) = x(0)e^{\int_0^t A(s)ds} + \int_0^t e^{\int_u^t A(s)ds}\left[\int_0^u B(u,s)\phi(s)ds + a(u)\right]du.$$

The reader readily shows it is a contraction.

This result and the next one form what might be called boundaries for a whole set of results of the type which can be found in Burton (2005b; Section 1.6). In the last result we try to derive all the stability from $A(t)$. In the next result we borrow all of $B(t,s)$ to help with the stability. Frequently, we can get nice results by just borrowing some of $B(t,s)$.

Theorem 3.7.3.4. *Let $D(t) := A(t)+\int_0^\infty B(u+t,t)du$ be defined for $t \geq 0$ and suppose there is a $J > 0$ such that $\int_0^t D(u)du \leq J$, $\int_0^t e^{\int_u^t D(s)ds}du \leq J$. Suppose also that there is an $\alpha < 1$ with*

$$\sup_{t\geq 0}\left[\int_0^t \int_{t-s}^\infty |B(v+s,s)|dvds + \int_0^t e^{\int_u^t D(s)ds}|D(u)|\int_0^u \int_{u-s}^\infty |B(v+s,s)|dvdsdu\right] \leq \alpha.$$

Then every solution of (3.7.1) on $[0,\infty)$ is in $\mathcal{BC}$ for every $a \in \mathcal{BC}$.

Proof. We can write (3.7.1) as

$$\begin{aligned} x'(t) &= A(t)x(t) + \int_0^\infty B(u+t,t)dux(t) \\ &\quad - \frac{d}{dt}\int_0^t \int_{t-s}^\infty B(u+s,s)dux(s)ds + a(t) \\ &= D(t)x(t) - \frac{d}{dt}\int_0^t \int_{t-s}^\infty B(u+s,s)dux(s)ds + a(t). \end{aligned}$$

By the variation of parameters formula followed by integration by parts we have

$$\begin{aligned} &x(t) \\ &= x(0)e^{\int_0^t D(s)ds} - \int_0^t e^{\int_u^t D(s)ds}\frac{d}{du}\int_0^u \int_{u-s}^\infty B(v+s,s)dvx(s)dsdu \\ &+ \int_0^t e^{\int_u^t D(s)ds}a(u)du \\ &= x(0)e^{\int_0^t D(s)ds} - e^{\int_u^t D(s)ds}\int_0^u \int_{u-s}^\infty B(v+s,s)dvx(s)ds\Big|_0^t \\ &- \int_0^t e^{\int_u^t D(s)ds}D(u)\int_0^u \int_{u-s}^\infty B(v+s,s)dvx(s)dsdu \\ &+ \int_0^t e^{\int_u^t D(s)ds}a(u)du \\ &= x(0)e^{\int_0^t D(s)ds} - \int_0^t \int_{t-s}^\infty B(v+s,s)dvx(s)ds \\ &- \int_0^t e^{\int_u^t D(s)ds}D(u)\int_0^u \int_{u-s}^\infty B(v+s,s)dvx(s)dsdu \\ &+ \int_0^t e^{\int_u^t D(s)ds}a(u)du. \end{aligned}$$

This will define a mapping in the usual way and by the assumptions it will map $\mathcal{BC}$ into $\mathcal{BC}$ and will be a contraction.

The proof of the next result differs in just three places from that of Theorem 2.1.12. As the last step is significantly different, we give the complete proof.

Theorem 3.7.3.5. *Suppose there is an $\alpha > 0$ with*

$$A(t) + \int_0^\infty |B(u+t,t)|du \leq -\alpha.$$

Suppose also that there is a function $\Phi : [0,\infty) \to (0,\infty)$ which is differentiable, decreasing, and $L^1[0,\infty)$ with

$$\Phi(t-s) \geq \int_{t-s}^\infty |B(u+s,s)|du.$$

If $a \in \mathcal{BC}$ so is each solution of (3.7.1) on $[0,\infty)$.

Proof. Define the Liapunov functional

$$V(t,x(\cdot)) = |x(t)| + \int_0^t \int_{t-s}^\infty |B(u+s,s)|du|x(s)|ds$$

with derivative along a solution of (3.7.1) satisfying

$$\begin{aligned} V'(t,x(\cdot)) &\leq A(t)|x(t)| + \int_0^t |B(t,s)x(s)|ds + |a(t)| \\ &\quad + \int_0^\infty |B(u+t,t)|du|x(t)| - \int_0^t |B(t,s)x(s)|ds \\ &\leq -\alpha|x(t)| + |a(t)|. \end{aligned}$$

For a fixed solution, we write $V(t,x(\cdot)) = V(t)$. Suppose that there is a $t > 0$ with $V(t) = \max_{0\leq s\leq t} V(s)$. Then for $0 \leq s \leq t$ we have

$$\frac{dV(s)}{ds} \leq -\alpha|x(s)| + |a(s)|$$

and

$$\frac{dV(s)}{ds}\Phi(t-s) \leq -\alpha|x(s)|\Phi(t-s) + |a(s)|\Phi(t-s).$$

An integration by parts of the left-hand-side yields

$$\begin{aligned}
\int_0^t \frac{dV(s)}{ds}\Phi(t-s)ds &= V(s)\Phi(t-s)\Big|_0^t - \int_0^t V(s)\frac{d}{ds}\Phi(t-s)ds \\
&= V(t)\Phi(0) - V(0)\Phi(t) - \int_0^t V(s)\frac{d}{ds}\Phi(t-s)ds \\
&\geq V(t)\Phi(0) - V(0)\Phi(t) - V(t)\int_0^t \frac{d}{ds}\Phi(t-s)ds \\
&= V(t)\Phi(0) - V(0)\Phi(t) - V(t)[\Phi(0) - \Phi(t)] \\
&= V(t)\Phi(t) - V(0)\Phi(t).
\end{aligned}$$

Hence, for $|a(t)| \leq a_0$ we then have

$$\begin{aligned}
-|x(0)|\Phi(0) &= -V(0)\Phi(0) \\
&\leq -V(0)\Phi(t) \\
&\leq V(t)\Phi(t) - V(0)\Phi(t) \\
&\leq -\alpha\int_0^t \Phi(t-s)|x(s)|ds + a_0\int_0^t \Phi(t-s)ds \\
&\leq -\alpha\int_0^t \int_{t-s}^\infty |B(u+s,s)|du|x(s)|ds + a_0\int_0^\infty \Phi(s)ds.
\end{aligned}$$

From this we see that for that fixed solution there is a positive number K with

$$\int_0^t \int_{t-s}^\infty |B(u+s,s)|du|x(s)|ds \leq K.$$

This means that

$$|x(t)| \leq V(t) \leq |x(t)| + K$$

and

$$V'(t) \leq -\alpha|x(t)| + a_0.$$

From that last pair it follows trivially that there is an $L > 0$ with $|x(t)| \leq L$. Indeed, a simple geometrical argument in the $(V, |x|)$ plane with the lines $V = |x|$ and $V = |x| + K$ shows that $|x(t)| \leq (a_0/\alpha) + K + |x(0)|$ since V' is negative for $|x| > a_0/\alpha$. This completes the proof.

Virtually the same proof would allow us to show that if $a \in L^1[0,\infty)$, then x, $Z(t,0)$, and $\int_0^t Z(t,s)a(s)ds$ are also in $L^1[0,\infty)$. In our next theorem we use a quadratic Liapunov functional and for $a \in L^p$ we can get $x \in L^2$. The next result is interesting in that its main condition averages the two main conditions of Theorems 3.7.3.2 and 3.7.3.5.

Theorem 3.7.3.6. *Suppose there is an $\alpha > 0$ and a $k > 0$ with*

$$\sup_{t\geq 0}\left[2A(t) + \int_0^\infty |B(u+t,t)|du + \int_0^t |B(t,s)|ds\right] + k \leq -\alpha.$$

Suppose also that there is a function $\Phi : [0,\infty) \to (0,\infty)$ which is differentiable, decreasing, and $L^1[0,\infty)$ with

$$\Phi(t-s) \geq \int_{t-s}^\infty |B(u+s,s)|du.$$

If $a \in \mathcal{BC}$, so is the solution of (3.7.1) on $[0,\infty)$.

Proof. For a fixed $a \in \mathcal{BC}$ and a fixed $x(0)$, find a_0 with $a^2(t) \leq a_0$ and define

$$V(t) = x^2(t) + \int_0^t \int_{t-s}^\infty |B(u+s,s)|du x^2(s)ds$$

with derivative along the fixed solution of (3.7.1) satisfying

$$\begin{aligned}
V'(t) &= 2A(t)x^2(t) + 2x(t)\int_0^t B(t,s)x(s)ds + 2x(t)a(t) \\
&+ \int_0^\infty |B(u+t,t)|du x^2(t) - \int_0^t |B(t,s)|x^2(s)ds \\
&\leq 2A(t)x^2(t) + \int_0^t |B(t,s)|\Big(x^2(t) + x^2(s)\Big)ds + kx^2(t) \\
&+ \frac{1}{k}a^2(t) + \int_0^\infty |B(u+t,t)|du x^2(t) - \int_0^t |B(t,s)|x^2(s)ds \\
&= \left[2A(t) + \int_0^t |B(t,s)|ds + \int_0^\infty |B(u+t,t)|du + k\right]x^2(t) + \frac{a_0}{k} \\
&\leq -\alpha x^2(t) + a_0/k.
\end{aligned}$$

We now introduce $\Phi(t-s)$ exactly as in the proof of the last theorem and show that the integral term in V is bounded. We continue and show that $x^2(t)$ is bounded to complete the proof.

The next result and its corollary are proved exactly as in Theorems 3.2.7 and 3.2.8.

Theorem 3.7.3.7. *Suppose that*

$$B(t,s) \le 0,\ B_s(t,s) \le 0,\ B_{st}(t,s) \ge 0,\ B_t(t,0) \ge 0.$$

Then the derivative of

$$V(t) = x^2(t) - \int_0^t B_s(t,s)\left(\int_s^t x(u)du\right)^2 ds - B(t,0)\left(\int_0^t x(s)ds\right)^2$$

along any solution of (3.7.1) on $[0,\infty)$ *satisfies*

$$V'(t) = 2A(t)x^2(t) + 2x(t)a(t) - \int_0^t B_{st}(t,s)\left(\int_s^t x(u)du\right)^2 ds$$
$$- B_t(t,0)\left(\int_0^t x(s)ds\right)^2.$$

Corollary 3.7.3.8. *Let the conditions of Theorem 3.7.3.7 hold and suppose there is an* $\alpha > 0$ *with* $A(t) \le -\alpha$:

(i) If $a \in L^2[0,\infty)$, *so is any solution of (3.7.1) on* $[0,\infty)$, *as is* $Z(t,0)$ *and* $\int_0^t Z(t,s)a(s)ds$.

(ii)If $a(t)$ *is bounded and if there is an* $M > 0$ *with*

$$\int_0^t |B_s(t,s)|(t-s)^2 ds + t^2|B(t,0)| \le M$$

then (3.7.4) holds.

Theorem 3.7.3.9. *Let the conditions of Theorem 3.7.3.7 hold. Suppose there is an* $\alpha > 0$ *and a function* $k : [0,\infty) \to (0,\infty)$ *with* $\int_0^t e^{-\int_u^t k(s)ds}du$ *bounded and*

$$2A(t) \le -k(t) - \alpha,\ B_{st}(t,s) \ge -k(t)B_s(t,s),\ B_t(t,0) \ge -k(t)B(t,0).$$

If $a \in \mathcal{BC}$, *so is any solution of (3.7.1) on* $[0,\infty)$.

Proof. Under these conditions we see that for V defined in Theorem 3.7.3.7 we have

$$V'(t) \le -k(t)V(t) + (1/\alpha)a^2(t)$$

so that

$$x^2(t) \le V(t) \le V(0)e^{-\int_0^t k(s)ds} + (1/\alpha)\int_0^t e^{-\int_u^t k(s)ds}a^2(u)du.$$

and that is bounded by assumption.

In our earlier proofs using Liapunov functionals we could have greatly simplified the proofs by asking a bit more from $A(t)$, using a coefficient larger than 1 in front of the integral, and getting a differential inequality. Such details are left to the reader.

3.7.4 The periodic case

We have been struggling to show that $\int_0^t |Z(t,s)|ds$ is bounded, while the Liapunov theory is showing that $\int_s^\infty |Z(u,s)|du$ is bounded. This very brief section is just a reminder that we had solved this in the periodic case in Section 3.5. Here are the details for quick reference, as it is a very important property.

We are supposing that

$$B(t+T, s+T) = B(t,s), \qquad A(t+T) = A(t), \quad \exists T > 0.$$

Our main result was the following.

Theorem 3.5.1. *Let A and B be continuous, $Z(t,s)$ satisfy (3.5.2), and let (3.5.5) hold.*

(i) *If there is a $J > 0$ such that*

$$\int_s^t |B(u,s)|\, du \le J \quad \text{for} \quad 0 \le s \le t < \infty \tag{3.5.8}$$

and

$$\int_0^T \int_s^\infty |Z(t,s)|\, dt\, ds < \infty\,, \tag{3.5.9}$$

then

$$\sup_{t\ge 0} \int_0^t |Z(t,s)|\, ds = M \quad \text{for some} \quad M > 0\,. \tag{3.5.10}$$

(ii) *If there is a $K > 0$ such that*

$$\int_0^t |B(t,s)|\, ds \le K \quad \text{for} \quad t \ge 0\,, \tag{3.5.11}$$

then (3.5.10) implies (3.5.9).

The main utility of the theorem is now stated.

Corollary. *Let the conditions of Theorem 3.5.1 hold. If there is an $E > 0$ with*

$$\int_s^\infty |Z(t,s)|dt \le E, \quad 0 \le s \le T,$$

then

$$\sup_{t \geq 0} \int_0^t |Z(t,s)| ds < \infty.$$

With this result we avoided all the difficulties encountered in the last subsection concerning a bounded and continuous function $a(t)$. The function Φ in Theorem 3.7.3.5 is not needed. This was emphasized in our Example 3.5.1 which logically belongs here and will be stated below as Theorem 3.7.5.2.

3.7.5 L^p results

If we do not have the periodicity, we can still have some results in the above mentioned direction. Suppose that $a(t) \in L^1[0, \infty)$ and consider the solution of (3.7.1) on $[0, \infty)$ given by

$$x(t) = Z(t,0)x(0) + \int_0^t Z(t,s)a(s)ds.$$

Could it be that $x \in L^1[0, \infty)$? In the convolution case we would have

$$x(t) = Z(t,0)x(0) + \int_0^t Z(t-s)a(s)ds$$

and if $Z \in L^1[0, \infty)$, then the convolution of two L^1 functions is an L^1-function. Thus, we would want to show that $Z \in L^1$. But in the nonconvolution case all of this breaks down, while something interesting happens. The easy case is to prove, not (3.7.4), but that there is an $M > 0$ with

$$\sup_{0 \leq s \leq t < \infty} \int_s^\infty |Z(t,s)| dt \leq M. \tag{3.7.6}$$

And that is exactly what is required here.

Theorem 3.7.5.1. *If (3.7.6) holds, then $a \in L^1[0, \infty)$ implies $x \in L^1[0, \infty)$.*

Proof. From the variation of parameters formula we have

$$\begin{aligned}
\int_0^t |x(s)|ds &\leq \int_0^t |Z(s,0)x(0)|ds + \int_0^t \int_0^u |Z(u,s)a(s)|dsdu \\
&= \int_0^t |Z(s,0)x(0)|ds + \int_0^t \int_s^t |Z(u,s)a(s)|duds \\
&\leq \int_0^t |Z(s,0)x(0)|ds + \int_0^t |a(s)| \int_s^\infty |Z(u,s)|duds \\
&\leq \int_0^t |Z(s,0)x(0)|ds + M \int_0^t |a(s)|ds.
\end{aligned}$$

Shortly we will present some results ensuring that (3.7.6) holds. However, this leads us to one of the most tantalizing ideas in all of Volterra equations. We know that if (3.7.4) holds then every solution of (3.7.1) is bounded for every bounded continuous $a(t)$. Using the sequence of inequalities just displayed we can say that if (3.7.6) holds then for every bounded and continuous $a(t)$ it is true that the average of every solution of (3.7.1) with $x(0) = 0$ is bounded:

$$\frac{1}{t}\int_0^t |x(s)|ds \tag{3.7.7}$$

is bounded. This suggests that (3.7.4) and (3.7.6) are very closely related. In many problems it may be that (3.7.7) is as useful as knowing that $x(t)$ is bounded. Here are the details. Note that if (3.7.6) holds and if $|a(t)| \leq K$ for some $K > 0$, then

$$\int_0^t |x(s)|ds \leq \int_0^t |Z(s,0)x(0)|ds + MKt = MKt.$$

We will now give a selection of results proving that (3.7.6) holds, as well as some related forms. Notice how much easier 18 the arguments are when we use the same Liapunov functionals, but take $a(t) = 0$ and strive for (3.7.6) instead of (3.7.4). The natural first result here is Example 3.5.1 which we restate here for reference. Review its proof for the form of the Liapunov functional.

Notice again that Theorem 3.7.5.2 uses afirst order Liapunov functional and integrates one coordinate, while Theorem 3.7.5.3 uses z^2 and integrates two coordinates.

Theorem 3.7.5.2. *Consider (3.7.1) and suppose that* $\int_{t-v}^{\infty} |B(u+v,v)|du$ *is continuous and that* $a(t) \equiv 0$. *If there is an* $\alpha > 0$ *with*

$$\sup_{t \geq 0} \left[A(t) + \int_0^{\infty} |B(u+t,t)|du \right] \leq -\alpha$$

for $0 \leq s \leq t$, *then the resolvent,* $Z(t,s)$, *satisfies* $\int_s^t Z(u,s)du \leq 1/\alpha$ *for* $0 \leq s \leq t < \infty$.

There is, of course, the parallel result showing the integrability of Z^2 with respect to t.

Theorem 3.7.5.3. *Consider (3.7.1) and suppose that* $\int_{t-v}^{\infty} |B(u+v,v)|du$ *is continuous and that* $a(t) \equiv 0$. *If there is an* $\alpha > 0$ *with*

$$\sup_{t \geq 0} \left[2A(t) + \int_0^{\infty} |B(u+t,t)|du + \int_s^t |B(t,u)|du \right] \leq -\alpha$$

for $0 \leq s \leq t$, *then the resolvent,* $Z(t,s)$, *satisfies* $\int_s^t Z^2(u,s)du \leq 1/\alpha$ *for* $0 \leq s \leq t < \infty$.

Proof. Use Becker's resolvent equation again and take $z(t) = Z(t,s)$. Define the Liapunov functional

$$V(t) = z^2(t) + \int_s^t \int_{t-v}^{\infty} |B(u+v,v)|duz^2(v)dv$$

with derivative satisfying

$$\begin{aligned} V'(t) &= 2A(t)z^2(t) + 2z(t) \int_s^t B(t,u)z(u)du \\ &+ \int_0^{\infty} |B(u+t,t)|duz^2(t) - \int_s^t |B(t,v)|z^2(v)dv \\ &\leq 2A(t)z^2(t) + \int_s^t |B(t,u)| \Big(z^2(u) + z^2(t) \Big) du \\ &+ \int_0^{\infty} |B(u+t,t)|duz^2(t) - \int_s^t |B(t,v)|z^2(v)dv \\ &= [2A(t) + \int_0^{\infty} |B(u+t,t)|du + \int_s^t |B(t,u)|du]z^2(t) \\ &\leq -\alpha z^2(t). \end{aligned}$$

Hence,

$$0 \leq V(t) \leq V(s) - \alpha \int_s^t z^2(u)du = 1 - \alpha \int_s^t z^2(u)du$$

from which the result follows.

The interested reader could borrow $B(t,s)$ and form $A(t)+\int_t^\infty B(u,t)du$ in place of $A(t)$ to get a stability result. Details, which are quite lengthy, may be found in Burton (2005c; p. 146).

Next, we revisit the Liapunov functional of Theorem 3.7.3.7. Here, the differential inequality needed in the proof of Theorem 3.7.3.9 is totally unnecessary.

Theorem 3.7.5.4. *Consider (3.7.1) with $a(t) \equiv 0$. Suppose there is an $\alpha > 0$ with*

$$B(t,s) \leq 0,\ B_s(t,s) \leq 0,\ B_{st}(t,s) \geq 0,\ B_t(t,s) \geq 0,\ 2A(t) \leq -\alpha.$$

Then the derivative of

$$V(t) = z^2(t) - \int_s^t B_u(t,u)\left(\int_u^t z(v)dv\right)^2 du - B(t,s)\left(\int_s^t z(u)du\right)^2$$

along the solution $Z(t,s)$ with $Z(s,s)=1$ of the resolvent equation

$$z'(t) = A(t)z(t) + \int_s^t B(t,u)z(u)du$$

satisfies

$$V'(t) \leq 2A(t)z^2(t) \leq -\alpha Z^2(t,s).$$

Thus,

$$0 \leq V(t) \leq 1 - \alpha \int_s^t Z^2(u,s)du$$

and

$$\int_s^t Z^2(u,s)du \leq 1/\alpha$$

for $0 \leq s \leq t < \infty$.

Proof. We have

$$\begin{aligned} V'(t) &= 2A(t)z^2(t) + 2z(t)\int_s^t B(t,u)z(u)du \\ &- \int_s^t B_{ut}(t,u)\left(\int_u^t z(v)dv\right)^2 du \\ &- 2z(t)\int_s^t B_u(t,u)\int_u^t z(v)dvdu \\ &- B_t(t,s)\left(\int_s^t z(u)du\right)^2 - 2z(t)B(t,s)\int_s^t z(u)du. \end{aligned}$$

We integrate the third-to-last term by parts and obtain

$$\begin{aligned} &-2z(t)\left[B(t,u)\int_u^t z(v)dv\Big|_s^t + \int_s^t B(t,u)z(u)du\right] \\ &\quad = -2z(t)\left[-B(t,s)\int_s^t z(v)dv + \int_s^t B(t,u)z(u)du\right]. \end{aligned}$$

Cancel terms and obtain

$$\begin{aligned} V'(t) = 2A(t)z^2(t) &- \int_s^t B_{ut}(t,u)\left(\int_u^t z(v)dv\right)^2 du \\ &- B_t(t,s)\left(\int_s^t z(u)du\right)^2. \end{aligned}$$

The conclusion follows from this.

Corollary 3.7.5.5. *Let the conditions of Theorem 3.7.5.4 hold and suppose that there is a continuous function $k : [0,\infty) \to [0,\infty)$ with $k(t) \leq \alpha$,*

$$-B_{ut}(t,u) \leq k(t)B_u(t,u)$$

and

$$-B_t(t,s) \leq k(t)B(t,s).$$

Then for V defined in Theorem 3.7.5.4 we have $V' \leq -k(t)V$ and $Z^2(t,s) \leq Z^2(s,s)e^{-\int_s^t k(u)du}$.

We turn now to the fixed point methods of Theorem 3.7.3.3, set $a(t) = 0$, and develop a weighted norm which will yield exponential decay on $Z(t,s)$. The reader will recognize that there are many alternatives to our presentation. Basically, one chooses a norm of the form $|\phi|_r = \sup_{t\geq s}|\phi(t)|/r(t)$

where $r : [0, \infty) \to (0, \infty)$ and $r \in L^2[0, \infty)$. The task then is to show that a certain mapping, which we define below, will map the resulting space into itself and be a contraction. It is a much more flexible technique than some of our Liapunov arguments because it relies on averages instead of pointwise conditions. It is often the case that exponential decay of a solution requires exponential decay of the kernel. Here, we do require such decay on the kernel.

Theorem 3.7.5.6. *Suppose there is a $\gamma > 0$ and an $\alpha < 1$ with $A(v)+\gamma \leq 0$ and*

$$\sup_{0\leq s\leq t<\infty} \int_s^t e^{\int_u^t [A(v)+\gamma]dv} \int_s^u |B(u,v)e^{\gamma(u-v)}|dvdu \leq \alpha.$$

Then the unique solution, $Z(t,s)$, of

$$z'(t) = A(t)z(t) + \int_s^t B(t,u)z(u)du, \quad z(s) = 1,$$

tends to zero exponentially.

Proof. Fix $s \geq 0$ and define a norm on the Banach space $(U, |\cdot|_\gamma)$ of continuous functions $\phi : [s, \infty) \to \mathbb{R}$ for which

$$\sup_{t\geq s} |\phi(t)|e^{\gamma(t-s)} =: |\phi|_\gamma$$

exists.

By the variation-of-parameters formula we have

$$z(t) = e^{\int_s^t A(u)du} + \int_s^t e^{\int_u^t A(v)dv} \int_s^u B(u,v)z(v)dvdu$$

which we use in the usual way to define a mapping P on U. Then, if $\phi \in U$ we have

$$\begin{aligned}(P\phi)(t)e^{\gamma(t-s)} &= e^{\int_s^t [A(u)+\gamma]du} \\ &\quad + \int_s^t e^{\int_u^t [A(v)+\gamma]dv} \int_s^u B(u,v)e^{\gamma(u-v)}\phi(v)e^{\gamma(v-s)}dvdu\end{aligned}$$

and

$$\begin{aligned}&|(P\phi)(t)e^{\gamma(t-s)}| \\ &\leq 1 + \int_s^t e^{\int_u^t [A(v)+\gamma]dv} \int_s^u |B(u,v)e^{\gamma(u-v)}||\phi(v)|e^{\gamma(v-s)}dvdu\end{aligned}$$

so that $P\phi \in U$. Next, if $\phi, \eta \in U$ then

$$\begin{aligned}
&|(P\phi)(t) - (P\eta)(t)|e^{\gamma(t-s)} \\
&\leq \int_s^t e^{\int_u^t [A(v)+\gamma]dv} \int_s^u |B(u,v)e^{\gamma(u-v)}||\phi(v) - \eta(v)|e^{\gamma(v-s)}dvdu \\
&\leq \alpha|\phi - \eta|_\gamma
\end{aligned}$$

a contraction. Thus, there is a fixed point and it is in U, completing the proof.

3.8 Nonlinearities

This section involves a fairly simple substitution of $g(x)$ in the integral for x and then suggests some changes in the Liapunov functional. Our work will center on the scalar equation

$$x(t) = a(t) - \int_0^t C(t,s)g(x(s))ds \tag{3.8.1}$$

where $a : [0,\infty) \to \Re$, $C : [0,\infty) \times [0,\infty) \to \Re$, $g : \Re \to \Re$, with a, C, and g continuous and

$$xg(x) > 0 \text{ for } x \neq 0. \tag{3.8.2}$$

These conditions are sufficient to ensure that (3.8.1) has at least one solution which, if it remains bounded, can be continued to $[0,\infty)$. We will now offer two results to the general effect that if $a' \in L^p$ then any solution of (3.8.1) satisfies $g(x(t)) \in L^p$. Here, C_1 or C_t means the partial of $C(t,s)$ with respect to t. The first has appeared many places in the linear case, but notice how (3.8.3) integrates only the first coordinate and how well $a(t)$ and $x(t)$ separate in V'; that will be a major concern here.

Notice below that there is no growth condition on g.

Theorem 3.8.1. *Suppose that $\int_{t-s}^\infty |C_1(u+s,s)|du$ is continuous, (3.8.2) holds, $a' \in L^1[0,\infty)$, and that there is an $\alpha > 0$ with*

$$-C(t,t) + \int_0^\infty |C_1(u+t,t)|du \leq -\alpha. \tag{3.8.3}$$

If x is a solution of (3.8.1) then $g(x) \in L^1[0,\infty)$ and x is bounded so $g(x) \in L^p$ for $p \geq 1$.

Proof. If $x(t)$ solves (3.8.1) then it also solves

$$x'(t) = a'(t) - C(t,t)g(x(t)) - \int_0^t C_1(t,s)g(x(s))ds \tag{3.8.4}$$

and we define a Liapunov functional by

$$V(t) = |x(t)| + \int_0^t \int_{t-s}^{\infty} |C_1(u+s,s)|du|g(x(s))|ds. \tag{3.8.5}$$

We find that

$$\begin{aligned} V'(t) &\leq |a'(t)| - C(t,t)|g(x(t))| + \int_0^t |C_1(t,s)g(x(s))|ds \\ &+ \int_0^{\infty} |C_1(u+t,t)|du|g(x(t))| - \int_0^t |C_1(t,s))g(x(s))|ds \\ &\leq |a'(t)| - \alpha|g(x(t))|. \end{aligned}$$

Thus, so long as the solution is defined then

$$|x(t)| \leq V(t) \leq V(0) + \int_0^t |a'(s)|ds - \alpha \int_0^t |g(x(s))|ds.$$

Hence, $|x(t)| \leq V(0) + \int_0^{\infty} |a'(s)|ds$ so x can be continued on $[0,\infty)$. Moreover, $g(x) \in L^1[0,\infty)$.

In the next result note that it is $a' \in L^{2n}$, resulting in $x \in L^{2n}$. Notice also how both coordinates of C are integrated, but the burden falls on the second coordinate as n becomes large in marked contrast to the last result. These properties are so important to help us understand the richness of the nonconvolution case.

Theorem 3.8.2. *Let $\int_{t-s}^{\infty} |C_1(u+s,s)|du$ be continuous and (3.8.2) hold. Suppose there is a positive integer n with $a'(t) \in L^{2n}[0,\infty)$, a constant $\alpha > 0$, and a constant $N > 0$ with*

$$\begin{aligned} &\frac{2n-1}{2nN^{\frac{2n}{2n-1}}} - C(t,t) + \frac{2n-1}{2n}\int_0^t |C_1(t,s)|ds \\ &\qquad + \frac{1}{2n}\int_0^{\infty} |C_1(u+t,t)|du \leq -\alpha. \end{aligned} \tag{3.8.6}$$

Then the solution x of (3.8.1) is defined on $[0,\infty)$ and $g(x) \in L^{2n}[0,\infty)$. If $\int_0^x g^{2n-1}(s)ds \to \infty$ as $|x| \to \infty$ then any solution x of (3.8.1) is bounded.

Proof. Obtain (3.8.4) and define

$$V(t) = \int_0^{x(t)} g^{2n-1}(s)ds + \frac{1}{2n}\int_0^t \int_{t-s}^{\infty} |C_t(u+s,s)|du g^{2n}(x(s))ds.$$

Then we have

$$\begin{aligned}
V'(t) &= a'(t)g^{2n-1}(x(t)) \\
&- C(t,t)g^{2n}(x(t)) - g^{2n-1}(x(t))\int_0^t C_1(t,s)g(x(s))ds \\
&+ \frac{1}{2n}\int_0^{\infty} |C_1(u+t,t)|du g^{2n}(x(t)) - \frac{1}{2n}\int_0^t |C_1(t,s)|g^{2n}(x(s))ds \\
&\le \frac{(2n-1)g^{2n}(x(t))}{2nN^{\frac{2n}{2n-1}}} + \frac{(Na'(t))^{2n}}{2n} - C(t,t)g^{2n}(x(t)) \\
&+ \int_0^t |C_t(t,s)| \left[\frac{(2n-1)g^{2n}(x(t))}{2n} + \frac{g^{2n}(x(s))}{2n}\right] ds \\
&+ \frac{1}{2n}\int_0^{\infty} |C_1(u+t,t)|du g^{2n}(x(t)) - \frac{1}{2n}\int_0^t |C_1(t,s)|g^{2n}(x(s))ds \\
&\le \frac{(Na'(t))^{2n}}{2n} - \alpha g^{2n}(x(t)).
\end{aligned}$$

If we integrate the first and last terms we see that there is a positive number, K, such that so long as $x(t)$ is defined, say on $[0,T)$, then $\int_0^t g^{2n}(x(s))ds \le K$. Let $2n = p$ and $(1/p) + (1/q) = 1$. Then for such t we have

$$\begin{aligned}
|x(t)| &\le |a(t)| + \int_0^t |C(t,s)||g(x(s))|ds \\
&\le |a(t)| + \left(\int_0^t |C(t,s)|^q ds\right)^{1/q} \left(\int_0^t g^{2n}(x(s))ds\right)^{1/2n} \\
&\le |a(t)| + \left(\int_0^t |C(t,s)|^q ds\right)^{1/q} K^{1/2n};
\end{aligned}$$

thus, $x(t)$ is bounded if $T < \infty$, and this means that we can continue the solution to ∞. We now have

$$\int_0^{x(t)} g^{2n-1}(s)ds \le V(t) \le V(0) + \int_0^t \frac{(Na'(s))^{2n}}{2n} ds - \alpha \int_0^t g^{2n}(x(s))ds.$$

As the solution can be defined for all future time, $g(x) \in L^{2n}$. If the integral on the left diverges with $|x|$, then x is bounded.

This result shows that when (3.8.6) holds then for $0 < \beta < 1$ and

$$a(t) = (t+1)^{\beta} + \sin(t+1)^{\beta} + (t+1)^{1/2}\sin(t+1)^{1/3} \tag{3.8.7}$$

then $g(x) \in L^{2n}[0,\infty)$ for some $n > 0$. Now, under a compatible set of conditions on C, we let $\beta = 1$ and notice that the solution is asymptotically periodic. The big functions are largely absorbed into the equation, while the little function takes permanent control.

The fact that no condition on divergence of the integral of g is needed for boundedness in Theorem 3.8.1, but divergence is needed in Theorem 3.8.2, and is a common problem usually traceable to method of proof rather than a property of the equation. Even if $\int_0^x g(s)ds$ fails to diverge, one may use $g \in L^{2n}$ in several ways to obtain boundedness of the solution. For example, by squaring (3.8.1) we get $(1/2)x^2(t) \le a^2(t) + \int_0^t C^2(t,s)ds \int_0^t g^2(x(s))ds$.

Let $\mathcal{P}_T$ be the set of continuous $T-$periodic scalar functions and let Q be the set of continuous functions $q : [0,\infty) \to R$ such that $q(t) \to 0$ as $t \to \infty$. Denote by $(Y, \|\cdot\|)$ the Banach space of functions $\phi : [0,\infty) \to R$ where $\phi \in Y$ implies that $\phi = p + q$ with $p \in \mathcal{P}_T$ and $q \in Q$ and where $\|\cdot\|$ is the supremum norm.

In our work here, we will use a contraction mapping and that will require that g have a bounded derivative, as required in (3.8.10) below. But that is a requirement based only on the method of proof and it can likely be removed using a different kind of proof, possibly by the method of *a priori* bounds.

Suppose that $C : \Re \times \Re \to \Re$ is continuous with

$$C(t+T, s+T) = C(t,s), \tag{3.8.8}$$

that

$$\int_{-\infty}^{t} C(t,s)ds \text{ is bounded and continuous,} \tag{3.8.9}$$

and that g has a continuous derivative, denoted by g^*, with

$$|g^*(x)| \le 1. \tag{3.8.10}$$

For a fixed $\phi = p + q \in Y$ by the mean value theorem for derivatives we have

$$g(\phi(t)) = g(p(t)) + g(p(t)+q(t)) - g(p(t)) = g(p(t)) + g^*(\xi(t))q(t)$$

where $\xi(t)$ is between $p(t)+q(t)$ and $q(t)$. This means that $g(\phi) \in Y$. Note that this did not require (3.8.10) since g^* would be bounded along the fixed function, ϕ. Thus, the ideas seem fully nonlinear.

We will also require

$$\int_{-\infty}^{0} |C(t,s)|ds \to 0 \text{ as } t \to \infty \tag{3.8.11}$$

and for $q \in Q$ then

$$\int_{0}^{t} C(t,s)q(s)ds \to 0 \text{ as } t \to \infty. \tag{3.8.12}$$

In order to have a contraction we ask that there exists an $\alpha < 1$ with

$$\int_{0}^{t} |C(t,s)|ds \leq \alpha. \tag{3.8.13}$$

To prove that (3.8.1) has an asymptotically periodic solution, we begin by defining a mapping from (3.8.1) by $\phi = p + q \in Y$ implies that

$$(P\phi)(t) = a(t) - \int_{0}^{t} C(t,s)g(\phi(s))ds. \tag{3.8.14}$$

Notice that some growth of g is needed to bound the solution and that (3.8.2) is not needed.

Theorem 3.8.3. *If (3.8.8) - (3.8.13) hold and if $a \in Y$, so is the unique solution of (3.8.1) on $[0, \infty)$.*

Proof. Clearly, (3.8.14) is a contraction, but we must show that $P : Y \to Y$. Write $a = p^* + q^* \in Y$ and for $\phi = p + q \in Y$ define

$$(P\phi)(t) = p(t) - \int_{-\infty}^{t} C(t,s)g(p(s))ds - \int_{0}^{t} C(t,s)g'(\xi(s))q(s)ds$$
$$+ \int_{-\infty}^{0} C(t,s)g(p(s))ds.$$

Clearly, the first two terms on the right are periodic, while the remainder is in Q. Thus, $P : Y \to Y$ and there is a fixed point.

It may be seen that (3.8.3) and (3.8.13) are closely related, but for large n then (3.8.6) and (3.8.13) are very different. We come now to a Liapunov functional which will require much more about the behavior of C, but it is a naturally nonlinear functional. It is closely adapted from work of Levin (1963) and we have used it several times in our linear work.

Notice that growth of g is required for boundedness of the solution.

Theorem 3.8.4. *Let (3.8.2) hold, $H(t,s) := C_t(t,s)$, and suppose there is an $\alpha > 0$ with $C(t,t) \geq \alpha$ and*

$$H(t,s) \geq 0,\ H_t(t,s) \leq 0,\ H_s(t,s) \geq 0,\ H_{st}(t,s) \leq 0. \tag{3.8.15}$$

(i) If $a' \in L^2[0,\infty)$, then any solution of (3.8.1) or (3.8.4) defined on $[0,\infty)$ satisfies $g(x) \in L^2[0,\infty)$.

(ii) If a' is bounded, if $g^2(x) \to \infty$ as $|x| \to \infty$, and if there is an $M > 0$ with

$$\int_0^t H_s(t,s)(t-s)\int_s^t |a'(u)|^2 du ds + H(t,0)t\int_0^t |a'(u)|^2 du \leq M, \tag{3.8.16}$$

then any solution of (3.8.1) or (3.8.4) defined on $[0,\infty)$ has $\int_0^x g(s)ds$ bounded.

Proof. Define

$$\begin{aligned}
V(t) &= \int_0^{x(t)} g(s)ds + (1/2)\int_0^t H_s(t,s)\left(\int_s^t g(x(u))du\right)^2 ds \\
&+ (1/2)H(t,0)\left(\int_0^t g(x(s))ds\right)^2 \\
&= \int_0^{x(t)} g(s)ds + (1/2)\int_0^t C_{st}(t,s)\left(\int_s^t g(x(u))du\right)^2 ds \\
&+ (1/2)C_t(t,0)\left(\int_0^t g(x(s))ds\right)^2
\end{aligned} \tag{3.8.17}$$

so that if $V(t)$ is bounded, so is $\int_0^x g(s)ds$. Next, write

$$x' = a'(t) - C(t,t)g(x(t)) - \int_0^t H(t,s)g(x(s))ds.$$

Then the derivative of V along a solution is

$$\begin{aligned}V'(t) &= a'(t)g(x) - C(t,t)g^2(x) - g(x)\int_0^t C_t(t,s)g(x(s))ds \\ &+ (1/2)\int_0^t H_{st}(t,s)\Big(\int_s^t g(x(u))du\Big)^2 ds \\ &+ g(x(t))\int_0^t H_s(t,s)\int_s^t g(x(u))duds \\ &+ (1/2)H_t(t,0)\Big(\int_0^t g(x(s))ds\Big)^2 \\ &+ H(t,0)\int_0^t g(x(s))dsg(x(t)).\end{aligned}$$

We integrate the fifth term on the right by parts and obtain

$$\begin{aligned}&g(x(t))[H(t,s)\int_s^t g(x(u))du\Big|_0^t + \int_0^t H(t,s)g(x(s))ds] \\ &= -g(x(t))H(t,0)\int_0^t g(x(u))du + \int_0^t H(t,s)g(x(s))dsg(x(t)).\end{aligned}$$

Canceling terms and taking into account sign conditions yields

$$\begin{aligned}V'(t) &\le a'(t)g(x) - C(t,t)g(x)^2 \\ &\le (1/2\alpha)|a'(t)|^2 + (\alpha/2)g^2(x) - \alpha g^2(x) \\ &\le (1/2)(|a'(t)|^2)/\alpha - \alpha g^2(x)).\end{aligned}$$

Hence,

$$2\int_0^x g(s)ds \le 2V(t) \le 2V(0) + (1/\alpha)\int_0^t |a'(s)|^2 ds - \alpha\int_0^t g^2(x(s))ds$$

so (i) follows. Note that $\int_0^x g(s)ds$ is bounded if V is bounded.

Now, assume $a'(t)$ bounded and let (3.8.16) hold; we will bound V and, hence, $\int_0^x g(s)ds$. From V' we see that there is a $\mu > 0$ such that if $V'(t) > 0$ then $|x(t)| < \mu$. Suppose, by way of contradiction, that V is not bounded. Then there is a sequence $\{t_n\} \uparrow \infty$ with $V'(t_n) \ge 0$ and $V(t_n) \ge V(s)$ for $0 \le s \le t_n$; thus, $|x(t_n)| \le \mu$. If $0 \le s \le t_n$ then

$$0 \le 2V(t_n) - 2V(s) \le -\alpha\int_s^{t_n} g^2(x(u))du + (1/\alpha)\int_s^{t_n} |a'(u)|^2 du.$$

Using these values in the formula for V, taking $|x(t_n)| \le \mu$, $t = t_n$, and applying the Schwarz inequality yields

$$V(t) \le \int_0^{\pm\mu} g(s)ds + \int_0^t H_s(t,s)(t-s)\int_s^t (1/\alpha^2)|a'(u)|^2 duds$$
$$+ H(t,0)t(1/\alpha^2)\int_0^t |a'(u)|^2 du \le \int_0^{\pm\mu} g(s)ds + (1/\alpha^2)M.$$

Thus, $V(t)$ and $\int_0^x g(s)ds$ are bounded.

We have constructed a very intricate Liapunov functional and it has exactly the derivative we sought. But it is crippled by the inconclusive requirement that the solution be defined on $[0,\infty)$. That can be cured, of course, by asking that $\int_0^x g(s)ds \to \infty$ as $|x| \to \infty$.

Note The material in Section 3.8 is taken from Burton (2007d). It was first published by the Mathematical Institute of the Slovak Academy of Sciences in the journal *Tatra Mountains Mathematical Publications* with complete details in the bibliography. It was also presented to the Conference on Differential and Difference Equations and Applications 2006 (CDDEA '06)in Rajecke Teplice, Slovakia in June of 2006.

3.9 Singular Integrodifferential equations

In this section we study a scalar nonlinear integrodifferential equation of the form

$$x'(t) = f(t) - h(t,x(t)) - \int_0^t C(t,s)q(s,x(s))ds \tag{3.9.1}$$

and also a linear vector equation, together with its resolvent. The objective is to determine qualitative properties of solutions when there is a $p \in [1,\infty)$ with

$$f \in L^p[0,\infty), \quad xh(t,x) \ge 0, \quad xq(t,x) \ge 0, \tag{3.9.2}$$

and C has a weak singularity at $t = s$.

It is to be emphasized that we introduce Liapunov functionals for integrodifferential equations with singular kernels and compare the results with parallel work done in 1975 by Grimmer and Seifert using a Razumikhin technique.

Two very different classes of problems are studied. Either C is a small kernel with no sign specified, or C is a convex kernel in the following sense. There is an $\epsilon > 0$ and for $0 \leq s \leq t - \epsilon$ we have

$$C(t,s) \geq 0, \quad C_2(t,s) \geq 0, \quad C_{2,1}(t,s) \leq 0, \quad C_1(t,0) \leq 0, \tag{3.9.3}$$

where $C_1(t,s) = C_t(t,s)$, $C_2(t,s) = C_s(t,s)$ and $C_{2,1}(t,s) = C_{s,t}(t,s)$.

The first problem is to continue the work of Section 3.7.5 to include the singular kernels. The second is to complete a project of Grimmer and Seifert (1975) in which they developed a Razumikhin technique to deal with a vector equation

$$x'(t) = Ax(t) + \int_0^t B(t,s)x(s)ds + f(t) \tag{3.9.4}$$

where A is a constant matrix which is negative definite, B is a matrix satisfying

$$\lim_{h \to 0} \int_0^t |B(t,s) - B(t+h,s)|ds = 0 \tag{3.9.5}$$

and

$$\lim_{h \to 0} \int_t^{t+h} |B(t+h,s)|ds = 0, \quad t \geq 0, \tag{3.9.6}$$

as well as a number of other conditions, some of which are listed below.

Here is the development of their question. They give conditions under which solutions of (3.9.4) will have certain qualitative properties in case f is bounded and continuous. All of that work is based on a Razumikhin technique which utilizes a Liapunov function instead of a Liapunov functional. Its central requirement is that for a constant matrix K satisfying

$$A^T K + KA = -I \text{ then } \int_0^t |KB(t,s)|ds \leq M \tag{3.9.7}$$

where M is related to the eigenvalues of K and, generally, M is small.

On the last page of their paper, Grimmer and Seifert express the desire to show that the solution of (3.9.4) is in L^p when f is in L^q for some positive integers p and q. Their conditions allow for B to have weak singularities. To solve that problem is to understand two fundamental differences between Liapunov functions and Razumikhin functions. Here are two loose principles. For a kernel $C(t,s)$, the first coordinate is the Liapunov coordinate and it relates to L^p solutions. The second coordinate is the Razumikhin coordinate and relates to L^∞ solutions.

Our second project here is to construct Liapunov functionals for (3.9.4) when B has weak singularities which will give the desired L^p properties of the solutions of (3.9.4).

In both projects, we also consider linear equations and associated resolvents.

Existence: Direct Fixed Point Mappings

The first part of this section concerns the existence of a solution of (3.9.1) with continuous derivative when C has some discontinuities. In our subsequent work we will only allow discontinuities of C at $t = s$, typified by $C(t-s) = (t-s)^{-1/2}$ which occurs so often in the literature. Our existence result here will be more general and it will rest on ideas from Burton and Zhang (1998).

The classical method for proving existence involves writing (3.9.1) as an integral equation, assuming h and g Lipschitz, introducing a weighted norm to bring the Lipschitz constants down to a contraction constant, and use the integral equation to define a fixed point mapping. Then one fixes an arbitrary interval $[0, T]$, adjusts the weighted norm for that interval, and shows that this is a contraction mapping with unique fixed point, $x(t)$ on $[0, T]$, a continuous solution of the integral equation. When C is continuous we simply differentiate the integral equation and have a solution of (3.9.1) with continuous derivative. When C is not continuous there are stern questions about that differentiation and its consequences. While those questions may be answered, it is often difficult and troubling. There is a simple way around it.

In this section we show that by using (3.9.1) itself as the mapping equation we totally avoid those questions and have a solution of (3.9.1) with a continuous derivative. There is a pleasant surprise along the way. We naturally look at h and see no way to introduce that weighted norm to bring the Lipschitz constant down to a contraction constant. The reader will see an unexpected way to avoid that problem.

For (3.9.1) we suppose that $f : [0, \infty) \to \Re^n$ is continuous, $h, g : [0, \infty) \times \Re^n \to \Re^n$ are both continuous and both satisfy a global Lipschitz condition for the same constant K.

Theorem 3.9.1. *In addition to these continuity conditions, let $C(t, s)$ be weakly singular on Ω in the sense of Definition 1.2.1. Suppose also that for each $T > 0$ and each $k \in (0, 1)$, there is a constant $\gamma_1 > 0$ with*

$$\int_0^t e^{-\gamma_1(t-s)}|C(t, s)|ds \le k$$

for $t \in [0, T]$. Then for every $x_0 \in \Re^n$ (3.9.1) has a unique solution $x(t)$ with a continuous derivative and satisfying $x(0) = x_0$.

Proof. Let $T > 0$ and $x_0 \in \Re^n$ be given and let $(Y, \|\cdot\|)$ be the Banach space of continuous functions $\phi : [0, T] \to \Re^n$ with the supremum norm. Define $P : Y \to Y$ by $\phi \in Y$ implies that

$$(P\phi)(t) = f(t) - h(t, x_0 + \int_0^t \phi(s)ds) - \int_0^t C(t,s)q(s, x_0 + \int_0^s \phi(u)du)ds.$$

By the continuity assumptions and the weak singularity, $P\phi \in Y$. As the existence of γ_1 implies that for any $\gamma > \gamma_1$ we also have $\int_0^t e^{-\gamma(t-s)}|C(t,s)|ds \leq k$ (see Lemma 3.9.1 below), we will define a weighted norm $\|\cdot\|_T$ by $\phi \in Y$ implies that

$$\|\phi\|_T = \sup_{0 \leq t \leq T} e^{-\gamma t}|\phi(t)|,$$

where $\gamma \geq \gamma_1$ is yet to be chosen. Note that $(Y, \|\cdot\|_T)$ is a Banach space.

If $\phi, \eta \in Y$ then

$$\begin{aligned}
&|(P\phi)(t) - (P\eta)(t)|e^{-\gamma t} \\
&\leq e^{-\gamma t}\left[\left|h(t, x_0 + \int_0^t \phi(s)ds) - h(t, x_0 + \int_0^t \eta(s))ds\right|\right. \\
&\left. + \int_0^t |C(t,s)|\left|q(s, x_0 + \int_0^s \phi(u)du) - q(s, x_0 + \int_0^s \eta(u)du)\right|ds\right] \\
&\leq e^{-\gamma t}K\int_0^t |\phi(s) - \eta(s)|ds \\
&+ e^{-\gamma t}K\int_0^t |C(t,s)|\int_0^s |\phi(u) - \eta(u)|duds \\
&= K\int_0^t e^{-\gamma(t-s)}e^{-\gamma s}|\phi(s) - \eta(s)|ds \\
&+ K\int_0^t |C(t,s)|e^{-\gamma(t-s)}e^{-\gamma s}\int_0^s |\phi(u) - \eta(u)|duds.
\end{aligned}$$

Now the last line yields

$$\begin{aligned}
&K\int_0^t |C(t,s)|e^{-\gamma(t-s)}e^{-\gamma s}\int_0^s |\phi(u)-\eta(u)|duds\\
&\leq K\int_0^t |C(t,s)|e^{-\gamma(t-s)}\int_0^s e^{-\gamma(s-u)}e^{-\gamma u}|\phi(u)-\eta(u)|duds\\
&\leq K\|\phi-\eta\|_T\int_0^t T|C(t,s)|e^{-\gamma(t-s)}ds\\
&= \|\phi-\eta\|_T KT\int_0^t |C(t,s)|e^{-\gamma(t-s)}ds\\
&\leq \|\phi-\eta\|_T KTk.
\end{aligned}$$

We then have

$$\begin{aligned}
&|(P\phi)(t)-(P\eta)(t)|e^{-\gamma t}\\
&\qquad\leq K\|\phi-\eta\|_T\int_0^t e^{-\gamma(t-s)}ds+\|\phi-\eta\|_T KTk\\
&\qquad\leq \|\phi-\eta\|_T\left[K\frac{e^{-\gamma(t-s)}}{\gamma}\bigg|_0^t+KTk\right]\\
&\qquad\leq \|\phi-\eta\|_T\Big[(K/\gamma)+KTk\Big].
\end{aligned}$$

Now, take k so small and γ so large that $(K/\gamma)+KTk\leq 1/2$. Thus, we have a contraction and a unique $\phi\in Y_T$ with $P\phi=\phi$ and, clearly, $\left[x_0+\int_0^t\phi(s)ds\right]'=\phi(t)$. That unique continuous ϕ is the continuous derivative of that unique solution.

Definition 1.2.1 is far more general than we will be needing here. We will allow a singularity only at $t=s$ and we will have a corresponding condition. Our next result shows that the new condition is satisfied if condition (*) is satisfied.

Lemma 3.9.1. *Let $C(t,s)$ be a weakly singular kernel on the set Ω and fix $T>0$. Moreover, suppose that for any $k\in(0,1)$ there exists an $\epsilon:=\epsilon(k,T)>0$ such that*

$$\int_{t-\epsilon}^t |C(t,s)|ds\leq k \text{ for all } t\in[0,T], \tag{*}$$

where we have set $C(t,s) = 0, (t,s) \in \Re - \Omega$. *Then there always exists a* $\gamma_{k,T} > 0$ *such that for any* $\gamma \geq \gamma_{k,T}$ *we have*

$$\int_0^t e^{-\gamma(t-s)}|C(t,s)|ds \leq k \text{ for all } t \in [0,T].$$

Proof. Let T be given and let $k \in (0,1)$ be chosen. Clearly, it follows immediately from (*) that for $t \in [0,\epsilon]$ our result is true for any $\gamma > 0$, so for the rest of the proof we assume that $t \geq \epsilon$. As the function

$$D(t) = \int_0^t |C(t,s)|ds, \ t \in [0,T]$$

is continuous, there exists an $M := M_T$ with

$$\int_0^t |C(t,s)|ds \leq M, \ t \in [0,T].$$

In view of (*) let

$$\epsilon = \epsilon(\frac{k}{2}, T)$$

and (provided that $k < 2M$, otherwise take any $\gamma_{k,T} > 0$) set

$$\gamma_{k,T} := \frac{1}{\epsilon} \ln\left(\frac{k}{2M}\right).$$

For $t \in [\epsilon, T]$ and $\gamma \geq \gamma_{k,T}$ we have

$$\begin{aligned}
&\int_0^t e^{-\gamma(t-s)}|C(t,s)|ds \\
&= \int_0^{t-\epsilon} e^{-\gamma(t-s)}|C(t,s)|ds + \int_{t-\epsilon}^t e^{-\gamma(t-s)}|C(t,s)|ds \\
&\leq e^{-\gamma t}\int_0^{t-\epsilon} e^{\gamma s}|C(t,s)|ds + \int_{t-\epsilon}^t |C(t,s)|ds \\
&\leq e^{-\gamma t}\int_0^{t-\epsilon} e^{\gamma(t-\epsilon)}|C(t,s)|ds + k/2 \\
&= e^{-\gamma t+\gamma(t-\epsilon)}\int_0^{t-\epsilon} |C(t,s)|ds + k/2 \\
&= e^{-\gamma\epsilon}\int_0^{t-\epsilon} |C(t,s)|ds + k/2 \\
&\leq e^{-\gamma\epsilon}\int_0^{t} |C(t,s)|ds + k/2 \\
&\leq e^{-\gamma\epsilon}M + k/2.
\end{aligned}$$

That is,

$$\int_0^t e^{-\gamma(t-s)}|C(t,s)|ds \leq e^{-\gamma\epsilon}M + k/2. \qquad (**)$$

Noting that

$$\gamma_{k,T} = -\frac{1}{\epsilon}\ln\left(\frac{k}{2M}\right) \leq \gamma,$$
$$-\gamma\epsilon \leq \ln\left(\frac{k}{2M}\right),$$
$$e^{-\gamma\epsilon}M \leq k/2,$$

from (**) it then follows that

$$\int_0^t e^{-\gamma(t-s)}|C(t,s)|ds \leq k, \text{ for } t \in [0,T]$$

which proves our assertion.

Note For

$$C(t,s) = (t-s)^{-q}, \; q \in (0,1)$$

we have for an $\epsilon > 0$ and any $t \geq \epsilon$

$$\int_{t-\epsilon}^t |C(t,s)|ds = \int_{t-\epsilon}^t (t-s)^{-q}ds$$
$$= \left[-\frac{(t-s)^{1-q}}{1-q}\right]_{t-\epsilon}^t$$
$$= \frac{\epsilon^{1-q}}{1-q}$$

from which it follows that for kernels of type $(t-s)^{-q}$ the condition (*) is always satisfied taking

$$\frac{\epsilon^{1-q}}{1-q} = k \iff \epsilon = [k(1-q)]^{\frac{1}{1-q}}.$$

Note that ϵ is independent of T.

In the following material we will assume that the Liapunov results are being applied to problems in which existence has been established.

Theorem 3.9.2. *Let (3.9.2) hold with $p = 1$. Suppose there is a $\gamma > 0$ with $|h(t,x)| \geq \gamma |q(t,x)|$ on $[0,\infty) \times \Re$. Suppose also that there is a $\beta > 0$ so that for each $\epsilon > 0$ we have $\int_\epsilon^\infty |C(u+t,t)|du \leq \beta$ for all $t \geq 0$, where $\gamma - \beta =: \mu > 0$. Finally, if there is an $\eta < \mu$ and a fixed $\epsilon > 0$ with*

$$\int_s^t |C(u+\epsilon,s) - C(u,s)|du \leq \eta \text{ for } 0 \leq s \leq t < \infty,$$

then any solution $x(t)$ of (3.9.1) on $[0,\infty)$ satisfies $q(t,x(t)) \in L^1[0,\infty)$.

Proof. For the fixed $\epsilon > 0$, define a Liapunov functional

$$V(t,\epsilon) = |x(t)| + \int_0^t \left[\int_{t-s+\epsilon}^\infty |C(u+s,s)|du\right] |q(s,x(s))|ds$$

so that since

$$-|C(t+\epsilon,s)| \leq -|C(t,s)| + |C(t+\epsilon,s) - C(t,s)|$$

we have

$$\begin{aligned}
V'(t,\epsilon) &\leq |f(t)| - |h(t,x(t))| + \int_0^t |C(t,s)g(s,x(s))|ds \\
&+ \int_\epsilon^\infty |C(u+t,t)|du|q(t,x(t))| - \int_0^t |C(t+\epsilon,s)||q(s,x(s))|ds \\
&\leq |f(t)| - \gamma|q(t,x(t))| + \int_0^t |C(t,s)q(s,x(s))|ds \\
&+ \beta|q(t,x(t))| - \int_0^t |C(t,s)q(s,x(s))|ds \\
&+ \int_0^t |C(t+\epsilon,s) - C(t,s)||q(s,x(s))|ds \\
&= |f(t)| - \mu|q(t,x(t))| + \int_0^t |C(t+\epsilon,s) - C(t,s)||q(s,x(s))|ds.
\end{aligned}$$

In preparation for integration of this expression we calculate

$$\begin{aligned}
&\int_\epsilon^t \int_0^u |C(u+\epsilon,s) - C(u,s)||q(s,x(s))|dsdu \\
&\quad \leq \int_0^t \int_s^t |C(u+\epsilon,s) - C(u,s)|du|q(s,x(s))|ds \\
&\quad \leq \int_0^t \eta|q(s,x(s)|ds.
\end{aligned}$$

With this conclusion in hand, we now integrate V' obtaining

$$\begin{aligned}
V(t,\epsilon) &\leq V(\epsilon,\epsilon) + \int_\epsilon^t |f(u)|du - \mu \int_\epsilon^t |q(s,x(s))|ds \\
&+ \int_\epsilon^t \int_0^u |C(u+\epsilon,s) - C(u,s)||q(s,x(s))|dsdu \\
&\leq V(\epsilon,\epsilon) + \int_0^t |f(u)|du \\
&- (\mu - \eta)\int_\epsilon^t |q(s,x(s))|ds + \eta \int_0^\epsilon |q(s,x(s))|ds.
\end{aligned}$$

This completes the proof.

This case with $p = 1$ is very simple and the proof is very short. It illustrates the type of results obtained in Burton and Purnaras (2012) which is not covered in this book. We offered it here as a contrast to the work in the rest of this section. Here, we consider kernels which would be convex, but they involve a singularity at $t = s$. There is no limit to their magnitude. Our work here is an extension to the singular case of a line of research by Levin (1963) which continued for many years. The Liapunov functional was suggested by Volterra (1928). Our next result does not contain a singularity, but it does introduce a new differential inequality relation and it is used primarily to show that the singularity causes us to add a term to the equation very much like the Ax of Grimmer and Seifert and for the same reason. Moreover, we streamline the proof so that the reader can see with ease exactly what techniques are involved. The techniques of these first two proofs include many of the ideas needed in the rest of the section. In the case of small kernels we can allow many values for p, but in the convex case we are restricted to $p = 2$. The results for the convex case are extendable to vector equations, as may be verified by consulting Burton (1980) for the nonsingular case. However, the details are very lengthy.

We begin by showing that with a nonsingular convex kernel then we can obtain $x \in L^\infty$ when $f \in L^1[0,\infty)$ and it allows $A = 0$. There is nothing linear about the result. It can be proved for $x' = f(t) - \int_0^t C(t,s)g(x(s))ds$ where $xg(x) > 0$ if $x \neq 0$, but we must extract a differential inequality from a relation $V' \leq |f(t)||g(x)|$ of a type which will be clearly seen below and that is difficult unless $g(x) = x^n$ with n an odd positive integer. The details of differentiation of V are not simple, but are parallel to those given in full in the proof of Theorem 3.9.4.

Theorem 3.9.3. *Let*

$$x'(t) = f(t) - \int_0^t C(t,s)x(s)ds$$

with C convex (this means that (3.9.3) holds and $\epsilon = 0$) and with $f \in L^1[0,\infty)$. Then $x \in L^\infty[0,\infty)$.

Proof. Define

$$V(t) = x^2(t) + \int_0^t C_2(t,s)\left(\int_s^t x(u)du\right)^2 ds + C(t,0)\left(\int_0^t x(s)ds\right)^2$$

and using the chain rule and Leibnitz's rule for differentiation we obtain

$$V'(t) = 2x(t)f(t) + \int_0^t C_{2,1}(t,s)\left(\int_s^t x(u)du\right)^2 ds \tag{3.9.8}$$
$$+ C_1(t,0)\left(\int_0^t x(s)ds\right)^2 \le 2x(t)f(t) \le 2\sqrt{V(t)}|f(t)|.$$

Separation of variables in (3.9.8) yields

$$V^{-1/2}(t)V'(t) \le 2|f(t)|$$

and so

$$2|x(t)| \le 2V^{1/2}(t) \le 2V^{1/2}(0) + 2\int_0^t |f(s)|ds.$$

The singular convex case

There is a pleasant surprise as we proceed from integral equations to integrodifferential equations. In parallel work on integral equations with weakly singular kernels in Chapter 2 we required that $|q(t,x)| \le |x|$ for integral equations, and that is not needed here.

Theorem 3.9.4. *Let x be a continuous solution of (3.9.1) on $[0,\infty)$ and let (3.9.2) and (3.9.3) be satisfied. If $\epsilon > 0$ is chosen so that (3.9.3) is satisfied and if $V(t,\epsilon)$ is defined for $t \geq \epsilon$ by*

$$V(t,\epsilon) = 2\int_0^{x(t)} q(t,s)ds + \int_0^{t-\epsilon} C_2(t,s)\Big(\int_s^t q(u,x(u))du\Big)^2 ds + C(t,0)\Big(\int_0^t q(u,x(u))du\Big)^2 \tag{3.9.9}$$

we have

$$\frac{dV(t,\epsilon)}{dt} \leq 2\int_0^{x(t)} q_t(t,s)ds \tag{3.9.10}$$
$$+ 2q(t,x(t))\Big[f(t) - h(t,x(t)) + C(t,t-\epsilon)\int_{t-\epsilon}^t q(u,x(u))du - \int_{t-\epsilon}^t C(t,s)q(s,x(s))ds\Big] + C_2(t,t-\epsilon)\Big(\int_{t-\epsilon}^t q(u,x(u))du\Big)^2.$$

Proof. Let x be a continuous solution of (3.9.1) on $[0,\infty)$. For any $t \geq \epsilon$ we have $C_1(t,0) \leq 0$ and $C_{2,1}(t,s) \leq 0$ when $0 \leq s \leq t-\epsilon$ so by Leibnitz's rule and the chain rule we have

$$\begin{aligned}
V'(t,\epsilon) \leq{}& 2\int_0^{x(t)} q_t(t,s)ds + C_2(t,t-\epsilon)\Big(\int_{t-\epsilon}^t q(u,x(u))du\Big)^2 \\
&+ 2q(t,x(t))\Big[f(t) - h(t,x(t)) - \int_0^t C(t,s)q(s,x(s))ds\Big] \\
&+ 2q(t,x(t))\int_0^{t-\epsilon} C_2(t,s)\int_s^t q(u,x(u))duds \\
&+ 2q(t,x(t))C(t,0)\int_0^t q(u,x(u))du.
\end{aligned}$$

Integrating the next-to-last term by parts yields

$$
\begin{aligned}
&2q(t,x(t))\int_0^{t-\epsilon} C_2(t,s)\int_s^t q(u,x(u))duds \\
&= 2q(t,x(t))\left[C(t,s)\int_s^t q(u,x(u))du\Big|_0^{t-\epsilon} \right.\\
&\left. + \int_0^{t-\epsilon} C(t,s)q(s,x(s))ds\right] \\
&= 2q(t,x(t))\left[C(t,t-\epsilon)\int_{t-\epsilon}^t q(u,x(u))du \right.\\
&\left. - C(t,0)\int_0^t q(u,x(u))du + \int_0^{t-\epsilon} C(t,s)q(s,x(s))ds\right].
\end{aligned}
$$

We cancel two terms and obtain

$$
\begin{aligned}
&V'(t,\epsilon) \le 2\int_0^{x(t)} q_t(t,s)ds \\
&+ 2q(t,x(t))\left[f(t) - h(t,x(t)) - \int_0^t C(t,s)q(s,x(s))ds\right] \\
&+ C_2(t,t-\epsilon)\left(\int_{t-\epsilon}^t q(u,x(u))du\right)^2 \\
&+ 2q(t,x(t))\left[C(t,t-\epsilon)\int_{t-\epsilon}^t q(u,x(u))du \right.\\
&\left. + \int_0^{t-\epsilon} C(t,s)q(s,x(s))ds\right].
\end{aligned}
$$

Now, write that last integral as

$$
\begin{aligned}
\int_0^{t-\epsilon} C(t,s)q(s,x(s))ds = &\int_0^t C(t,s)q(s,x(s))ds \\
&- \int_{t-\epsilon}^t C(t,s)q(s,x(s))ds
\end{aligned}
$$

and cancel two terms. This will yield

$$\begin{aligned}
V'(t,\epsilon) &\le 2\int_0^{x(t)} q_t(t,s)ds + C_2(t,t-\epsilon)\left(\int_{t-\epsilon}^t q(u,x(u))du\right)^2 \\
&+ 2q(t,x(t))\left[C(t,t-\epsilon)\int_{t-\epsilon}^t q(u,x(u))du \right. \\
&- \left.\int_{t-\epsilon}^t C(t,s)q(s,x(s))ds\right] \\
&+ 2q(t,x(t))[f(t)-h(t,x(t))],
\end{aligned}$$

as required.

Three relations will be needed for us to parlay this Liapunov functional derivative into a qualitative result for a solution of (3.9.1).

In our opening theorem we saw that when the kernel is nonsingular, then we could take $h(t,x) \equiv 0$. Thus, we can expect that the larger the singularity, the more we will need from $h(t,x)$. Our assumption is that there is a $\gamma > 0$ with

$$|h(t,x)| \ge \gamma |q(t,x)|, \quad 0 \le t < \infty, \quad x \in \Re. \tag{3.9.11}$$

This condition bears some loose relation to the Grimmer-Seifert condition (3.9.7) and Theorem 3.9.2 shows that it is needed only because of the singularity. But notice the weakness of (3.9.2) in that $xq(t,x) \ge 0$ so that h can be zero for any value of x. We have not lost the essential properties of Theorem 3.9.2.

We now verify a certain relation which will be needed in the middle of the proof of the next theorem. It will assist in the flow of logic of the argument.

Claim 1 *If (3.9.11) and the sign assumptions in (3.9.2) hold, then for $(t,x) \in [0,\infty) \times \Re$ we have*

$$2q(t,x(t))[f(t)-h(t,x(t))] \le \frac{1}{\gamma}f^2(t) - \gamma q^2(t,x(t)).$$

Proof. From the sign properties in (3.9.2) we have

$$\begin{aligned}
&|h(t,x(t))| \ge \gamma|q(t,x(t))| \\
&|h(t,x(t))||q(t,x(t))| \ge \gamma|q(t,x(t))||q(t,x(t))| \\
&h(t,x(t))q(t,x(t)) \ge \gamma q^2(t,x(t)) \\
&-h(t,x(t))q(t,x(t)) \le -\gamma q^2(t,x(t))
\end{aligned}$$

and so

$$\begin{aligned}2q(t,x(t))[f(t)-h(t,x(t))] &= 2q(t,x(t))f(t) - 2q(t,x(t))h(t,x(t))\\ &\le \gamma q^2(t,x(t)) + \frac{1}{\gamma}f^2(t) - 2\gamma q^2(t,x(t))\\ &= \frac{1}{\gamma}f^2(t) - \gamma q^2(t,x(t)).\end{aligned}$$

Next, we ask for $\alpha > 0, \quad \beta > 0$ with $\alpha + \beta < \gamma$ and

$$\int_s^{s+\epsilon} [\epsilon C_2(u,u-\epsilon) + C(u,u-\epsilon) + |C(u,s)|]du < \alpha \tag{3.9.12}$$

for $0 \le s < \infty$. For technical reasons we ask $C(t,s) = 0$ for $s < 0$ and $t \ge 0$. Also, let

$$C(t,t-\epsilon)\epsilon + \int_{t-\epsilon}^t |C(t,s)|ds < \beta \tag{3.9.13}$$

for $\epsilon \le t < \infty$. Note in (3.9.3) that we do not specify the sign of $C(s,s)$ so the absolute value is needed in these relations.

Remark Conditions (3.9.12) and (3.9.13) allow for gross singularities to occur at $s = t$. We have already noticed that $C_2(u, u-\epsilon)$ occurring in the derivative of V is always well away from the singularity and (3.9.2) holds for it. While (3.9.12) and (3.9.13) occur as technical necessities in a later computation, repeated again and again in this work here, one would really like at least a theoretical rational for them. To begin with, for weak singularities such as $C(t,s) = (t-s)^{-p}$ for $0 < p < 1$, those conditions are too lenient to make any sense; for in that case $\alpha + \beta \to 0$ as $\epsilon \to 0$.

The next claim will again assist in the flow of the logic in the proof of the next result.

Claim 2 *Using (3.9.12) and the Hobson-Tonelli test we can verify the relation*

$$\begin{aligned}\int_\epsilon^t \int_{u-\epsilon}^u [\epsilon C_2(u,u-\epsilon) + C(u,u-\epsilon) + |C(u,s)|]q^2(s,x(s))dsdu&\\ \le \int_0^t \alpha q^2(s,x(s))ds.&\end{aligned} \tag{3.9.14}$$

Proof. We have

$$\begin{aligned}
&\int_{\epsilon}^{t}\int_{u-\epsilon}^{u}[\epsilon C_2(u,u-\epsilon)+C(u,u-\epsilon)+|C(u,s)|]q^2(s,x(s))dsdu\\
&\leq \int_{0}^{t}\int_{s}^{s+\epsilon}[\epsilon C_2(u,u-\epsilon)+C(u,u-\epsilon)+|C(u,s)|]q^2(s,x(s))duds\\
&\leq \int_{0}^{t}\alpha q^2(s,x(s))ds.
\end{aligned}$$

When we treated the parallel problem for integral equations in Chapter 2 using Liapunov functionals and assumptions very much like the ones given here the function $q(t,x)$ was allowed to depend on t in a very natural way and caused no difficulty. However, investigators going all the way back to Levin (1963) have been forced to require that q be independent of t. That is a definite defect and one which we try to rectify here. Several things need to be said about the term in the Liapunov function $\int_0^{x(t)} q(t,s)ds$ and, in order to not break the flow of ideas here, we will discuss this in the example at the end of the section. This will include a very instructive example of q. In this result, if $q(t,x)$ is independent of t and if $\int_0^x q(t,s)ds \to \infty$ as $|x| \to \infty$ then it does yield a bounded solution, just as was the case in Levin's original theorem.

Theorem 3.9.5. *Let x be a continuous solution of (3.9.1) on $[0,\infty)$ and let (3.9.2), (3.9.3), (3.9.11)-(3.9.13) hold. Moreover, assume that*

$$xq_t(t,x) \leq 0 \text{ for } t \in [0,\infty),\ x \in \Re, \tag{q_1}$$

$$\left|\int_0^{\pm\infty} q_t(t,s)ds\right| < \infty \text{ for each fixed } t \in [0,\infty)\ , \tag{q_2}$$

and that for the function

$$Q(t) = \max\left\{\left|\int_0^{\pm\infty} q_t(t,s)ds\right|\right\}, \quad t \in [0,\infty) \tag{Q}$$

we have $Q \in L^1(0,\infty)$. If, in addition, $f \in L^2[0,\infty)$ so are $q(t,x(t))$ and $q(t,x(t)) - f(t)$.

Proof. We begin by organizing the derivative of V which we computed in (3.9.10). First, by the Schwarz inequality we have

$$C_2(t,t-\epsilon)\left(\int_{t-\epsilon}^{t} q(u,x(u))du\right)^2 \le \epsilon C_2(t,t-\epsilon)\int_{t-\epsilon}^{t} q^2(u,x(u))du.$$

Next,

$$\begin{aligned}
&|2q(t,x(t))C(t,t-\epsilon)\int_{t-\epsilon}^{t} q(u,x(u))du| \\
&\qquad \le C(t,t-\epsilon)\int_{t-\epsilon}^{t} [q^2(t,x(t))+q^2(u,x(u))]du
\end{aligned}$$

and

$$|2q(t,x(t))\int_{t-\epsilon}^{t} C(t,s)q(s,x(s))ds| \le \int_{t-\epsilon}^{t} C(t,s)[q^2(t,x(t))+q^2(s,x(s))]ds.$$

These three relations along with the result of Claim 1 in (3.9.10) yield

$$\begin{aligned}
V'(t,\epsilon) &\le 2\left|\int_0^{x(t)} q_t(t,s)ds\right| + C_2(t,t-\epsilon)\left(\int_{t-\epsilon}^{t} q(u,x(u))du\right)^2 \\
&+ 2q(t,x(t))\left[C(t,t-\epsilon)\left(\int_{t-\epsilon}^{t} q(u,x(u))du\right)\right. \\
&\left. - \int_{t-\epsilon}^{t} C(t,s)q(s,x(s))ds\right] + 2q(t,x(t))[f(t)-h(t,x(t))] \\
&\le 2Q(t) + \epsilon C_2(t,t-\epsilon)\left(\int_{t-\epsilon}^{t} q^2(u,x(u))du\right) \\
&+ C(t,t-\epsilon)\int_{t-\epsilon}^{t} [q^2(t,x(t))+q^2(s,x(s))]ds \\
&+ \int_{t-\epsilon}^{t} |C(t,s)|[q^2(t,x(t))+q^2(s,x(s))]ds + \frac{1}{\gamma}f^2(t) - \gamma q^2(t,x(t)) \\
&= 2Q(t) + q^2(t,x(t))\int_{t-\epsilon}^{t} [C(t,t-\epsilon)+|C(t,s)|]ds \\
&+ \int_{t-\epsilon}^{t} [\epsilon C_2(t,t-\epsilon)+C(t,t-\epsilon)+|C(t,s)|]q^2(s,x(s))ds \\
&+ \frac{1}{\gamma}f^2(t) - \gamma q^2(t,x(t)).
\end{aligned}$$

That is,

$$V'(t,\epsilon) \le 2Q(t) + q^2(t,x(t))\beta + \frac{1}{\gamma}f^2(t) - \gamma q^2(t,x(t))$$
$$+ \int_{t-\epsilon}^{t} [\epsilon C_2(t,t-\epsilon) + C(t,t-\epsilon) + |C(t,s)|]q^2(s,x(s))ds.$$

Integrating from ϵ to t and invoking Claim 2 yields

$$\begin{aligned} V(t,\epsilon) - V(\epsilon,\epsilon) &\le 2\int_{\epsilon}^{t} Q(s)ds + \frac{1}{\gamma}\int_{\epsilon}^{t} f^2(s)ds \\ &- [\gamma - \beta]\int_{s}^{t} q^2(s,x(s))ds + \int_{0}^{t} \alpha q^2(s,x(s))ds \\ &= 2\int_{\epsilon}^{t} Q(s)ds + \frac{1}{\gamma}\int_{\epsilon}^{t} f^2(s)ds \\ &- [\gamma - \beta - \alpha]\int_{\epsilon}^{t} q^2(s,x(s))ds \end{aligned}$$

from which we have

$$\begin{aligned} V(t,\epsilon) + [\gamma - \beta - \alpha]&\int_{\epsilon}^{t} q^2(s,x(s))ds \\ &\le V(\epsilon,\epsilon) + 2\int_{\epsilon}^{t} Q(s)ds + \frac{1}{\gamma}\int_{\epsilon}^{t} f^2(s)ds. \end{aligned}$$

As x is continuous and ϵ is a positive number, it follows that $x(\epsilon)$ is finite so by the continuity of $q(t,s)$ we see that $\int_0^{x(\epsilon)} q(\epsilon,s)ds < \infty$, thus

$$V(\epsilon,\epsilon) = 2\int_{0}^{x(\epsilon)} q(\epsilon,s)ds + C(\epsilon,0)\left(\int_{0}^{\epsilon} q(u,x(u))du\right)^2 < \infty.$$

Hence, for any $t \ge \epsilon$ we have

$$\int_{\epsilon}^{t} q^2(s,x(s))ds \le \frac{1}{\gamma - \beta - \alpha}\left[V(\epsilon,\epsilon) + 2\int_{\epsilon}^{t} Q(s)ds + \frac{1}{\gamma}\int_{0}^{\infty} f^2(s)ds\right]$$

which proves our assertion.

The Resolvent Let C be an $n \times n$ matrix of functions with weak singularities and consider

$$x'(t) = Ax(t) - \int_0^t C(t,s)x(s)ds + f(t), \quad x(0) = x_0, \tag{3.9.15}$$

where A is an $n \times n$ constant matrix all of whose characteristic roots have negative real parts. There is then an $n \times n$ symmetric matrix B with

$$A^T B + BA = -I. \tag{3.9.16}$$

Associated with (3.9.15) is the resolvent equation

$$\frac{d}{dt} Z(t,s) = AZ(t,s) - \int_s^t C(t,u)Z(u,s)du, \quad Z(s,s) = I, \tag{3.9.17}$$

whose columns are the vector equations

$$z'(t,s) = Az(t,s) - \int_s^t C(t,u)z(u,s)du. \tag{3.9.18}$$

There is then the variation of parameters formula

$$x(t) = Z(t,0)x_0 + \int_0^t Z(t,s)f(s)ds. \tag{3.9.19}$$

We focus on three fundamental results.

(i) If we can show that there is an $M > 0$ with $\int_0^t |Z(t,s)|ds \leq M$, then for $f \in L^\infty$, we see that there is the bounded solution of (3.9.15), $x(t) = \int_0^t Z(t,s)f(s)ds$.

(ii) If we can show that there is an $M > 0$ with $\int_s^t |Z(u,s)|du \leq M$ and if $f \in L^1[0,\infty)$ then for $x(0) = 0$ we have $|x(t)| \leq \int_0^t |Z(t,s)f(s)|ds$ so

$$\begin{aligned}
\int_0^t |x(s)|ds &\leq \int_0^t \int_0^u |Z(u,s)||f(s)|dsdu \\
&= \int_0^t \int_s^t |Z(u,s)|du|f(s)|ds \\
&\leq M \int_0^t |f(s)|ds
\end{aligned}$$

is bounded; $f \in L^1$ yields $x \in L^1$.

(iii) If C is scalar, if there is an $M > 0$ with $\int_s^t Z^2(u,s)du \leq M$ and if $f \in L^1[0,\infty)$ then for $x(0) = 0$ we have

$$|x(t)|^2 \leq \left(\int_0^t |Z(t,s)f(s)|ds\right)^2 \leq \int_0^t |f(s)|ds \int_0^t Z^2(t,s)|f(s)|ds$$

and $x \in L^2[0,\infty)$ by the argument in (ii).

Convex kernels: scalar equations

We now revisit (3.9.15)–(3.9.19) and ask that x be a scalar function so that (3.9.16) will simplify to the statement that A is a negative constant. We also assume that (3.9.3) holds. Conditions (3.9.12) and (3.9.13) will be modified here.

Theorem 3.9.6. *Let (3.9.15) be a scalar equation, A be a negative constant, and C have a singularity at $t = s$. We assume that there exist an $\epsilon > 0$ such that for $0 \leq s \leq t - \epsilon$ then (3.9.3) holds. Suppose there exist $\alpha^* > 0$ and $\beta^* > 0$ with*

$$\int_s^{s+\epsilon} |\epsilon C_2(u, u-\epsilon) + |C(u, u-\epsilon) - C(u, s)||du < \alpha^*, \quad 0 \leq s < \infty, \tag{3.9.12*}$$

$$\int_{t-\epsilon}^{t} |C(t, t-\epsilon) - C(t, s)|ds < \beta^*, \quad \epsilon \leq t < \infty, \tag{3.9.13*}$$

and

$$A < -\frac{\alpha^* + \beta^*}{2}. \tag{3.9.20}$$

If $V(t, \epsilon)$ is defined by

$$V(t, \epsilon) = z^2(t, s) + \int_s^{t-\epsilon} C_2(t, u)\left(\int_u^t z(v, s)dv\right)^2 du + C(t, s)\left(\int_s^t z(v, s)dv\right)^2$$

for $t \geq \epsilon$ and $s \leq t - \epsilon$ and if $z(t, s)$ solves (3.9.18) then

$$z^2(t, s) + \mu \int_0^t z^2(u, s)du \leq V(s + \epsilon, \epsilon) + |2A + \beta^*| \int_0^s z^2(u, s)du.$$

That is, $z(t, s) \in L_t^2[0, \infty) \cap L_t^\infty[0, \infty)$.

Proof. We have

$$V'(t,\epsilon) = 2z(t,s)\Big[Az(t,s) - \int_s^t C(t,u)z(u,s)du\Big]$$

$$+ C_2(t,t-\epsilon)\Big(\int_{t-\epsilon}^t z(v,s)dv\Big)^2$$

$$+ \int_s^{t-\epsilon} C_{2,1}(t,u)\Big(\int_u^t z(v,s)dv\Big)^2 du$$

$$+ C_1(t,s)\Big(\int_s^t z(v,s)dv\Big)^2$$

$$+ 2z(t,s)C(t,s)\int_s^t z(v,s)dv$$

$$+ 2z(t,s)\int_s^{t-\epsilon} C_2(t,u)\int_u^t z(v,s)dvdu$$

Integration of the last term by parts yields

$$2z(t,s)\Big[C(t,u)\int_u^t z(v,s)dv\Big|_s^{t-\epsilon} + \int_s^{t-\epsilon} C(t,u)z(u,s)du\Big]$$

$$= 2z(t,s)\Big[C(t,t-\epsilon)\int_{t-\epsilon}^t z(v,s)dv - C(t,s)\int_s^t z(v,s)dv$$

$$+ \int_s^{t-\epsilon} C(t,u)z(u,s)du\Big].$$

Thus,

$$V'(t,\epsilon) \le C_2(t,t-\epsilon)\epsilon\int_{t-\epsilon}^t z^2(v,s)dv$$

$$+ 2z(t,s)\Big[Az(t,s) - \int_s^t C(t,u)z(u,s)du\Big]$$

$$+ 2z(t,s)\Big[C(t,t-\epsilon)\int_{t-\epsilon}^t z(v,s)dv + \int_s^t C(t,u)z(u,s)du$$

$$- \int_{t-\epsilon}^t C(t,u)z(u,s)du\Big]$$

$$= 2z(t,s)\Big[Az(t,s) + C(t,t-\epsilon)\int_{t-\epsilon}^t z(v,s)dv$$

$$- \int_{t-\epsilon}^t C(t,u)z(u,s)du\Big] + C_2(t,t-\epsilon)\epsilon\int_{t-\epsilon}^t z^2(v,s)dv$$

$$\leq C_2(t,t-\epsilon)\epsilon\int_{t-\epsilon}^{t} z^2(v,s)dv + 2Az^2(t,s)$$
$$+\int_{t-\epsilon}^{t} [|C(t,u) - C(t,t-\epsilon)|][z^2(t,s) + z^2(u,s)]du$$
$$= \left[2A + \int_{t-\epsilon}^{t} |C(t,u) - C(t,t-\epsilon)|du\right] z^2(t,s)$$
$$+\int_{t-\epsilon}^{t} [|C(t,t-\epsilon) - C(t,u)| + C_2(t,t-\epsilon)\epsilon)]z^2(u,s)du.$$

For $0 \leq s \leq t-\epsilon$ integrating the last term on the interval $[s+\epsilon, t]$ and taking (3.9.12*) into consideration, we obtain

$$\int_{s+\epsilon}^{t}\int_{v-\epsilon}^{v} [|C(v,u) - C(v,v-\epsilon)| + C_2(v,v-\epsilon)\epsilon]z^2(u,s)dudv$$
$$\leq \int_{s}^{t}\int_{u}^{u+\epsilon} [|C(v,u) - C(v,v-\epsilon)| + C_2(v,v-\epsilon)\epsilon]dvz^2(u,s)du$$
$$\leq \int_{0}^{t}\int_{u}^{u+\epsilon} [|C(v,u) - C(v,v-\epsilon)| + C_2(v,v-\epsilon)\epsilon]dvz^2(u,s)du$$
$$\leq \int_{0}^{t} \alpha^* z^2(u,s)du.$$

Then, taking into account (3.9.12*) and (3.9.13*), by a change in the order of integration we have for $s+\epsilon \leq t$

$$V(t,\epsilon) - V(s+\epsilon,\epsilon)$$
$$\leq \int_{s+\epsilon}^{t} [2A + \int_{v-\epsilon}^{v} [|C(v,u) - C(v,v-\epsilon)|du]z^2(v,s)dv$$
$$+\int_{s+\epsilon}^{t}\left[\int_{v-\epsilon}^{v} [|C(v,v-\epsilon) - C(v,u)| + C_2(v,v-\epsilon)\epsilon]z^2(u,s)du\right]dv$$
$$\leq \int_{s+\epsilon}^{t} (2A+\beta^*)z^2(v,s)dv$$
$$+\int_{s}^{t}\left[\int_{u}^{u+\epsilon} [|C(v,v-\epsilon) - C(v,u)| + C_2(v,v-\epsilon)\epsilon]z^2(u,s)dv\right]du$$
$$\leq \int_{s+\epsilon}^{t} (2A+\beta^*)z^2(v,s)dv + \int_{s}^{t} \alpha^* z^2(u,s)du$$
$$\leq \int_{s+\epsilon}^{t} (2A+\beta^*)z^2(v,s)dv + \int_{0}^{t} \alpha^* z^2(u,s)du$$

and so

$$
\begin{aligned}
&V(t,\epsilon) - V(s+\epsilon,\epsilon) \\
&\leq \int_0^t (2A+\beta^*+\alpha^*)z^2(v,s)dv - \int_0^{s+\epsilon} (2A+\beta^*)z^2(u,s)du
\end{aligned}
$$

Hence for $t \geq s+\epsilon$ we obtain

$$
\begin{aligned}
&z^2(t,s) - (2A+\beta^*+\alpha^*)\int_0^t z^2(u,s)du \\
&\leq V(s+\epsilon,\epsilon) - (2A+\beta^*)\int_0^s z^2(u,s)du.
\end{aligned}
$$

By (3.9.20) we have $2A+\beta^*+\alpha^* < 0$ and $2A+\beta^* < 0$.

Since for any fixed s the right-hand-side of the above inequality is a positive constant which does not depend on t, taking into consideration the fact that the solution $z(u,s)$ is continuous on $[0,\epsilon]\times\{s\}$ for all $s \geq 0$, it follows that for any $s \geq 0$ there exists an $M_z(s) > 0$ with

$$
z^2(t,s) \leq M_z(s), \quad t \geq 0
$$

and

$$
\int_0^t z^2(u,s)du \leq M_z(s), \qquad \textit{for all } t \geq 0,
$$

as required.

Continuation of this work for small kernels is found in Burton and Purnaras (2012) which is to appear in Journal of Integral Equations and Applications.

Notes The material in this section was taken from Burton and Purnaras (2011). It was first published by Elsevier in the *Journal of Mathematical Analysis and Applications.* Full details are given in the bibliography.

Chapter 4

Strategy

Once more we go back to the statement of Miller that if we can differentiate an integral equation, obtaining an integrodifferential equation, then we can apply Liapunov's direct method using the chain rule to unite the Liapunov functional with the integrodifferential equation. We begin in that way, but quickly change direction. It turns out that if we start with the integral equation, differentiate it, and then form $x' + kx$ for some positive constant k then several unexpected events occur which lead us to many new results. The first section is very brief and shows the basic ideas by means of an example. That work was taken much further in Burton-Haddock (2009) and the interested reader will find many ideas which are new and have proved useful in concrete problems. The second section applies the ideas to a number of problems, while the third and fourth sections apply the techniques to obtain bounded and periodic solutions of several integral equations.

Sections 3 and 4 focus on a classical problem occurring throughout Liapunov theory. If we begin with a uniformly asymptotically stable equation, perturb it with a bounded function, and apply a Liapunov functional, we frequently arrive at a relation

$$V'(t) \leq -W(|x|) + M$$

where W is positive definite and M is a positive constant. A number of arguments are then formulated which will show that solutions are uniformly bounded or uniformly ultimately bounded. Such arguments are found throughout Yoshizawa (1966) and Burton (2005b,c), for example. The problems become much more difficult when we have a perturbed relaxation-oscillation problem in which there is positive damping for small x. Most of the work in Sections 3 and 4 deal with this latter type.

4.1 Adding x and x'

Consider the scalar equation

$$x(t) = a(t) - \int_0^t (1 + t - s)^{-2} x(s)ds \tag{4.1.1}$$

where we first suppose that

$$a \in L^1[0, \infty) \text{ and } a' \in L^1[0, \infty). \tag{4.1.2}$$

Observe that

$$\int_0^\infty (1 + u)^{-2} du = 1$$

and that

$$\int_0^\infty 2(1 + u)^{-3} du = 1 = C(t, t).$$

A study of these relations with the techniques of Chapter 2 suggests that our Liapunov theorems from Chapter 2 will not yield $x \in L^1$ and neither will our Razumikhin techniques. We show how to rectify the situation. Notice also that, while the kernel is completely monotone, the crucial property that the kernel have infinite integral fails; thus, we can not claim that critical property of (2.5.1.22) stating that the resolvent has an integral equal to 1.

From (4.1.1) we have

$$x'(t) = a'(t) - x(t) - \int_0^t -2(1 + t - s)^{-3} x(s)ds$$

and for $k > 0$ we have

$$kx(t) = ka(t) - \int_0^t k(1 + t - s)^{-2} x(s)ds$$

so that

$$\begin{aligned} x'(t) + kx(t) &= a'(t) + ka(t) - x(t) \\ &\quad - \int_0^t \{k(1 + t - s)^{-2} - 2(1 + t - s)^{-3}\} x(s)ds \end{aligned}$$

or

$$\begin{aligned} x'(t) &= a'(t) + ka(t) - (1 + k)x(t) \\ &\quad - \int_0^t \{k(1 + t - s)^{-2} - 2(1 + t - s)^{-3}\} x(s)ds. \end{aligned} \tag{4.1.3}$$

Two advantages of $x' + kx$

While there are other ways to proceed, it is easiest here to observe that if we select $k = 2$ then the kernel does not change sign; the second term in the new kernel continually subtracts from the first term, decreasing the integral of the kernel and, at the same time, the ode part of $-x$ is increased to $-3x$ so that the fact that we multiplied by 2 is balanced in the ode term. In view of our Liapunov theory, we win in both crucial places.

Thus, for $k = 2$ we define a Liapunov functional for (4.1.3) as

$$\begin{aligned} V(t) &= |x(t)| + \int_0^t \int_{t-s}^{\infty} |2C(u+s,s) + C_t(u+s,s)|du|x(s)|ds \\ &= |x(t)| + \int_0^t \int_{t-s}^{\infty} \{2(1+u)^{-2} - 2(1+u)^{-3}\}du|x(s)|ds \end{aligned}$$

and that integrand is non-negative. We have

$$\begin{aligned} V'(t) &\leq |2a(t) + a'(t)| - 3|x(t)| \\ &\quad + \int_0^t \{2(1+t-s)^{-2} - 2(1+t-s)^{-3}\}|x(s)|ds \\ &\quad + \int_0^{\infty} \{2(1+u)^{-2} - 2(1+u)^{-3}\}du|x(t)| \\ &\quad - \int_0^t \{2(1+t-s)^{-2} - 2(1+t-s)^{-3}\}|x(s)|ds \\ &= |2a(t) + a'(t)| - 3|x(t)| + |x(t)|. \end{aligned}$$

Theorem 4.1.1. *If (4.1.2) holds then the solution of (4.1.1) and any solution of (4.1.3) for $k = 2$ satisfies x is bounded, $x \in L^1[0,\infty)$, and $x' \in L^1[0,\infty)$.*

Proof. A solution of (4.1.1) also solves (4.1.3). From V and V' we have

$$|x(t)| \leq V(t) \leq V(0) + \int_0^t |2a(s) + a'(s)|ds - 2\int_0^t |x(s)|ds.$$

This yields the first two conclusions. As $x \in L^1$, it is readily verified from (4.1.3) that $x' \in L^1$.

Theorem 4.1.2. *If*

$$2a(t) + a'(t) \text{ is bounded} \tag{4.1.4}$$

then the solution of (4.1.1) and any solution of (4.1.3) on $[0,\infty)$ for $k = 2$ is bounded.

Proof. We use the Razumikhin function $V(t) = |x|$ and notice that if $x(t)$ is any fixed unbounded solution of (4.1.3) then there is a sequence $\{t_n\} \uparrow \infty$ with $|x(t)| \leq |x(t_n)|$ for $0 \leq t \leq t_n$ and $|x(t_n)| \uparrow \infty$. Let $|2a(t)+a'(t)| \leq M$ and note that for $0 \leq t \leq t_n$ we have

$$V'(t) \leq |2a(t) + a'(t)| - 3|x(t)| + |x(t_n)|$$

and at $t = t_n$ we must have $V'(t_n) \geq 0$. But $V'(t_n) \leq M - 3|x(t_n)| + |x(t_n)| < 0$ if $M < 2|x(t_n)|$. This is a contradiction and completes the proof.

For further results both with $x' + kx$ and $x'' + kx'$ see Burton and Haddock (2009). The technique is also effective when the kernel is convex, especially in the nonlinear case. A fourth application involves the existence of periodic solutions.

Notice! This process is producing a uniformly asymptotically ode with an integral perturbation.

Here is a very enlightening exercise. Take $d > 0$ and

$$C(t, s) = de^{-(t-s)} \cos(t - s).$$

Form $x' + kx$ with $k = 2$ and divide $[0, \infty)$ into $[0, \pi]$ and $[\pi, \infty)$. Approximate the new kernel on $[\pi, \infty)$ by the exponential part. Determine how large d can be chosen so that our Liapunov functional given above will satisfy

$$V'(t) \leq |a'(t) + 2a(t)| - \beta|x(t)|$$

for some $\beta > 0$. This will show how understated our theorems are unless the kernel is prepared by careful considerations.

4.2 Liapunov functionals and convex kernels

Using the simple device of forming the new equation from $x' + kx$ many of the difficulties of the convex case with which we struggled in earlier chapters will be substantially reduced. The reader might wish to review Theorem 2.1.10 and Theorem 2.9.7 before starting this section.

We are concerned here with an integral equation

$$x(t) = a(t) - \int_0^t C(t, s)g(x(s))ds \tag{4.2.1}$$

where $a : [0, \infty) \to \Re$ is continuous, while C is continuous for $0 \leq s \leq t < \infty$, and $g : \Re \to \Re$ is continuous with $xg(x) > 0$ if $x \neq 0$. Later on this condition will change to $xg(x) > 0$ for large $|x|$. Continuity of a, C, g will ensure the existence of a solution. If the solution remains bounded, then it can be continued on $[0, \infty)$.

It is always assumed that the kernel, $C(t, s)$, is convex in the sense that

$$C(t,s) \geq 0, C_s(t,s) \geq 0, C_{st}(t,s) \leq 0, C_t(t,s) \leq 0. \tag{4.2.2}$$

Convolution problems of this type are seen in Levin (1965) and Londen (1972), for example. It is an enduring problem.

Volterra (1928) noticed that a great many real world problems were being modeled by integral and integrodifferential equations with convex kernels which inherently suggested a fading memory. He conjectured that there is a Liapunov functional for such kernels which would yield much qualitative information about solutions and which would allow very large kernels; Levin (1963) constructed the first such functional. Today we see problems in biology, nuclear reactors, viscoelasticity, and neural networks being modeled using convex kernels.

For the linear form of (4.2.1) there is also a Liapunov functional for the resolvent equation and we discussed this in some detail in Theorem 2.7.1 when $\sup_{0\leq s\leq t<\infty} \int_s^t C^2(u,s)du =: \Gamma < \infty$. This section seeks to extend some of that work to the case $\Gamma = \infty$ using $x' + kx$. In the nonlinear integral equation there is a severe technical problem in dealing with the derivative of the Liapunov functional and the investigator must make some undesirable assumptions about the nonlinearity. We studied this in Theorem 2.9.7. Use the Liapunov functional

$$V_1(t) = \int_0^t C_s(t,s)\left(\int_s^t g(x(u))du\right)^2 ds + C(t,0)\left(\int_0^t g(x(u))du\right)^2,$$

and obtain a derivative satisfying

$$V_1'(t) \leq 2g(x)[a(t) - x(t)].$$

In order to relate $g(x)$ to $a(t)$ we need to be able to separate that relation into

$$V_1'(t) \leq |p(a(t))| - |q(x(t))|$$

for some functions p and q with q positive definite with respect to x or $g(x)$ and p positive definite with respect to $a(t)$ so that

$$0 \leq V_1(t) \leq V_1(0) + \int_0^t |p(a(s))|ds - \int_0^t |q(x(s))|ds.$$

That separation has proved to be very cumbersome and investigators (See Zhang (2009) and Section 2.1.1 of Chapter 2.) have resorted to *ad hoc* assumptions, as well as stringent conditions on g in order to use Young's inequality (See Section 2.1.1, Lemma 2.1.1.1 and Theorem 2.1.1.1.). A definite example will show the need for the theory which is to follow.

Example 4.2.1. *Consider the scalar equation*

$$x(t) = a(t) - \int_0^t [1+t-s]^{-1/4} g(x(s))ds$$

where g is an arbitrary continuous function satisfying $xg(x) > 0$ if $x \neq 0$. For $a \in L^2[0,\infty)$ if x is a solution on $[0,\infty)$ then we know of no result or technique in the literature that will yield $g(x) \in L^p[0,\infty)$. The fact that the kernel is completely monotone seems to help very little in the general nonlinear case. The Liapunov functional mentioned above will yield the indicated derivative and we find no way to perform the required separation. The difficulty will vanish using $x' + kx$ in Theorem 4.2.1 and (4.2.14). We will immediately find $g(x) \in L^2[0,\infty)$ without further restriction on g.

In the linear case (Theorem 2.1.10), $g(x) = x$ and a convex kernel, we have

$$V_1'(t) \leq a^2(t) - x^2(t)$$

so that

$$\int_0^t x^2(s)ds \leq \int_0^t a^2(s)ds,$$

a very useful relation. Moreover, it extends to the resolvent equation

$$R(t,s) = C(t,s) - \int_s^t C(t,u)R(u,s)du$$

as

$$V_2(t) = \int_s^t C_v(t,v)\left(\int_s^t R(u,s)du\right)^2 dv + C(t,s)\left(\int_s^t R(u,s)du\right)^2,$$

with a derivative satisfying

$$V_2'(t) \leq -R^2(t,s) + C^2(t,s)$$

as may be seen Theorem 2.6.4.1 in Chapter 2, Section 2.6.4. This yields

$$\int_s^t R^2(u,s)du \leq \int_s^t C^2(u,s)du$$

which is so useful in the variation of parameters formula

$$x(t) = a(t) - \int_0^t R(t,s)a(s)ds.$$

But we have a difficulty here also. If $\sup_{0\leq s\leq t<\infty} \int_s^t C^2(u,s)du =: \Gamma < \infty$ then we have a very useful parallel property for R. On the other hand, if $\Gamma = \infty$ then the property is lost and we are left with the obvious fact that if $a \in L^2$ then $x \in L^2$ so by default $\int_0^t R(t,s)a(s)ds \in L^2$ and $x - a \in L^2$, but we can not extract from that any essential properties of R itself. From the complete monotonicity we know that $\int_0^\infty R(s)ds = 1$, but properties of $\int_0^t R^2(s)ds$ are unknown at this point of our study.

Example 4.2.2. *We can continue Example 4.2.1 with $g(x) = x$ and study $x(t) = a(t) - \int_0^t [1+t-s]^{-1/4}x(s)ds$. The Liapunov functional will yield $\int_0^t x^2(s)ds \leq \int_0^t a^2(s)ds$ and $\int_0^t R(t,s)a(s)ds \in L^2[0,\infty)$ when $a \in L^2$ without any independent property of R. Our next goal is to obtain basic properties of a resolvent independent of $a(t)$. That resolvent will not be R but it will serve in a parallel way to R.*

Thus, we encounter fundamental problems in both the nonlinear and linear cases. These two unsolved problems will drive this section.

In an effort to avoid the difficulties just mentioned differentiate (4.2.1) to obtain

$$x'(t) = a'(t) - C(t,t)x(t) - \int_0^t C_t(t,s)x(s)ds$$

which seems promising since for $C(t,t) \geq \alpha > 0$ we have a perturbation of the uniformly asymptotically stable equation

$$x' + C(t,t)x = 0.$$

However, that gain pales in comparison to our great loss in that $C_t(t,s)$ is no longer convex; hence, we would require some restrictions on the magnitude of $C_t(t,s)$ in order to use standard results on qualitative properties. To avoid all of those problems we develop a strategy which yields very good results.

Moreover, there is an added benefit, uncommon in the theory of convex kernels. If we can find a function $f : [0,\infty) \to [0,\infty)$ with $\int_0^t \frac{ds}{f(s)}$ continuous for $t \geq 0$,

$$|g(x)| \leq f\Big(\int_0^x g(s)ds\Big),\ x \in \Re,\ f \in \nearrow, \tag{4.2.3}$$

and

$$\int_0^\infty \frac{ds}{f(s)} = \infty, \tag{4.2.4}$$

then we prove that the solution has certain integral properties.

The work is based on four Liapunov functionals, a differential inequality, and a strategy for finding a strongly stable equation which has a solution of (4.2.1) as one of its solutions.

Differentiate (4.2.1) obtaining

$$x'(t) = a'(t) - C(t,t)g(x) - \int_0^t C_t(t,s)g(x(s))ds. \tag{4.2.5}$$

Under some general conditions, if $C(t,s)$ is convex and if k is a sufficiently large positive constant, then it is true that

$$D(t,s) := kC(t,s) + C_t(t,s) \tag{4.2.6}$$

is convex. For example, it is readily verified that if r is a positive constant, then $k = r+3$ is a suitable constant for $C(t,s) = [1+t-s]^{-r}$ (this pertains to Example 4.2.1), while $k = r+1$ is suitable for $C(t,s) = e^{-r(t-s)}$.

If we form $x' + kx$ then we have

$$x'(t) = a'(t) + ka(t) - [kx + C(t,t)g(x(t))] - \int_0^t D(t,s)g(x(s))ds. \tag{4.2.7}$$

This is a one-parameter family of totally different equations having exactly one property in common: a solution of (4.2.1) satisfies every one of those equations. If all solutions of (4.2.7) satisfy a certain property, so does a solution of (4.2.1). Two things have happened. Since C is convex, $C(t,t) \geq 0$ and, hence, $x' + kx + C(t,t)g(x) = 0$ is uniformly asymptotically stable. If $a'(t) + ka(t) \in L^2[0,\infty)$ and $C(t,t) \geq \alpha > 0$, then Levin's original Liapunov functional will yield $g(x(t)) \in L^2[0,\infty)$. In addition, if (4.2.3) and (4.2.4) hold and if $a' + ka \in L^1[0,\infty)$, then we will obtain an L^2 result for x.

We have used (4.2.5) to introduce a differential equation, but we have overwhelmed it with the integral equation by taking k large.

What is, perhaps, more interesting is the fact that when $g(x) = x$, then Becker's (2006) resolvent equation, (1.2.1.8), for (4.2.7) is

$$Z_t(t,s) = -[k + C(t,t)]Z(t,s) - \int_s^t D(t,u)Z(u,s)du \tag{4.2.8}$$

and a slight modification of Levin's (1963) Liapunov functional will yield

$$\sup_{0 \leq s \leq t < \infty} \int_s^t Z^2(u,s)du < \infty. \tag{4.2.9}$$

This is a critical result in the variation of parameters formula

$$x(t) = Z(t,0)x(0) + \int_0^t Z(t,s)[a'(s) + ka(s)]ds \tag{4.2.10}$$

where x solves (4.2.1) if $x(0) = a(0)$.

Now Levin's (1963) equation was a convolution form of

$$x' = -\int_0^t D(t,s)g(x(s))ds \tag{4.2.11}$$

with $xg(x) > 0$ for $x \neq 0$ and D convex. He constructed the Liapunov functional

$$\begin{aligned} V_3(t) = \int_0^x g(s)ds + \frac{1}{2}\int_0^t D_s(t,s)\left(\int_s^t g(x(u))du\right)^2 ds \\ + \frac{1}{2}D(t,0)\left(\int_0^t g(x(u))du\right)^2 \end{aligned} \tag{4.2.12}$$

and found that $V_3'(t) \leq 0$ along a solution of (4.2.11). This means that

$$\int_0^{x(t)} g(s)ds \leq V_3(t) \leq V_3(0) = \int_0^{x(0)} g(s)ds$$

so that if $\int_0^{\pm\infty} g(s)ds = \infty$, then every solution of (11) is bounded.

Main Remark We are going to use the same Liapunov functional on (4.2.7). In (4.2.16) below recall that $xg(x) > 0$ if $x \neq 0$ and that $C(t,t) \geq 0$ so kx and $C(t,t)g(x)$ have the same sign. It is very mild to ask that $|kx + C(t,t)g(x)| \geq \mu|g(x)|$. The obvious and usual condition is that $C(t,t)$ be greater than a positive constant, entirely consistent with the convexity. Indeed, in the convolution case $C(t) \geq 0$ and $C'(t) \leq 0$ so if $C(0) = 0$ then $C(t) \equiv 0$. Even if this fails, in the next step of the proof we get x bounded. None of the *ad hoc* assumptions on g needed in Young's inequality used in Section 2.1.1 are needed here.

Theorem 4.2.1. *Suppose that D is convex,*

$$D(t,s) \geq 0, D_s(t,s) \geq 0, D_{st}(t,s) \leq 0, D_t(t,s) \leq 0, \tag{4.2.13}$$

that $xg(x) > 0$ if $x \neq 0$, and that V_3 is defined in (4.2.12). Then along a solution of (4.2.7) we have

$$V_3'(t) \leq g(x(t))[a'(t) + ka(t)] - g(x(t))[kx(t) + C(t,t)g(x(t))]. \tag{4.2.14}$$

If, in addition, $|kx + C(t,t)g(x)| \geq \mu|g(x)|$, *for some* $\mu > 0$, *then* $a' + ka \in L^2[0,\infty)$ *implies* $g(x(t)) \in L^2[0,\infty)$. *In particular, any solution* x *in Example 4.2.1 satisfies* $g(x) \in L^2[0,\infty)$.

If (4.2.14) and (4.2.3) hold then along a solution of (4.2.7) we have

$$V_3'(t) \leq f(V_3(t))|a'(t) + ka(t)|. \tag{4.2.15}$$

If, in addition, (4.2.4) holds, $\int_0^{\pm\infty} g(s)ds = \infty$, *and* $a' + ka \in L^1[0,\infty)$, *then every solution of (4.2.7) is bounded and*

$$\int_0^\infty g(x(s))[kx(s) + C(s,s)g(x(s))]ds < \infty. \tag{4.2.16}$$

Proof. Along a solution of (4.2.7) we have

$$\begin{aligned} V_3'(t) &\leq g(x)[a'(t) + ka(t)] - g(x)[kx + C(t,t)g(x)] \\ &\quad - g(x)\int_0^t D(t,s)g(x(s)ds \\ &\quad + g(x)D(t,0)\int_0^t g(x(u))du + g(x)\int_0^t D_s(t,s)\int_s^t g(x(u))duds. \end{aligned}$$

Integrating the last term by parts yields

$$\begin{aligned} g(x)[D(t,s)\int_s^t g(x(u))du\Big|_0^t + \int_0^t D(t,s)g(x(s))ds] \\ = g(x)[-D(t,0)\int_0^t g(x(u))du + \int_0^t D(t,s)g(x(s))ds] \end{aligned}$$

so that

$$V_3'(t) \leq g(x)[a'(t) + ka(t)] - g(x)[kx + C(t,t)g(x)].$$

Now, if $|kx + C(t,t)g(x)| \geq \mu|g(x)|$ then

$$V_3'(t) \leq \alpha[a'(t) + ka(t)]^2 - \beta g^2(x)$$

for some positive α and β, from which $a' + ka \in L^2$ implies $g(x) \in L^2$.

Next, if (4.2.14) and (4.2.3) hold, then

$$V_3'(t) \le f\left(\int_0^x g(s)ds\right)|a'(t) + ka(t)| \le f(V_3)|a'(t) + ka(t)|$$

so

$$\int_{V_3(0)}^{V_3(t)} \frac{du}{f(u)} \le \int_0^t |a'(s) + ka(s)|ds.$$

If (4.2.4) holds and $a' + ka \in L^1$, then $V_3(t)$ is bounded and, hence, $x(t)$ is bounded. This means that $g(x)[a' + ka] \in L^1$ so from (4.2.14) we see that (4.2.16) follows. The proof is complete.

These results raise questions for the linear case. For we then see that $a' + ka \in L^1$ yields $x \in L^2$, but $a' + ka \in L^2$ also yields $x \in L^2$. Linear theory shows that $x \in L^1[0, \infty)$ is intimately related to uniform asymptotic stability. The next result shows that for certain choices of g we approximate $x \in L^1[0, \infty)$.

Theorem 4.2.2. *Suppose that D is convex and that $g(x) = x^{1/n}$ where n is an odd positive integer. If $a' + ka \in L^1[0, \infty)$ then $\int_0^\infty |x(s)|^{\frac{1+n}{n}} ds < \infty$.*

Proof. Note that there is a positive number p with

$$|g(x)| = |x^{1/n}| = \left(|x^{1/n}|^{(n+1)}\right)^{\frac{1}{n+1}} = p\left(\int_0^x s^{1/n} ds\right)^{\frac{1}{n+1}}.$$

Hence $f(r) = p(r)^{\frac{1}{n+1}}$. We then have

$$V_3'(t) \le p(V_3(t))^{\frac{1}{n+1}}|a'(t) + ka(t)|$$

and

$$\frac{1}{p}\int_{V_3(0)}^{V_3(t)} \frac{ds}{s^{\frac{1}{n+1}}} = \frac{1}{p} s^{(1-\frac{1}{n+1})}\Big|_{V_3(0)}^{V_3(t)}$$

so

$$\frac{1}{p}V_3(t)^{\frac{n}{n+1}} \le \frac{1}{p}V_3(0)^{\frac{n}{n+1}} + \int_0^t |a'(s) + ka(s)|ds.$$

Hence, $V_3(t)$ is bounded so $x(t)$ is bounded and $x(t)[a'(t) + ka(t)] \in L^1[0, \infty)$. But

$$V_3'(t) \le |g(x)||a'(t) + ka(t)| - kxg(x)$$

with $xg(x) = xx^{1/n} = x^{1+\frac{1}{n}}$ so $\int_0^\infty |x(s)|^{\frac{1+n}{n}} ds < \infty$.

Notice that $V_3'(t) \leq -\beta g^2(x)+\alpha(a'(t)+ka(t))^2$ with $a'+ka \in L^2$ would yield $\int_0^\infty x^{2/n}(s)ds < \infty$, an entirely different property. Suppose now that $n > 2$ and that $C(t,t) \geq \alpha > 0$ so that both of our integral relations hold. Note that if $|x(t)| \geq 1$ then $|x(t)| \leq |x(t)|^{1+\frac{1}{n}}$. If $|x(t)| < 1$ then $|x(t)| \leq |x(t)|^{2/n}$. Hence, we conclude that $\int_0^\infty |x(s)|ds < \infty$.

If $g(x) = x$ then (4.2.7) becomes

$$x'(t) = a'(t) + ka(t) - [k + C(t,t)]x(t) - \int_0^t D(t,s)x(s)ds \qquad (4.2.17)$$

and Becker's (2006) resolvent equation (see (1.2.1.8)) is

$$Z_t(t,s) = -[k+C(t,t)]Z(t,s) - \int_s^t D(t,u)Z(u,s)du,\ Z(s,s) = 1, \quad (4.2.18)$$

(where $Z_t = \frac{\partial Z}{\partial t}$) with variation of parameters formula

$$x(t) = Z(t,0)x(0) + \int_0^t Z(t,s)[a'(s) + ka(s)]ds. \qquad (4.2.19)$$

The Grossman-Miller (1970) (see (1.2.1.5)) resolvent equation is

$$H_s(t,s) = H(t,s)[k+C(s,s)] + \int_s^t H(t,u)D(u,s)du,\ H(t,t) = 1, \quad (4.2.20)$$

and it is true that

$$H(t,s) = Z(t,s). \qquad (4.2.21)$$

With D convex, a Liapunov functional for (4.2.18) is

$$V_4(t) = Z^2(t,s) + \int_s^t D_u(t,u)\left(\int_u^t Z(v,s)dv\right)^2 du + D(t,s)\left(\int_s^t Z(v,s)dv\right)^2. \qquad (4.2.22)$$

Theorem 4.2.3. *If D is convex and $k > 0$ then the derivative of V_4 along a solution of (4.2.18) satisfies*

$$V_4'(t) \leq -2[k+C(t,t)]Z^2(t,s), \text{ and } \sup_{0\leq s\leq t<\infty} \int_s^t Z^2(u,s)du < \infty. \quad (4.2.24)$$

Proof. We have

$$V_4'(t) \leq -2[k + C(t,t)]Z^2(t,s) - 2Z(t,s)\int_s^t D(t,u)Z(u,s)du$$

$$+ 2Z(t,s)D(t,s)\int_s^t Z(v,s)dv + 2Z(t,s)\int_s^t D_u(t,u)\int_u^t Z(v,s)dvdu.$$

An integration of the last term by parts yields

$$2Z(t,s)\left[D(t,u)\int_u^t Z(v,s)dv\bigg|_s^t + \int_s^t D(t,u)Z(u,s)du\right]$$

$$= 2Z(t,s)\left[-D(t,s)\int_s^t Z(v,s)dv + \int_s^t D(t,u)Z(u,s)du\right].$$

Cancellation of terms yields the required conclusion.

We then see that

$$Z^2(t,s) \leq V_4(t) \leq V_4(s) - 2k\int_s^t Z^2(u,s)du$$

with $Z^2(s,s) = 1$ yielding

$$Z^2(t,s) + 2k\int_s^t Z^2(u,s)du \leq 1. \tag{4.2.25}$$

This is a significant difference from the integral equation resolvent which requires $\int_s^t C^2(u,s)du$ bounded in order to get the parallel conclusion for the resolvent. Notice that $\int_0^t Z^2(u,0)du \leq 1/(2k)$; as $k \to \infty$, the integral tends to zero.

It is most direct to obtain $x \in L^2[0,\infty)$ in the linear case from (4.2.1) with the Liapunov functional

$$V_1(t) = \int_0^t C_s(t,s)\left(\int_s^t x(u)du\right)ds + C(t,0)\left(\int_0^t x(u)du\right)^2,$$

yielding

$$V_1'(t) \leq -x^2(t) + a^2(t).$$

We are coming to one of our central issues. From $a \in L^2$ we obtain $x \in L^2$ and, hence, from (4.2.29) we have by default that

$$\int_0^t R(t,s)a(s)ds \in L^2 \text{ and } x - a \in L^2.$$

However, we have no independent property of R which can be used without $a(t)$. We seek integral properties on R alone and the following is a typical way in which we would use them. Recall that we found that for C convex, then

$$\int_s^t R^2(u,s)ds \leq \int_s^t C^2(u,s)du \leq \Gamma \leq +\infty.$$

We just noted that $a \in L^2$ yields $x - a \in L^2$ by default. But $\Gamma < \infty$ yields $x - a \in L^2$ by direct computation, not by default, and that is such a desirable property in other contexts.

Proposition 4.2.1. *If $\Gamma < \infty$ then $a \in L^1[0,\infty)$ implies $x - a \in L^2[0,\infty)$.*

We will give a proof of a parallel result below, but it is sketched as follows.

$$(x(t) - a(t))^2 = \left(- \int_0^t R(t,s)a(s)ds \right)^2$$

so integration, followed by the Schwarz inequality and interchange of the order of integration will yield the result.

Our focus here is on the case of $\Gamma = +\infty$ and we attempt to obtain an integrability property of a resolvent. The first step is to note that (4.2.25) did not require $\Gamma < \infty$.

Proposition 4.2.2. *If (4.2.25) holds, then $a' + ka \in L^1[0,\infty)$ implies $x \in L^2[0,\infty)$ and $x - Z(t,0)x(0) \in L^2[0,\infty)$.*

Proof. Let $a'(t) + ka(t) =: p(t)$ and from (4.2.19) we have

$$(1/2)x^2(t) \leq Z^2(t,0)x^2(0) + \left(\int_0^t Z(t,s)p(s)ds \right)^2.$$

The last term is in L^1, not by default, but by the nonconvolution extension of the result that the convolution of an L^1–function with an L^2–function is an L^2–function. Here are the details. We have

$$(1/2)\int_0^t x^2(u)du \le \int_0^t Z^2(u,0)x^2(0)du$$

$$+\int_0^t \left(\int_0^u Z(u,s)p(s)ds\right)^2 du$$

$$\le \int_0^t Z^2(u,0)x^2(0)du + \int_0^t \int_0^u |p(s)|ds \int_0^u Z^2(u,s)|p(s)|dsdu$$

$$\le \int_0^t Z^2(u,0)x^2(0)du + \int_0^\infty |p(s)|ds \int_0^t \int_s^t Z^2(u,s)du|p(s)|ds$$

$$\le \int_0^t Z^2(u,0)x^2(0)du + \left(\int_0^\infty |p(s)|ds\right)^2 (1/2k).$$

We will now see how this applies to $R(t,s)$.

If we begin with C convex and

$$x(t) = a(t) - \int_0^t C(t,s)x(s)ds, \tag{4.2.26}$$

we have the resolvent equation

$$R(t,s) = C(t,s) - \int_s^t C(t,u)R(u,s)du \tag{4.2.27}$$

and the variation of parameters formula

$$x(t) = a(t) - \int_0^t R(t,s)a(s)ds. \tag{4.2.28}$$

For (4.2.26) there is the Liapunov functional

$$V_1(t) = \int_0^t C_s(t,s)\left(\int_s^t x(u)du\right)^2 ds + C(t,0)\left(\int_0^t x(u)du\right)^2 \tag{4.2.29}$$

and a calculation yields

$$V_1'(t) \le -x^2(t) + a^2(t) \tag{4.2.30}$$

and

$$\int_0^t x^2(s)ds \le \int_0^t a^2(s)ds. \tag{4.2.31}$$

In a parallel manner we have a Liapunov functional for the resolvent equation given by

$$V_2(t) = \int_s^t C_2(t,u)\left(\int_u^t R(v,s)dv\right)^2 du + C(t,s)\int_s^t R(u,s)du\right)^2 \quad (4.2.32)$$

and a calculation will yield

$$V_2'(t) \leq -R^2(t,s) + C^2(t,s) \quad (4.2.33)$$

with

$$\int_s^t R^2(u,s)du \leq \int_s^t C^2(u,s)du. \quad (4.2.34)$$

We explored consequences of these relations in Theorem 2.6.4.1 of Chapter 2 for the case

$$\sup_{0\leq s\leq t<\infty} \int_s^t C^2(u,s)du < \infty. \quad (4.2.35)$$

Here, we look at the case where

$$\sup_{0\leq s\leq t<\infty} \int_s^t C^2(u,s)du = \infty \quad (4.2.36)$$

so that V_2 yields nothing about R. We find a substitute for

$$\sup_{0\leq s\leq t<\infty} \int_s^t R^2(u,s)du < \infty \quad (4.2.37)$$

when (4.2.36) holds.

Theorem 4.2.5. *If D is defined in (4.2.6), D convex, and if $\frac{d}{ds}C(s,s)$ is continuous, then*

$$R(t,s) = Z_s(t,s) - kZ(t,s). \quad (4.2.38)$$

Proof. From (4.2.17), (4.2.18), and (4.2.24) we see that for (4.2.1)

$$x(t) = a(t) - \int_0^t R(t,s)a(s)ds$$

and

$$\begin{aligned}
x(t) &= Z(t,0)a(0) + \int_0^t Z(t,s)[a'(s) + ka(s)]ds \\
&= Z(t,0)a(0) + Z(t,s)a(s)\Big|_0^t - \int_0^t Z_s(t,s)a(s)ds \\
&+ k\int_0^t Z(t,s)a(s)ds \\
&= Z(t,0)a(0) + a(t) - Z(t,0)a(0) - \int_0^t [Z_s(t,s) + kZ(t,s)]a(s)ds \\
&= a(t) - \int_0^t [Z_s(t,s) - kZ(t,s)]a(s)ds.
\end{aligned}$$

This means that for any $a(t)$ with $a'(t)$ continuous then

$$\int_0^t R(t,s)a(s)ds = \int_0^t [Z_s(t,s) - kZ(t,s)]a(s)ds. \tag{4.2.39}$$

Looking back at the Grossman-Miller (1970) resolvent (4.2.20) and noting that $H(t,s) = Z(t,s)$ we see that if $C(s,s)$ has a continuous derivative, then $H_{ss} = Z_{ss}$ is continuous. We should also note from (4.2.27) that R_s is continuous. Thus, for any fixed t we can pick $a(s) = Z_s(t,s) - kZ(t,s) - R(t,s)$ and have from (4.2.39) with t fixed that

$$\int_0^t [Z_s(t,s) - kZ(t,s) - R(t,s)]^2 ds = 0. \tag{4.2.40}$$

Thus, the integrand is identically zero and (4.2.38) holds. This completes the proof.

The variation of parameters formula for (4.2.1) now becomes

$$x(t) = a(t) - \int_0^t [Z_s(t,s) - Z(t,s)]a(s)ds. \tag{4.2.41}$$

We have independent properties of Z, as well as Z_s through (4.2.20) and through integration by parts.

Notes The material for Section 4.2 is found in Burton (2010). It was first published by InforMath Publishing Group in the journal *Nonlinear Dynamics and Systems Theory*. Full details are given in the bibliography.

4.3 Periodic Solutions

In this section and the next we consider several nonlinear scalar integral equations of the form

$$x(t) = a(t) - \int_{\alpha(t)}^{t} C(t,s)g(x(s))ds$$

where $\alpha(t)$ may be zero, $-\infty$, or $t-h$ for some constant $h>0$. In each of the problems the kernel need not be convex, but the assumption is that there is a constant $k>0$ with

$$D(t,s) := C_t(t,s) + kC(t,s)$$

convex.

Volterra (1928) noted that many physical problems were being modeled by integral and integrodifferential equations with convex kernels. Such kernels are natural representations of fading memory. Today we see such models in problems in biology, neural networks, viscoelasticity, nuclear reactors, and many other places. See, for example, Burton (1993b) and (2008c), Burton-Dwiggins (2010), Levin (1965), Londen (1972), Volterra (1928), Zhang (2009a,b) for work on integral equations with convex kernels.

In each case, the idea is to form $x' + kx$. For $\alpha = 0$ we have

$$x' + kx = a'(t) + ka(t) - C(t,t)g(x(t)) - \int_0^t D(t,s)g(x(s))ds$$

with D defined above. There are six important observations.

Advantages of $x' + kx$ continued

(i) $x' + kx$ is a uniformly asymptotically stable operator for $k>0$.

(ii) If $C(t,t) \geq 0$ and if $xg(x) > 0$ for $x \neq 0$, then $x' + kx + C(t,t)g(x)$ is an operator of the same, but stronger, type.

(iii) If C and C_t differ in sign then $D(t,s)$ is smaller than the larger of the two terms.

(iv) Under general conditions if C is convex and k is large then D is convex, while the kernel for x' alone has lost its convexity.

(v) If $C(t,s)$ is not convex while $D(t,s)$ is, then the combined equation $x' + kx$ is the right form to apply Liapunov functionals.

(vi) The utility of a Liapunov functional often depends on the separation of its derivative into a difference, say $|p(t)| - |h(x)|$. Using C alone, that can require strong conditions on g, but when using D there is a natural separation seen in the last section.

That is the background and we now move along with some new problems. First we prove the existence of a periodic solution when $\alpha = \infty$. We then study the case of $\alpha = 0$ proving boundedness properties. Finally, we take $\alpha = t - h$ and prove both boundedness and periodicity.

Let $R^+ = [0, \infty), R = (-\infty, \infty)$, and $C(X, Y)$ denote the space of continuous functions $\phi : X \to Y$. We also denote by $(P_T, \|\cdot\|)$ the Banach space of continuous T-periodic functions $\phi : R \to R$ with the supremum norm. For the existence of periodic solutions, we apply Schaefer's fixed point theorem with $F(x)$ being the right-hand side of our integral equation so that if F has a fixed point, then this fixed point is a periodic solution of that integral equation. We repeat Schaefer's theorem for convenience.

Theorem. *(Schaefer (1955)). Let $(P, \|\cdot\|)$ be a normed space, F a continuous mapping of P into P which is compact on each bounded subset of P. Then either*

(i) the equation $\phi = \lambda F\phi$ has a solution for $\lambda = 1$, or

(ii) the set of all such solutions ϕ, for $0 < \lambda < 1$, is unbounded.

Boundedness and Periodicity

We consider the equation

$$x(t) = \lambda[a(t) - \int_{-\infty}^{t} C(t,s)g(x(s))ds], \quad 0 \le \lambda \le 1 \tag{4.3.1}$$

where $a : R \to R$, $C : R \times R \to R$, $g : R \to R$ are all continuous. Suppose that there is a positive constant k so that

$$D(t,s) := C_t(t,s) + kC(t,s) \text{ is convex.} \tag{4.3.2}$$

We first want to show that there exists a constant $\gamma > 0$ such that $|x(t)| \le \gamma$ whenever x is a T-periodic solution of (4.3.1) for all $0 \le \lambda \le 1$. We then show the existence of a T-periodic solution of (4.3.1) for $\lambda = 1$ by applying Schaefer's fixed point theorem. Our main assumptions are that there is a $T > 0$ and $J > 0$ such that

$$a(t+T) = a(t), \quad C(t+T, s+T) = C(t,s) \tag{4.3.3}$$

for all $s \le t$ with a' continuous, that

$$\sup_{0 \le t \le T} \int_{-\infty}^{t} D(t,s)(t-s)ds \le J, \tag{4.3.4}$$

and that D satisfies

$$D(t,s) \geq 0, \quad D_s(t,s) \geq 0, \quad D_{st}(t,s) \leq 0. \tag{4.3.5}$$

Differentiate (4.3.1) and form

$$x' + kx = \lambda[a' + ka - C(t,t)g(x) - \int_{-\infty}^{t} D(t,s)g(x(s))ds]. \tag{4.3.6}$$

Now, we define the Liapunov functional

$$V(t) = 2\int_0^{x(t)} g(s)ds + \lambda \int_{-\infty}^{t} D_s(t,s)\left(\int_s^t g(x(v))dv\right)^2 ds \tag{4.3.7}$$

for $x \in (P_T, \|\cdot\|)$.

Theorem 4.3.1. *Suppose that (4.3.3), (4.3.4), and (4.3.5) hold. If $x(t)$ is a T-periodic solution of (4.3.1), then the derivative of V along that solution satisfies*

$$V'(t) \leq 2\lambda g(x)[a'(t) + ka(t)] - 2g(x)[kx + \lambda C(t,t)g(x)]. \tag{4.3.8}$$

If there is an $L > 0$ with

$$xg(x) \geq 0 \text{ for } |x| \geq L \tag{4.3.9}$$

and if, in addition, there is a $\mu > 0$ with

$$g(x)[kx + C(t,t)g(x)] \geq \mu g^2(x) \tag{4.3.10}$$

for $|x| \geq L$, then there is an $M > 0$ with

$$V'(t) \leq -|g(x(t))| + M. \tag{4.3.11}$$

Proof. We first define some constants to simplify notation. Integrating by parts, we obtain

$$\begin{aligned}\int_b^t D_s(t,s)(t-s)^2 ds &= D(t,s)(t-s)^2\Big|_b^t + 2\int_b^t D(t,s)(t-s)ds \\ &= -D(t,b)(t-b)^2 + 2\int_b^t D(t,s)(t-s)ds\end{aligned}$$

for each $b < t$. Since $D(t,s) \geq 0$ and $D_s(t,s) \geq 0$, letting $b \to -\infty$, we see that

$$\int_{-\infty}^{t} D_s(t,s)(t-s)^2 ds + \lim_{s\to-\infty} D(t,s)(t-s)^2$$
$$= 2\int_{-\infty}^{t} D(t,s)(t-s)ds \leq 2J. \tag{4.3.12}$$

Observe also that

$$D(t,b)(t-b)^2 \leq 2\int_{b}^{t} D(t,s)(t-s)ds \leq 2J$$

for all $b \leq t$. This then implies that $D(t,s)(t-s) \leq 2J/(t-s)$ for all $s < t$, and so we arrive at

$$\lim_{s\to-\infty}(t-s)D(t,s) = 0 \ \text{ for fixed } t \tag{4.3.13}$$

and obtain

$$\int_{-\infty}^{t} D_s(t,s)ds = \lim_{b\to-\infty}[D(t,t) - D(t,b)] \leq \sup_{0\leq t\leq T} D(t,t) =: B. \tag{4.3.14}$$

Now let x be a T-periodic solution of (4.3.1) and $V(t)$ be defined in (4.3.7). It follows from (4.3.12) that $V(t)$ is well-defined and T-periodic. We then find

$$V'(t) = 2g(x)x'(t) + \lambda\int_{-\infty}^{t} D_{st}(t,s)\left(\int_{s}^{t} g(x(v))dv\right)^2 ds$$
$$+ 2\lambda g(x)\int_{-\infty}^{t} D_s(t,s)\int_{s}^{t} g(x(v))dvds.$$

Integration of the last term by parts and use of (4.3.13) in the lower limit for that periodic solution yields

$$\int_{-\infty}^{t} D_s(t,s)\int_{s}^{t} g(x(v))dvds = \int_{-\infty}^{t} D(t,s)g(x(s))ds.$$

Since $D_{st}(t,s) \leq 0$, the second term of V' is not positive, and thus, if we use (4.3.6), we obtain

$$V'(t) \leq 2g(x)\Bigg[-kx + \lambda(a' + ka) - \lambda C(t,t)g(x)$$
$$-\lambda\int_{-\infty}^{t} D(t,s)g(x(s))ds\Bigg] + 2\lambda g(x)\int_{-\infty}^{t} D(t,s)g(x(s))ds$$
$$= 2\lambda g(x)[a'(t) + ka(t)] - 2g(x)[kx + \lambda C(t,t)g(x)]$$

verifying (4.3.8).

Next we choose $N > 1$ so large that $-\mu(N-1) < C_* = \min\{C(t,t) : 0 \le t \le T\}$, where $\mu > 0$ is defined in (4.3.10). If $|x| \ge L$, then $xg(x) \ge 0$, and by (4.3.10), we obtain for $|x(t)| \ge L$ that

$$\begin{aligned} V'(t) &\le 2\lambda|g(x)|[\|a'\| + k\|a\|] - 2g(x)[kx + \lambda C(t,t)g(x)] \\ &\le 2|g(x)|[\|a'\| + k\|a\|] - 2k\frac{1}{N}xg(x) \\ &- 2\lambda\left(1 - \frac{1}{N}\right) g(x)[kx + C(t,t)g(x)] - 2\lambda\frac{1}{N}C(t,t)g^2(x) \\ &\le 2|g(x)|[\|a'\| + k\|a\|] - 2k\frac{1}{N}xg(x) \\ &- 2\lambda\left(1 - \frac{1}{N}\right) \mu g^2(x) - 2\lambda\frac{1}{N}C(t,t)g^2(x) \\ &\le -2|g(x)|\left[\frac{1}{N}k|x| - (\|a'\| + k\|a\|)\right]. \end{aligned}$$

We may assume that $L \ge N(\|a'\| + k\|a\| + 1)/k$. Thus, if $|x(t)| \ge L$, then $V'(t) \le -|g(x(t))|$. It is clear that $V'(t) \le M$ for $0 \le |x(t)| \le L$, where

$$M = 2g_L[\|a'\| + k\|a\|] + 2g_L[kL + C^* g_L]$$

with $g_L = \sup\{|g(x)| : |x| \le L\}$ and $C^* = \sup\{|C(t,t)| : 0 \le t \le T\}$, and hence,

$$V'(t) \le -|g(x(t))| + M$$

for all $t \ge 0$. This completes the proof.

To establish an *a priori* bound for all possible T-periodic solutions of (4.3.1), we assume that

$$\lim_{s \to -\infty} (t-s)C(t,s) = 0 \ \text{ and } \ \int_{-\infty}^{t} |C_s(t,s)|(t-s)ds \le J_1 \qquad (4.3.15)$$

for $J_1 > 0$.

Theorem 4.3.2. *Suppose that (4.3.3)-(4.3.5), (4.3.9)-(4.3.0), and (4.3.15) hold. Then there exists a constant $\gamma > 0$ such that $\|x\| < \gamma$ whenever x is a T-periodic solution of (4.3.1).*

Proof. Let x be a T-periodic solution of (4.3.1) and $V(t)$ be defined in (4.3.7). Then (4.3.11) holds. Since $V(t)$ is T-periodic, $V(t)$ has a global maximum at $q \in [0, T]$ and, hence, at $t_n = q + nT$. So for $s \leq t_n$, we have

$$0 \leq V(t_n) - V(s) \leq -\int_s^{t_n} |g(x(v))|dv + M(t_n - s).$$

and so

$$\int_s^{t_n} |g(x(v))|dv \leq M(t_n - s).$$

Then $(x(t) - \lambda a(t))^2$ has a global maximum at $h_n := t_n + p$, where $0 \leq p \leq T$, and for $s \leq h_n$ we have

$$\int_s^{h_n} |g(x(v))|dv \leq \int_s^{t_{n+1}} |g(x(v))|dv \leq M(t_{n+1} - s).$$

It follows from (4.3.1) that

$$\begin{aligned}
\Big(x(h_n) - \lambda a(h_n)\Big)^2 &\leq \left(\int_{-\infty}^{h_n} C(h_n, s) g(x(s))ds\right)^2 \\
&= \Big(- C(h_n, s) \int_s^{h_n} g(x(v))dvds\Big|_{-\infty}^{h_n} \\
&+ \int_{-\infty}^{h_n} C_s(h_n, s) \int_s^{h_n} g(x(v))dvds\Big)^2 \\
&= \left(\int_{-\infty}^{h_n} C_s(h_n, s) \int_s^{h_n} g(x(v))dvds\right)^2 \\
&\leq \left(\int_{-\infty}^{h_n} |C_s(h_n, s)| \int_s^{t_{n+1}} |g(x(v))|dvds\right)^2 \\
&\leq M^2 \left(\int_{-\infty}^{h_n} |C_s(h_n, s)|(h_n + T - s)ds\right)^2.
\end{aligned}$$

Since $\int_{-\infty}^t |C_s(t, s)|ds$ is T-periodic, we see from (4.3.15) that

$$\sup_{0 \leq t \leq T} \int_{-\infty}^t |C_s(t, s)|ds \leq J_0 \tag{4.3.16}$$

for $J_0 > 0$, and hence,

$$\left(x(h_n) - \lambda a(h_n)\right)^2 \leq M^2 \left(TJ_0 + J_1\right)^2.$$

Noticing that M is a function of L, we find that

$$|x(h_n)| < \|a\| + M\left(TJ_0 + J_1\right) + 1 := \gamma.$$

This implies that $\|x\| < \gamma$ whenever x is a T-periodic solution of (4.3.1) for $0 \leq \lambda \leq 1$, and the proof is complete.

We now define a mapping F on P_T by

$$F(\phi)(t) = a(t) - \int_{-\infty}^{t} C(t,s)g(\phi(s))ds \text{ for } \phi \in P_T. \tag{4.3.17}$$

Theorem 4.3.3. *If (4.3.3)-(4.3.5), (4.3.9)-(4.3.10), and (4.3.15) hold, then (4.3.1) has a T-periodic solution for $\lambda = 1$.*

Proof. It is clear that $F(\phi) \in P_T$. We show that F is continuous on P_T and is compact on each bounded subset of P_T. If $\tilde{\phi}$, $\phi \in P_T$, then

$$|F(\phi)(t) - F(\tilde{\phi})(t)| = \left|\int_{-\infty}^{t} C(t,s)g(\phi(s))ds - \int_{-\infty}^{t} C(t,s)g(\tilde{\phi}(s))ds\right|$$

$$= \left|\int_{-\infty}^{t} C_s(t,s)\left(\int_{s}^{t} g(\phi(v))dv - \int_{s}^{t} g(\tilde{\phi}(v))dv\right)ds\right|. \tag{4.3.18}$$

Since g is uniformly continuous on $\{x \in R : |x| \leq \|\tilde{\phi}\| + 1\}$, for any $\varepsilon > 0$, there exists $0 < \delta < 1$ such that $\|\phi - \tilde{\phi}\| < \delta$ implies $|g(\phi(s)) - g(\tilde{\phi}(s))| < \varepsilon$ for
all $s \in [0,T]$. It follows from (4.3.18) that $\|F(\phi) - F(\tilde{\phi})\| \leq J_1\varepsilon$. Thus, F is continuous on P_T.

We now show that F is compact on each bounded subset of P_T. Let $\eta > 0$ and define

$$\Gamma = \{F(\phi) : \phi \in P_T,\ \|\phi\| \le \eta\}. \tag{4.3.19}$$

Since

$$\begin{aligned}
\frac{d}{dt}F(\phi)(t) &= a'(t) - C(t,t)g(\phi(t)) - \int_{-\infty}^{t} C_t(t,s)g(\phi(s))ds \\
&= a'(t) - C(t,t)g(\phi(t)) - \int_{-\infty}^{t} D(t,s)g(\phi(s))ds \\
&\quad + k\int_{-\infty}^{t} C(t,s)g(\phi(s))ds \\
&= a'(t) - C(t,t)g(\phi(t)) - \int_{-\infty}^{t} D(t,s)g(\phi(s))ds \\
&\quad + k\int_{-\infty}^{t} C_s(t,s)\int_{s}^{t} g(\phi(v))dvds
\end{aligned}$$

we have

$$\begin{aligned}
\left|\frac{d}{dt}F(\phi)(t)\right| &\le \|a'\| \\
&\quad + g_\eta \sup_{0\le t\le T}\left(|C(t,t)| + \int_{-\infty}^{t} D(t,s)ds + k\int_{-\infty}^{t} |C_s(t,s)|(t-s)ds\right)
\end{aligned}$$

where $g_\eta = \{|g(x)| : |x| \le \eta\}$, and thus, Γ is equi-continuous. The uniform boundedness of Γ follows from the inequality

$$|F(\phi)(t)| \le \|a\| + \int_{-\infty}^{t} |C_s(t,s)| \int_{s}^{t} |g(\phi(v))|dv \le \|a\| + J_1 g_\eta$$

for all $\phi \in \Gamma$. So, by the Ascoli-Arzela theorem, Γ lies in a compact subset of P_T. By combining Schaefer's theorem with Theorem 4.3.2, we see that F has a fixed point which is a T-periodic solution of (4.3.1) for $\lambda = 1$. This completes the proof.

Corollary. *Suppose that (4.3.3)-(4.3.5) hold. If there is an $L > 0$ and $\mu > 0$ with*

$$xg(x) \ge 0 \ \text{ for } \ |x| \ge L \ \text{ and } \ C(t,t) \ge \mu, \tag{4.3.20}$$

then there is an $M > 0$ with

$$V'(t) \le -|g(x)| + M$$

whenever x is a T-periodic solution of (4.3.1). If, in addition, (4.3.15) is satisfied, then (4.3.1) has a T-periodic solution for $\lambda = 1$.

Boundedness

We turn now to

$$x(t) = a(t) - \int_0^t C(t,s)g(x(s))ds \tag{4.3.21}$$

where $a : R^+ \to R$, $C : R^+ \times R^+ \to R$, $g : R \to R$ are all continuous with a, a' bounded. The project here is to show that solutions of (4.3.1) are bounded. We define

$$D(t,s) := C_t(t,s) + kC(t,s) \tag{4.3.22}$$

for a constant $k > 0$. Our main assumptions are that

$$D(t,s) \text{ is convex for } t \geq s \geq 0 \tag{4.3.23}$$

and there exists $B > 0$ with

$$C(t,t) \geq -B \text{ and } \int_0^t D(t,s)(t-s)ds \leq B \tag{4.3.24}$$

for all $t \geq 0$. Differentiate (4.3.21) and form

$$x' + kx = [a' + ka - C(t,t)g(x) - \int_0^t D(t,s)g(x(s))ds]. \tag{4.3.25}$$

Now, define the Liapunov functional

$$V(t) = 2G(x(t)) + \int_0^t D_s(t,s)\left(\int_s^t g(x(u))du\right)^2 ds + D(t,0)\left(\int_0^t g(x(u))du\right)^2 \tag{4.3.26}$$

for $x \in C(R^+, R)$, where $G(x) = \int_0^x g(s)ds$.

Theorem 4.3.4. *Suppose that $D(t,s)$ is convex and $C(t,t) \geq -B$ for a constant $B > 0$. If $x(t)$ is a solution of (4.3.21), then the derivative of V along that solution satisfies*

$$V'(t) \leq 2g(x)[a'(t) + ka(t)] - 2g(x)[kx + C(t,t)g(x)]. \tag{4.3.27}$$

If there is an $L > 0$ with

$$xg(x) \geq 0 \text{ for } |x| \geq L \tag{4.3.28}$$

and if, in addition, there is a $\mu > 0$ with

$$g(x)[kx + C(t,t)g(x)] \geq \mu g^2(x) \tag{4.3.29}$$

for $|x| \geq L$, then there is an $M > 0$ with

$$V'(t) \leq -|g(x)| + M. \tag{4.3.30}$$

Proof. First observe that if x is a solution of (4.3.21), then x is also a solution of (4.3.25). Now let x be a solution (4.3.21) and $V(t)$ be defined in (4.3.26). We then find

$$\begin{aligned} V'(t) &= 2g(x)x'(t) + \int_0^t D_{st}(t,s)\left(\int_s^t g(x(u))du\right)^2 ds \\ &\quad + 2g(x(t))\int_0^t D_s(t,s)\int_s^t g(x(u))duds \\ &\quad + D_t(t,0)\left(\int_0^t g(x(u))du\right)^2 + 2D(t,0)g(x(t))\int_0^t g(x(u))du. \end{aligned}$$

Integrate the third to last term by parts to obtain

$$\begin{aligned} &2g(x(t))\int_0^t D_s(t,s)\int_s^t g(x(u))duds \\ &= 2g(x(t))\left[D(t,s)\int_s^t g(x(u))du\Big|_{s=0}^{s=t} + \int_0^t D(t,s)g(x(s))ds\right] \\ &= 2g(x(t))\left[-D(t,0)\int_0^t g(x(s))ds + \int_0^t D(t,s)g(x(s))ds\right]. \end{aligned}$$

Cancel terms, use the sign conditions, and use (4.3.25) in the process to unite the Liapunov functional and the equation to obtain

$$\begin{aligned} V'(t) &\leq 2g(x)\left[-kx + (a' + ka) - C(t,t)g(x) - \int_0^t D(t,s)g(x(s))ds\right] \\ &\quad + 2g(x)\int_0^t D(t,s)g(x(s))ds \\ &\quad = 2g(x)[a'(t) + ka(t)] - 2g(x)[kx + C(t,t)g(x)] \end{aligned}$$

verifying (4.3.27).

Now assume that (4.3.28) and (4.3.29) hold. We may choose $N > 1$ so large that $\mu(N-1) > B$, where B and μ are defined in (4.3.24) and (4.3.29), respectively. If $|x| \geq L$, then $xg(x) \geq 0$, and by (4.3.29), we obtain for $|x(t)| \geq L$ that

$$\begin{aligned}
V'(t) &\leq 2|g(x)|[|a'| + k|a|] - 2g(x)[kx + C(t,t)g(x)] \\
&= 2|g(x)|[|a'| + k|a|] - 2k\frac{1}{N}xg(x) \\
&- 2\left(1 - \frac{1}{N}\right) g(x)[kx + C(t,t)g(x)] - 2\frac{1}{N}C(t,t)g^2(x) \\
&\leq 2|g(x)|[|a'| + k|a|] - 2k\frac{1}{N}xg(x) \\
&- 2\left(1 - \frac{1}{N}\right) \mu g^2(x) + 2\frac{1}{N}Bg^2(x) \\
&\leq -2|g(x)| \left[\frac{1}{N}k|x| - (|a'(t)| + k|a(t)|)\right].
\end{aligned}$$

We may assume that

$$L \geq N[\sup_{t \geq 0}(|a'(t)| + k|a(t)|) + 1]\Big/k.$$

Thus, if $|x(t)| \geq L$, then $V'(t) \leq -|g(x(t))|$. Since $-C(t,t) \leq B$, it is clear that $V'(t) \leq M$ for $0 \leq |x(t)| \leq L$, where the constant $M > 0$ is a function of L, and hence,

$$V'(t) \leq -|g(x(t))| + M$$

for all $t \geq 0$. This completes the proof.

To establish the boundedness of solutions, we assume that there is a $B_1 > 0$ with

$$|C(t,0)|\, t \leq B_1 \quad \textit{and} \quad \int_0^t |C_s(t,s)|(t-s+1)ds \leq B_1 \tag{4.3.32}$$

for $t \geq 0$. We also observe that

$$\begin{aligned}
\int_0^t D_s(t,s)(t-s)^2 ds &= D(t,s)(t-s)^2\Big|_0^t + 2\int_0^t D(t,s)(t-s)ds \\
&= -D(t,0)\, t^2 + 2\int_0^t D(t,s)(t-s)ds.
\end{aligned}$$

By (4.3.24), we now have

$$\int_0^t D_s(t,s)(t-s)^2 ds + D(t,0)\,t^2 = 2\int_0^t D(t,s)(t-s)ds \le 2B \quad (4.3.33)$$

for all $t \ge 0$.

Theorem 4.3.5. *If (4.3.23)-(4.3.24), (4.3.28)-(4.3.29), and (4.3.32) hold, then any solution of (4.3.21) is bounded.*

Proof. Let x be a solution of (4.3.21) and $V(t)$ be defined in (4.3.26). Then $V(t)$ is bounded below and satisfies (4.3.30). We now show that $V(t)$ is bounded above. If $V(t)$ is unbounded, then there exists a sequence $\{t_n\} \uparrow \infty$ with $V(t_n) \to \infty$ as $n \to \infty$ and

$$V(t_n) \ge V(s) \quad \text{for} \quad 0 \le s \le t_n.$$

It then follows from (4.3.30) that

$$0 \le V(t_n) - V(s) \le -\int_s^{t_n} |g(x(u))|du + M(t_n - s).$$

This implies that

$$\int_s^{t_n} |g(x(u))|du \le M(t_n - s). \quad (4.3.34)$$

Applying (4.3.34) to $V(t_n)$ and taking into account (4.3.33), we find that

$$\begin{aligned} V(t_n) &\le 2G(x(t_n)) + M^2\left[\int_0^{t_n} D_s(t_n,s)(t_n-s)^2 ds + D(t_n,0)\,t^2\right] \\ &\le 2G(x(t_n)) + 2BM^2. \end{aligned} \quad (4.3.35)$$

Now use (4.3.21), (4.3.32), and (4.3.34) to obtain

$$\begin{aligned} (x(t_n) - a(t_n))^2 &= \left(\int_0^{t_n} C(t_n,s)g(x(s))ds\right)^2 \\ &= \left(-C(t_n,s)\int_s^{t_n} g(x(u))du\Big|_{s=0}^{s=t_n} + \int_0^{t_n} C_s(t_n,s)\int_s^{t_n} g(x(u))duds\right)^2 \\ &= \left(C(t_n,0)\int_0^{t_n} g(x(s))ds + \int_0^{t_n} C_s(t_n,s)\int_s^{t_n} g(x(u))duds\right)^2 \\ &\le \left(|C(t_n,0)|M\,t_n + M\int_0^{t_n} |C_s(t_n,s)|(t_n-s)ds\right)^2 \le M^2(2B_1)^2. \end{aligned}$$

This implies that

$$|x(t_n)| \le \sup_{s\ge 0} |a(s)| + 2B_1 M := B_2$$

and that $|G(x(t_n))| \le B_3$ for a $B_3 > 0$. We now find that

$$V(t_n) \le 2G(x(t_n)) + 2BM^2 \le 2B_3 + 2BM^2 := B_4,$$

a contradiction. Thus, $V(t)$ is bounded. In fact, we have

$$2G(x(t)) \le V(t) \le 2G(x(0)) + B_4 \tag{4.3.37}$$

and hence

$$|V(t)| \le K \text{ for all } t \ge 0 \tag{4.3.38}$$

where $K := 2|\eta| + 2|G(x(0))| + B_4$ where $\eta = \inf\{G(u) : u \in R\}$. We also observe that $|C(t,0)| \le B_5$ for a $B_5 > 0$ whenever (4.3.32) holds.

Now integrate (4.3.30) from s to t and use (4.3.38) to obtain

$$\int_s^t |g(x(u))|du \le V(s) - V(t) + M(t-s) \le 2K + M(t-s). \tag{4.3.39}$$

Applying (4.3.39) to (4.3.21) we find

$$\begin{aligned}
|x(t)| &\le |a(t)| + \left|\int_0^t C(t,s)g(x(s))ds\right| \\
&\le |a(t)| + \left|C(t,0)\int_0^t g(x(s))ds + \int_0^t C_s(t,s)\int_s^t g(x(u))duds\right| \\
&\le |a(t)| + |C(t,0)|(Mt + 2K) + \int_0^t |C_s(t,s)|[M(t-s) + 2K]ds \\
&\le \sup_{s\ge 0}|a(s)| + B_1 M + 2KB_5 + B_1(M + 2K).
\end{aligned}$$

This implies that x is bounded. The proof is complete.

Corollary. *Suppose that (4.3.23)-(4.3.24) hold. If there is an $L > 0$ and $\mu > 0$ with*

$$xg(x) \ge 0 \text{ for } |x| \ge L \text{ and } C(t,t) \ge \mu, \tag{4.3.40}$$

then there is an $M > 0$ with

$$V'(t) \le -|g(x)| + M \tag{4.3.41}$$

whenever x is a solution of (4.3.21). If, in addition, (4.3.32) holds, then $V(t)$ satisfies (4.3.37) and any solution of (4.3.21) is bounded.

Remark Inequalities related to (4.3.37) and (4.3.41) are of fundamental importance in the study of boundedness and periodic solutions in differential equations by Liapunov's direct method (see Burton (2005c) and Yoshizawa (1966). Not only are these practical inequalities with many applications, but such combined relations are directly linked to the right-hand side of the equations, and hence, much of the qualitative properties of solutions can be derived by taking full advantage of the Liapunov functions.

The following example shows that if $C(t,s)$ is not convex while $D(t,s)$ is, then the combined equation $x' + kx$ is the right form to apply Liapunov functionals.

Example 4.3.1. *Consider the equation*

$$x(t) = a(t) - \int_0^t C(t,s)g(x(s))ds \tag{4.3.42}$$

where $a : R^+ \to R$ and $g : R \to R$ are continuous with a, a' bounded, and

$$C(t,s) = C(t-s) = -e^{-(t-s+3)^2} \quad \text{for } t \geq s \geq 0.$$

It is clear that $C(t,s)$ is not convex (it is not even positive). If we choose $k = 4$, then

$$D(t,s) = C_t(t,s) + kC(t,s) = 2(t-s+1)e^{-(t-s+3)^2}.$$

A straightforward calculation shows that $D(t,s)$ is convex and (4.3.24) holds. We also see that $C(t,s)$ satisfies (4.3.32). Thus, if there exist constants $L > 0$ and $\mu > 0$ with $xg(x) \geq 0$ for $|x| \geq L$ and

$$g(x)[kx + C(t,t)g(x)] = g(x)[4x - e^{-9}g(x)] \geq \mu g^2(x)$$

for $|x| \geq L$, then any solution of (4.3.42) is bounded by Theorem 4.3.5.

4.4 A Truncated Equation and Unification

We consider the finite delay equation

$$x(t) = a(t) - \int_{t-h}^t C(t,s)g(x(s))ds \tag{4.4.1}$$

in which $h > 0$ is a constant, $a : R^+ \to R$, $C : R^+ \times [-h, \infty) \to R$, $g : R \to R$ are all continuous with a, a' bounded. Write

$$D(t,s) := C_t(t,s) + kC(t,s) \tag{4.4.2}$$

for a positive constant k and assume that $D(t, s)$ is convex:

$$D(t,s) \geq 0,\ D_s(t,s) \geq 0,\ D_{st}(t,s) \leq 0,\ D_t(t,s) \leq 0 \tag{4.4.3}$$

for $t \geq s \geq -h$ and that

$$C(t,t-h) = 0,\ \ C_t(t,t-h) = 0,\ \ C(t,t) \geq -B \tag{4.4.4}$$

for all $t \geq 0$ and a constant $B > 0$, where $C_t(t,t-h)$ is the partial derivative of $C(t,s)$ with respect the first variable for $s = t - h$.

Before we get too far into the work, it is interesting to point out classical forms for C. Let

$$C(t,s) = C(t-s) = (-1)^n(t-s-h)^n,\ n > 2.$$

Not only does it satisfy (4.4.4), but it is a convex kernel for $0 \leq s \leq t \leq h$. Moreover, if we let $C(t) = 0$ for $t > h$ then that kernel will satisfy our work in Section 4.3. In Section 4.3 something very interesting happens. In the linear case we have

$$x(t) = a(t) - \int_0^t C(t-s)x(s)ds,$$

an equation about which there is a very straightforward theory. However, for $t \geq h$ it becomes

$$x(t) = a(t) - \int_{t-h}^t C(t-s)x(s)ds$$

and that belongs to a class of far more complex structure.

Differentiate (4.4.1) and take into account (4.4.4) to form

$$x' + kx = \left[a' + ka - C(t,t)g(x) - \int_{t-h}^t D(t,s)g(x(s))ds\right]. \tag{4.4.5}$$

Now, define the Liapunov functional

$$V(t) = 2G(x(t)) + \int_{t-h}^t D_s(t,s)\left(\int_s^t g(x(u))du\right)^2 ds \tag{4.4.6}$$

for $x \in C([-h,\infty),R)$, where $G(x) = \int_0^x g(s)ds$.

Theorem 4.4.1. *Suppose that (4.4.3) and (4.4.4) hold. If $x(t)$ is a solution of (4.4.1), then the derivative of V along that solution satisfies*

$$V'(t) \leq 2g(x)[a'(t) + ka(t)] - 2g(x)[kx + C(t,t)g(x)]. \tag{4.4.7}$$

If there is an $L > 0$ with

$$xg(x) \geq 0 \ \text{ for } \ |x| \geq L \tag{4.4.8}$$

and if, in addition, there is a $\mu > 0$ with

$$g(x)[kx + C(t,t)g(x)] \geq \mu g^2(x) \tag{4.4.9}$$

for $|x| \geq L$, then there is an $M > 0$ with

$$V'(t) \leq -|g(x)| + M \tag{4.4.10}$$

for all $t \geq 0$.

Proof. We first observe that if x is a solution of (4.4.1), then x is also a solution of (4.4.5). Now let x be a solution (4.4.1) and $V(t)$ be defined in (4.4.6). We then find that

$$\begin{aligned} V'(t) &= 2g(x)x'(t) + \int_{t-h}^{t} D_{st}(t,s)\left(\int_s^t g(x(u))du\right)^2 ds \\ &\quad - D_s(t,t-h)\left(\int_{t-h}^{t} g(x(u))du\right)^2 \\ &\quad + 2g(x(t))\int_{t-h}^{t} D_s(t,s)\int_s^t g(x(u))duds. \end{aligned}$$

Integration of the last term by parts and use of (4.4.4) yield

$$\begin{aligned} &2g(x(t))\int_{t-h}^{t} D_s(t,s)\int_s^t g(x(u))duds \\ &= 2g(x(t))\left[D(t,s)\int_s^t g(x(u))du\bigg|_{s=t-h}^{s=t} + \int_{t-h}^{t} D(t,s)g(x(s))ds\right] \\ &= 2g(x(t))\int_{t-h}^{t} D(t,s)g(x(s))ds. \end{aligned} \tag{4.4.11}$$

Since $D_{st}(t,s) \le 0$ and $D_s(t,t-h) \ge 0$, the middle two terms of V' are not positive, and if we use (4.4.5) and (4.4.11), we obtain

$$V'(t) \le 2g(x)\left[-kx + (a' + ka) - C(t,t)g(x) - \int_{t-h}^{t} D(t,s)g(x(s))ds\right]$$

$$+ 2g(x)\int_{t-h}^{t} D(t,s)g(x(s))ds$$

$$= 2g(x)[a'(t) + ka(t)] - 2g(x)[kx + C(t,t)g(x)]$$

verifying (4.4.7).

Now assume that (4.4.8) and (4.4.9) hold. We may choose $N > 1$ so large that $\mu(N-1) > -B$, where B and μ are defined in (4.4.4) and (4.4.9), respectively. If $|x| \ge L$, then $xg(x) \ge 0$, and by (4.4.9), we obtain for $|x(t)| \ge L$ that

$$V'(t) \le 2|g(x)|[|a'| + k|a|] - 2g(x)[kx + C(t,t)g(x)]$$

$$= 2|g(x)|[|a'| + k|a|] - 2k\frac{1}{N}xg(x)$$

$$-2\left(1 - \frac{1}{N}\right)g(x)[kx + C(t,t)g(x)] - 2\frac{1}{N}C(t,t)g^2(x)$$

$$\le 2|g(x)|[|a'| + k|a|] - 2k\frac{1}{N}xg(x)$$

$$-2\left(1 - \frac{1}{N}\right)\mu g^2(x) + 2\frac{1}{N}Bg^2(x)$$

$$\le -2|g(x)|\left[\frac{1}{N}k|x| - (|a'(t)| + k|a(t)|)\right].$$

We may assume that

$$L \ge N[\sup_{t\ge 0}(|a'(t)| + k|a(t)|) + 1]\big/k.$$

Thus, if $|x(t)| \ge L$, then $V'(t) \le -|g(x(t))|$. Since $-C(t,t) \le B$, it is clear that $V'(t) \le M$ for $0 \le |x(t)| \le L$, where the constant $M > 0$ is a function of L, and hence,

$$V'(t) \le -|g(x(t))| + M$$

for all $t \ge 0$. This completes the proof.

To establish boundedness of solutions, assume that there is a $B_1 > 0$ with

$$D(t,t) \leq B_1 \ \ \textit{and} \ \ \int_{t-h}^{t} |C_s(t,s)|ds \leq B_1 \tag{4.4.12}$$

for $t \geq 0$. We then see that

$$\int_{t-h}^{t} D_s(t,s)ds = D(t,t) \leq B_1 \ \ \textit{and} \ \ |C(t,t)| \leq \int_{t-h}^{t} |C_s(t,s)|ds \leq B_1.$$

Theorem 4.4.2. *If (4.4.3)-(4.4.4),(4.4.8)-(4.4.9), and (4.4.12) hold, then any solution of (4.4.1) is bounded.*

Proof. Let x be a solution of (4.4.1) and $V(t)$ be defined in (4.4.6). Then $V(t)$ is bounded below and satisfies (4.4.10). We now show that $V(t)$ is bounded above. If $V(t)$ is unbounded, then there exists a sequence $\{t_n\} \uparrow \infty$ with $V(t_n) \to \infty$ as $n \to \infty$ and

$$V(t_n) \geq V(s) \ \ \textit{for} \ \ 0 \leq s \leq t_n.$$

It then follows from (4.4.10) that

$$0 \leq V(t_n) - V(s) \leq -\int_{s}^{t_n} |g(x(u))|du + M(t_n - s).$$

This implies that

$$\int_{s}^{t_n} |g(x(u))|du \leq M(t_n - s) \tag{4.4.13}$$

and, in particular, that

$$\int_{s}^{t_n} |g(x(u))|du \leq M(t_n - s) \leq hM \tag{4.4.14}$$

for all $t_n - h \leq s \leq t_n$. Applying (4.4.14) to (4.4.1), we see that

$$\begin{aligned}
(x(t_n)-a(t_n))^2 &= \left(\int_{t_n-h}^{t_n} C(t_n,s)g(x(s))ds\right)^2 \\
&= \left(-C(t_n,s)\int_{s}^{t_n} g(x(s))ds\Big|_{s=t_n-h}^{s=t_n} + \int_{t_n-h}^{t_n} C_s(t_n,s)\int_{s}^{t_n} g(x(u))duds\right)^2 \\
&= \left(\int_{t_n-h}^{t_n} C_s(t_n,s)\int_{s}^{t_n} g(x(u))duds\right)^2 \\
&\leq \left(hM\int_{t_n-h}^{t_n} |C_s(t_n,s)|ds\right)^2 \leq h^2B_1^2M^2.
\end{aligned}$$

This implies that

$$|x(t_n)| \leq \sup_{s \geq 0} |a(s)| + hB_1 M := B_2 \tag{4.4.15}$$

and that $|G(x(t_n))| \leq B_3$ for a $B_3 > 0$. We now arrive at

$$\begin{aligned} V(t_n) &= 2G(x(t_n)) + \int_{t_n - h}^{t_n} D_s(t_n, s) \left(\int_s^{t_n} g(x(u))du \right)^2 ds \\ &\leq 2G(x(t_n)) + h^2 M^2 \int_{t_n - h}^{t_n} D_s(t_n, s)ds \\ &\leq 2B_3 + B_1 h^2 M^2 := B_4, \end{aligned} \tag{4.4.16}$$

a contradiction. Thus, $V(t)$ is bounded. In fact, we have

$$2G(x(t)) \leq V(t) \leq \max\{V(0), B_4\}$$

and hence

$$|V(t)| \leq K \text{ for all } t \geq 0 \tag{4.4.17}$$

where $K := 2|\eta| + 2|G(x(0))| + B_4$ with $\eta = \inf\{G(u) : u \in R\}$.

We now integrate (4.4.10) from s to t and use (4.4.17) to obtain

$$\int_s^t |g(x(u))|du \leq V(s) - V(t) + M(t - s) \leq 2K + M(t - s)$$

and hence

$$\int_s^t |g(x(u))|du \leq 2K + hM \quad \text{for} \quad t_n - h \leq s \leq t_n. \tag{4.4.18}$$

Applying (4.4.18) to (4.4.1) we find that

$$\begin{aligned} |x(t)| &\leq |a(t)| + \left| \int_{t-h}^t C(t, s)g(x(s))ds \right| \\ &\leq |a(t)| + \left| C(t, t - h) \int_{t-h}^t g(x(s))ds + \int_{t-h}^t C_s(t, s) \int_s^t g(x(u))duds \right| \\ &\leq |a(t)| + \int_{t-h}^t |C_s(t, s)|[2K + hM]ds \\ &\leq \sup_{s \geq 0} |a(s)| + B_1(2K + hM). \end{aligned} \tag{4.4.19}$$

This implies that x is bounded. The proof is complete.

Periodicity

We now study the existence of periodic solutions of (4.4.1). Assume that $a : R \to R$, $C : R \times R \to R$, and $g : R \to R$ are continuous and that there is a $T > 0$ with

$$a(t+T) = a(t),\ C(t+T, s+T) = C(t, s) \tag{4.4.20}$$

for all $t \geq s$. If (4.4.20) holds, then $C(t,t)$ and $\int_{t-h}^{t} |C_s(t,s)|ds$ are T-periodic, and so there are B and B_1 with

$$C(t,t) \geq -B, \quad \int_{t-h}^{t} |C_s(t,s)|ds \leq B_1;$$

then part of (4.4.4) and (4.4.12) are satisfied. We define a companion of (4.4.1) by

$$x(t) = \lambda \left[a(t) - \int_{t-h}^{t} C(t,s)g(x(s))ds \right], \quad 0 \leq \lambda \leq 1 \tag{4.4.21}$$

for $t \in R$ and form a differential equation

$$x' + kx = \lambda \left[a' + ka - C(t,t)g(x) - \int_{t-h}^{t} D(t,s)g(x(s))ds \right]. \tag{4.4.22}$$

To obtain an *a priori* bound for all T-periodic solutions of (4.4.21), define

$$V_1(t) = 2G(x(t)) + \lambda \int_{t-h}^{t} D_s(t,s) \left(\int_s^t g(x(u))du \right)^2 ds \tag{4.4.23}$$

for $t \in R$ and $x \in (P_T, \|\cdot\|)$.

Theorem 4.4.3. *If (4.4.3)-(4.4.4), (4.4.8)-(4.4.9), and (4.4.20) hold for $t \geq s$, then (4.4.1) has a T-periodic solution.*

Proof. Let x be a T-periodic solution of (4.4.21) and $V_1(t)$ be defined in (4.4.23). Then we have

$$V_1'(t) \leq -|g(x)| + M \tag{4.4.24}$$

for $t \geq 0$ and for an $M > 0$ independent of x and λ. Since $V_1(t)$ is T-periodic, $V_1(t)$ has a global maximum at $q \in [0,T]$, and hence, at $t_n = q + nT$. We then have

$$0 \leq V_1(t_n) - V_1(s) \leq -\int_s^{t_n} |g(x(u))|du + M(t_n - s)$$

for all $s \le t_n$. An argument similar to that of (4.4.13)-(4.4.16) shows that $V_1(t_n) \le B_4$ with B_4 defined in (4.4.16). Observing that

$$V_1(0) \le V_1(t_n) \le B_4,$$

we see that $|V_1(t)| \le K$ with $K = 2|\eta| + B_4$, where $\eta = \inf\{G(u) : u \in R\}$. We then follow the argument in (4.4.19) to arrive at

$$|x(t)| < \sup_{s\ge 0} |a(s)| + B_1(2K + hM)) + 1 := B^* \tag{4.4.25}$$

for all $t \in R$. This implies that $\|x\| < B^*$ whenever x is a T-periodic solution of (4.4.21) for $0 \le \lambda \le 1$.

Define a mapping F on P_T by

$$F(\phi)(t) = a(t) - \int_{t-h}^{t} C(t,s)g(\phi(s))ds \tag{4.4.26}$$

for each $\phi \in P_T$. It is clear that $F(\phi) \in P_T$. We will show that F is continuous on P_T and is compact on each bounded subset of P_T. If $\tilde{\phi}, \phi \in P_T$, then

$$\begin{aligned} |F(\phi)(t) - F(\tilde{\phi})(t)| &= \left| \int_{t-h}^{t} C(t,s)g(\phi(s))ds - \int_{t-h}^{t} C(t,s)g(\tilde{\phi}(s))ds \right| \\ &= \left| \int_{t-h}^{t} C_s(t,s) \left(\int_s^t g(\phi(v))dv - \int_s^t g(\tilde{\phi}(v))dv \right) ds \right|. \end{aligned} \tag{4.4.27}$$

Since g is uniformly continuous on $\{x \in R : |x| \le \|\tilde{\phi}\| + 1\}$, then for any $\varepsilon > 0$, there exists $0 < \delta < 1$ such that $\|\phi - \tilde{\phi}\| < \delta$ implies $|g(\phi(s)) - g(\tilde{\phi}(s))| < \varepsilon$ for all $s \in [0,T]$. It follows from (4.4.27) that $\|F(\phi) - F(\tilde{\phi})\| \le hB_1\varepsilon$. Thus, F is continuous on P_T.

We now show that F is compact on each bounded subset of P_T. Let $\eta > 0$ and define

$$\Gamma = \{F(\phi) : \phi \in P_T, \|\phi\| \le \eta\}. \tag{4.4.28}$$

Observe that

$$\begin{aligned}\frac{d}{dt}F(\phi)(t)=&a'(t) - C(t,t)g(\phi(t)) - \int_{t-h}^{t} C_t(t,s)g(\phi(s))ds\\ =&a'(t)-C(t,t)g(\phi(t))\\ &-\int_{t-h}^{t} D(t,s)g(\phi(s))ds + k\int_{t-h}^{t} C(t,s)g(\phi(s))ds\\ =&a'(t)-C(t,t)g(\phi(t))-\int_{t-h}^{t} D(t,s)g(\phi(s))ds\\ &+k\int_{t-h}^{t} C_s(t,s)\int_s^t g(\phi(v))dvds\end{aligned}$$

and that

$$\begin{aligned}&|\frac{d}{dt}F(\phi)(t)| \le \|a'\|\\ &+g^* \sup_{0\le t\le T}\left[|C(t,t)| + \int_{t-h}^{t} D(t,s)ds + k\int_{t-h}^{t} |C_s(t,s)|(t-s)ds\right]\\ &\le \|a'\| + g^*[\sup_{0\le t\le T} |C(t,t)| + hB_1 + hkB_1]\end{aligned}$$

where $g^* = \sup\{|g(u)| : |u| \le \eta\}$; thus, Γ is equi-continuous. The uniform boundedness of Γ follows from the inequality

$$|F(\phi)(t)| \le \|a\| + \int_{t-h}^{t} |C_s(t,s)| \int_s^t |g(\phi(v))|dv \le \|a\| + hB_1g^*$$

for all $\phi \in \Gamma$. So, by the Ascoli-Arzela theorem, Γ lies in a compact subset of P_T. By Schaefer's theorem, we see F has a fixed point which is a T-periodic solution of (4.4.1). The proof is complete.

We now give two examples which show a connection between this section and Section 4.3.

Example 4.4.1. *Consider the scalar equation*

$$x(t) = a(t) - \int_{-\infty}^{t} C(t-s)g(x(s))ds \tag{4.4.29}$$

where

$$\begin{aligned}C(t) &= (-1)^n(t-h)^n,\ 0 \le t \le h,\ n = 3, 4, ..\\ &= 0,\ t \ge h\end{aligned} \tag{4.4.30}$$

for some $h > 0$. It is readily verified that C'' is continuous for $0 \leq t < \infty$ and that $C(t-s)$ is convex. Moreover, using two changes of variable we find that

$$\begin{aligned} x(t) &= a(t) - \int_0^\infty C(u)g(x(t-u))du \\ &= a(t) - \int_0^h C(u)g(x(t-u))du \\ &= a(t) - \int_{t-h}^t C(t-s)g(x(s))ds \end{aligned} \tag{4.4.31}$$

and $C(h) = C'(h) = C''(h) = 0$. All of the work in Section 4.3 holds for this equation.

Example 4.4.2. . *Consider*

$$x(t) = a(t) - \int_0^t C(t-s)g(x(s))ds \tag{4.4.32}$$

with solution ϕ on $[0,h]$ where C satisfies (4.4.30). Then for $t \geq h$ we have

$$x(t) = a(t) - \int_{t-h}^t C(t-s)g(x(s))ds, \ t \geq h, \tag{4.4.33}$$

with initial function ϕ on $[0,h]$.

Theorem 4.4.4. *Let $\frac{dg(x)}{dx}$ be continuous, let $x(t)$ be the unique solution of (4.32), and let $y(t)$ be any continuous solution of (4.4.33). Suppose that there is an $L > 0$ with $\frac{dg(x)}{dx} \geq L$. Then $z(t) := x(t) - y(t) \in L^2[h,\infty)$. If, in addition, there is an $M > 0$ with $\frac{dg(x)}{dx} \leq M$, then $z(t) \to 0$ as $t \to \infty$.*

Proof. We have for $t \geq h$ that

$$z(t) = -\int_{t-h}^t C(t-s)[g(x(s)) - g(y(s))]ds \tag{4.4.34}$$

$$= -\int_{t-h}^t C(t-s)\frac{dg(\xi(s))}{dx}z(s)ds \tag{4.4.35}$$

where $\xi(s)$ is between $x(s)$ and $y(s)$. Define a Liapunov functional by

$$V(t) = \int_{t-h}^t C_s(t-s)\left(\int_s^t [g(x(u)) - g(y(u))]du\right)^2 ds \tag{4.4.36}$$

with derivative satisfying

$$V'(t) \leq -2[g(x(t)) - g(y(t))][x(t) - y(t)] = -2\frac{dg(\xi(t))}{dx}z^2(t) \leq -2Lz^2(t).$$

(4.4.37)

This yields the first conclusion. With the last assumption, note that

$$|z(t)| \leq M \int_{t-h}^{t} |C(t-s)||z(s)|ds \tag{4.4.38}$$

$$\leq M\sqrt{\int_{t-h}^{t} C^2(t-s)ds \int_{t-h}^{t} z^2(s)ds} \tag{4.4.39}$$

$$\leq M\sqrt{h^{2n+1}\int_{t-h}^{t} z^2(s)ds} \tag{4.4.40}$$

and

$$\int_{t-h}^{t} z^2(s)ds \to 0 \text{ as } t \to \infty. \tag{4.4.41}$$

Under the conditions here, with C defined by (4.4.30) we see that the solutions of the equations in Sections 4.3 and 4.4 all converge to the same function both pointwise and in L^2.

Notes The material for Sections 4.3 and 4.4 is found in Burton and Zhang (2012). It was first published by the Department of Mathematics of the Universidad de La Frontera Temuco-Chile and the Department of Mathematics of the Universidade Federal de Pernambuco, Recife-Brazil in the journal *CUBO A Mathematical Journal*. Full details are given in the bibliography.

Chapter 5

Appendix: Fixed Points & Fractional Equations

Our work with fractional differential equations reached a turning point with the work in Section 2.5.1 and, in particular, equation

$$^cD^q x = f(t) - k[x(t) + G(t, x(t))], \quad x(0) \in \Re, \quad 0 < q < 1,$$

inverted as

$$x(t) = -\frac{k}{\Gamma(q)} \int_0^t (t-s)^{q-1} \left[x(s) + G(s, x(s)) - \frac{f(s)}{k} \right] ds, \tag{2.5.1.7}$$

and transformed into

$$x(t) = y(t) - \int_0^t R(t-s) \left[G(s, x(s)) - \frac{f(s)}{k} \right] ds \tag{2.5.1.24}$$

with

$$y(t) = x(0) \left[1 - \int_0^t R(s) ds \right]$$

and

$$0 < R(t), \quad \int_0^\infty R(s) ds = 1.$$

It turns out that (2.5.1.24) is very fixed point friendly. Moreover, fixed point theory and Liapunov theory have always gone hand-in-glove.

First, we laid a foundation for fixed point theory using (2.5.1.24) setting up mappings and obtaining qualitative properties of solutions in the form of bounded continuous functions, asymptotically periodic solutions, and L^p solutions. These are detailed in Burton (2011), (2012a,b), Burton-Zhang (2012a,b,c,d). In Burton-Zhang (2012d) we obtained an *a priori* bound result which is just set up for Liapunov's direct method.

References

Banach, S. (1932). "Théorie des Opérations Linéairs" (reprint of the 1932 ed.). Chelsea, New York.

Barbashin, E. A. (1968). The construction of Lyapunov functions. *Differential Equations* **4**, 1097–1112.

Barroso, C. S. (2003). Krasnoselskii's fixed point theorem for weakly continuous maps. *Nonlinear Anal.* **55**, 25-31.

Barroso, C. S. and Teixeira, E. V. (2005). A topological and geometric approach to fixed point results for sum of operators and applications. *Nonlinear Anal.* **60**, 625-650.

Becker, L. C. (1979). Stability considerations for Volterra integro-differential equations. Ph.D. dissertation. Southern Illinois University, Carbondale, Illinois.

Becker, Leigh C. (2006). Principal matrix solutions and variation of parameters for a Volterra integro-differential equation and its adjoint. *E. J. Qualitative Theory of Diff. Equ.* **14**, pp. 1 - 22.

Becker, Leigh C. (2007). Function bounds for solutions of Volterra equations and exponential asymptotic stability. *Nonlinear Anal.* **67**, 382-397.

Becker, Leigh C. (2011). Resolvents and solutions of weakly singular linear Volterra integral equations. *Nonlinear Anal.*, **74**, 1892-1912.

Becker, Leigh C. (2012). Resolvents for weakly singular kernels and fractional differential equations. *Nonlinear Anal.*, **75**, 4839-4861.

Becker, L. C., Burton, T. A., and Krisztin, T. (1988). Floquet theory for a Volterra equation. *J. London Math. Soc.* **37**, 141–147.

Becker, L. C., Burton, T. A., and Purnaras, I. K. (2012). Singular integral equations, Liapunov functionals, and resolvents. *Nonlinear Anal.* **75**, pp. 3277-3291.

Boyd, D. W. and Wong, J. S. W. (1969). On nonlinear contractions. *Proc. Amer. Math. Soc.* **20**, 458-464.

Burton, T. A. (1980c). An integrodifferential equation. *Proc. Amer. Math. Soc.* **79**, 393–399.

Burton, T. A. (1983a). Volterra equations with small kernels. *J. Integral Equations* **5**, 271–285.

Burton, T. A. (1983b). "Volterra Integral and Differential Equations." Academic Press, Orlando.

Burton, T. A. (1984). Periodic solutions of linear Volterra equations. *Funkcial Ekvac.* **27**, 229-253.

Burton, T. A. (1985). "Stability and Periodic Solutions of Ordinary and Functional Differential Equations." Academic Press, Orlando.

Burton, T. A. (1991). The nonlinear wave equation as a Liénard equation. *Funkcialaj Ekvacioj* **34**, 529-545.

Burton, T. A., (1993). Boundedness and periodicity in integral and integro-differential equations. *Differential Equations and Dynamical Systems* **1**, 161-172.

Burton, T. A., (1993a). Averaged neural networks. *Neural Networks* **6**, 667-680.

Burton, T. A. (1994a). Liapunov functionals and periodicity in integral equations. *Tohoku Math. J.* **46**, 207-220.

Burton, T. A. (1994b). Differential inequalities and existence theory for differential, integral, and delay equations, pp. 35-56. In "Comparison Methods and Stability Theory." Xinzhi Liu and David Siegel, eds, Dekker, New York.

Burton, T. A. (1996a). Examples of Lyapunov functionals for non-differentiated equations. Proc. First World Congress of Nonlinear Analysts, 1992. V. Lakshmikantham, ed. Walter de Gruyter, New York. pp. 1203–1214.

Burton, T. A. (1996b). Integral equations, implicit functions, and fixed points. *Proc. Amer. Math. Soc.* **124**, 2383-2390.

Burton, T. A., (1997). Linear integral equations and periodicity. *Annals of Differential Equations* **13**, 313-326.

Burton, T. A. (1998a). A fixed point theorem of Krasnoselskii. *Appl. Math. Lett.* **11**, 85-88.

Burton, T. A. (2003). Stability by fixed point theory or Liapunov theory: a comparison. *Fixed Point Theory* **4**, 15-32.

Burton, T. A. (1998b). Basic neutral integral equations of advanced type. *Nonlinear Anal.* **31**, 295-310.

Burton, T. A. (2005b). "Stability and Periodic Solutions of Ordinary and Functional Differential Equations." Dover, New York. (This is a slightly corrected reprint of the 1985 edition published by Academic Press.)

Burton, T. A. (2005c). "Volterra Integral and Differential Equations, Second Edition." Elsevier, Amsterdam.

Burton, T. A. (2006a). Integral equations, Volterra equations, and the remarkable resolvent: contractions. *E. J. Qualitative Theory of Diff. Equ.* **No. 2**, 1-17. (http://www.math.u-szeged.hu/ejqtde/2006/200602.html)

Burton, T. A. (2006b). "Stability by Fixed Point Theory for Functional Differential Equations." Dover, New York.

Burton, T. A., (2007a). Integral equations, L^p-forcing, remarkable resolvent: Liapunov functionals. *Nonlinear Anal.* **68**, 35-46.

Burton, T. A., (2007b). Integral equations, large and small forcing functions: periodicity. *Math. Computer Modelling* **45**, 1363-1375.

Burton, T. A., (2007c). Integral equations, large forcing, strong resolvents. *Carpathian J. of Math.* **23**, 1-10.

Burton, T. A., (2007d). Scalar nonlinear integral equations. *Tatra Mountains Mathematical Publications* **38**, 41–56.

Burton, T. A., (2008a). Integral equations with contrasting kernels. *E. J. Qualitative Theory of Diff. Equ.* **No. 2**, pp. 1-22. *http:/www.math.u-szeged.hu/ejqtde/2008/200802.html*

Burton, T. A., (2008b). Integral equations, periodicity, and fixed points. *Fixed Point Theory* **9**, 47-65.

Burton, T. A., (2010). Liapunov functionals, convex kernels, and strategy. *Nonlinear Dynamics and Systems Theory* **10**, 325-337.

Burton, T. A., (2010a). A Liapunov functional for a linear integral equation. *E. J. Qualitative Theory of Diff. Equ.* **10**, 1-10. *http:/www.math.u-szeged.hu/ejqtde/*

Burton, T. A., (2010b). A Liapunov functional for a singular integral equation. *Nonlinear Anal.* **73**, 3873-3882.

Burton, T. A., (2010c). Six integral equations and a flexible Liapunov functional. *Trudy Instituta Matematiki i Mekhaniki UrO RAN* **16**(5), 241-252.

Burton, T. A., (2011). Fractional differential equations and Lyapunov functionals. *Nonlinear Anal.* **74**, 5648-5662.

Burton, T. A., (2011a). Kernel-resolvent relations for an integral equation. *Tatra Mountains Mathematical Publications*, **48**, 1-16.

Burton, T. A., (2012a). Fractional equations and a theorem of Brouwer-Schauder type. Preprint.

Burton, T. A., (2012b). Neutral differential equations and a Brouwer-Krasnoselskii type theorem. Preprint.

Burton, T. A. and Dwiggins, D. P., (2011). Resolvents of integral equations with continuous kernels. *Nonlinear Studies*, **18**, 293-305.

Burton, T. A. and Dwiggins, D. P., (2010). Resolvents, integral equations, and limit sets. *Mathematica Bohemica*, **135**, 337-354.

Burton, T. A., Eloe, P. W., and Islam, M. N. (1990). Periodic solutions of linear integrodifferential equations. *Math. Nach.* **147**, 175-184.

Burton, T. A. and Furumochi, Tetsuo (1994). A stability theory for integral equations. *J. Integral Equations Appl.* **6**, 445–477.

Burton, T. A. and Furumochi, Tetsuo (1995). Periodic solutions of a Volterra equation and robustness. *Nonlinear Anal.* **25**, 1199-1219.

Burton, T. A. and Furumochi, Tetsuo (1996). Periodic and asymptotically periodic solutions of Volterra integral equations. *Funkcialaj Ekvacioj* **39**, 87-107.

Burton, T. A. and Furumochi, Tetsuo (2001). A note on stability by Schauder's theorem. *Funkcialaj Ekvacioj* **44**, 73-82.
179, 193–209.

Burton, T. A. and Haddock, J. (2009). Qualitative properties of solutions of integral equations. *Nonlinear Anal.*, **71**, 5712–5723.

Burton, T. A. and J. R. Haddock, J. R. (2010). Parallel theories of integral equations, *Nonlinear Studies* **17**, 177-197.

Burton, T. A. and Kirk, Colleen (1998). A fixed point theorem of Krasnoselskii-Schaefer Type. *Math. Nachr.* **189**, 23-31.

Burton, T. A. and Makay, G. (2002). Continuity, compactness, fixed points, and integral equations. *E. J. Qualitative Theory of Diff. Equ.* **No. 14**, 1-13. (http://www.math.u-szeged.hu/ejqtde/2002/)

Burton, T. A. and Purnaras, I. K. (2011). L^p-solutions of singular integrodiffer-ntial equations. *J. Math. Anal. Appl.* **386**, 830-841.

Burton, T. A. and Purnaras, I. K. (2012). Singular integro-differential equations with small kernels. preprint.

Burton, T. A. and Somolinos, A. (2007). The Lurie control saisfies a Liénard equation. *Dyn. Contin., Discrete Impuls. Syst. Ser. B Appl. Algorithms* **14**, 625-640.

Burton, T. A. and Zhang, B. (1990). Uniform ultimate boundedness and periodicity in functional differential equations. *Tohoku Math. J.* **42**, 93–100.

Burton, T. A. and Bo Zhang, Bo (1998). Periodicity in delay equations by direct fixed point mappings, *Differential Equations and Dynamical Systems* **6**, 413-424.

Burton, T. A. and Zhang, B. (2004). Fixed points and stability of an integral equation: nonuniqueness. *Applied Math. Letters* **17**, 839-846.

Burton, T. A. and Zhang, Bo (2011). Periodic solutions of singular integral equations. *Nonlinear Dynamics and Systems Theory* **11(2)**, 113-123.

Burton, T. A. and Zhang, Bo (2012). Bounded and periodic solutions of integral equations. *Cubo: A Mathematical Journal* **14**(1), 55-79.

Burton, T. A. and Zhang, Bo (2012a). Asymptotically periodic solutions of fractional differential equations. Preprint.

Burton, T. A. and Zhang, Bo (2012b). L^p-solutions of fractional differential equations. *Nonlinear Studies* **19**(2), 307-324.

Burton, T. A. and Zhang, Bo (2012c). Fixed points and fractional differential equations: Examples. *Fixed Point Theory* to appear.

Burton, T. A. and Zhang,Bo (2012d). Fractional equations and generalizations of Schaefer's and Krasnoselskii's fixed point theorems, *Nonlinear Anal.:TMA* DOI 10.1016/j.na.2012.07.022.

Cao, Jinde, Li, H. X., and Ho, Daniel W. C. (2005). Synchronization criteria of Luré systems with time-delay feedback control. *Chaos, Solitons, and Fractals* **23:4**, 1285-1298.

Coddington, E.A. and Levinson, N. (1955). "Theory of Ordinary Differential Equations." McGraw-Hill, New York.

Consiglio, A. (1940). Risoluzione di una equazione integrale non lineare presentatasi in un problema di turbalenza, *Academia Gioenia di Scienze Naturali in Cantania*, **4**, No. XX., 1-13.

Corduneanu, C. (1971). "Principles of Differential and Integral Equations." Allyn and Bacon, Rockledge, New Jersey.

Corduneanu, C. (1973). "Integral Equations and Stability of Feedback Systems." Academic Press, New York.

Corduneanu, C. (1977). "Principles of Differential and Integral Equations." Chelsea Publishing Co., New York.

Corduneanu, C. (1991). "Integral Equations and Applications." Cambridge Univ. Press, Cambridge, U.K.

Diethelm, Kai (2004). "The Analysis of Fractional Differential Equations" Springer, New York.

Driver, R. D. (1962). Existence and stability of solutions of a delay-differential system. *Arch. Rational Mech. Anal.* **10**, 401–426.

Driver, R. D. (1963). Existence theory for a delay-differential system. *Contrib. Differential Equations* **1**, 317–335.

Driver, R. D. (1965). Existence and continuous dependence of solutions of a neutral functional-differential equation. *Archive Rat. Mech. Anal.* **19**, 149-186.

El'sgol'ts, L. E. (1966). "Introduction to the Theory of Differential Equations with Deviating Arguments." Holden-Day, San Francisco.

Feller, W. (1941). On the integral equation of renewal theory. *Ann. Math. Statist.* **12**, 243–267.

Gao, H., Li, Y. and Zhang, Bo (2011). A fixed point theorem of Krasnoselskii-Schaefer type and its applications in control and periodicity of integral equations. *Fixed Point Theory* **12**, 91-112.

Gopalsamy, K. (1992). "Stability and Oscillations in Delay Differential Equations of Population Dynamics." Kluwer, Dordrecht.

Gopalsamy, K. and Zhang, B.G. (1988). On a neutral delay logistic equation. *Dynamics Stability Systems* **2**, 183–195.

Graef, J. R., Qian, Chuanxi, and Zhang, Bo. (2004). Formulas of Liapunov Functions for systems of linear difference equations. *Proc. London Math. Soc.* **88**, 185-203.

Graves, L. M. (1946). "The Theory of Functions of Real Variables." McGraw-Hill, New York.

Grossman, S. I., and Miller, R. K. (1970). Perturbation theory for Volterra integrodifferential systems. *J. Differential Equations* **8**, 457–474.

Hartman, P. (1964). "Ordinary Differential Equations." Wiley, New York.

Henry, D. (1981). "Geometric Theory of Semilinear Parabolic Equations." Springer, Berlin.

Hewitt, E. and Stromberg, K. (1971). "Real and Abstract Analysis." Springer, Berlin.

Hurewicz, Witold (1958). "Lectures on Ordinary Differential Equations." The M.I.T. Press, Cambridge.

Islam, M. N. and Neugebauer, J. T. (2008). Qualitative properties of nonlinear Volterra integral equations. *E. J. Qualitative Theory of Diff Equ.* **12**, pp. 1-16 (http://www.math.u-szeged.hu/ejqtde/2008/200812.html)

Kaslik, Eva and Sivasundaram, Seenith (2012) Non-existence of periodic solutions in fractional-order dynamical systems and a remarkable difference between integer and fractional-order derivatives of periodic functions, *Nonlinear Analysis: RWA* **13**1489-1497.

Kato, J. (1994). Stability criteria for difference equations related with Liapunov functions for delay-differential equations. *Dynamic Systems and Applications* **3**, 75–84.

Kato, J., and Strauss, A. (1967). On the global existence of solutions and Liapunov functions. *Ann. Math. Pura. Appl.* **77**, 303–316.

Kirk, C. M. and Olmstead,W. E. (2000). The influence of two moving heat sources on blow-up in a reactive-diffusive medium. *Z. Angew. Math. Phys.* **51**, 1-16.

Kirk, C. M. and Olmstead, W. E. (2002). Blow-up in a reactive-diffusive medium with a moving heat source. *Z. Angew. Math. Phys.* **53**, 147-159.

Kirk, C. M. and Olmstead, W. E. (2005). Blow-up solutions of the two-dimensional heat equation due to a localized moving source. *Analysis and Appl.* **3**, 1-16.

Kirk, C. M. and Roberts, Catherine A. (2002). A quenching problem for the heat equation. *J. Integral Equations Appl.* **14**, 53-72.

Krasnoselskii, M. A. (1958). Some problems of nonlinear analysis. *Amer. Math. Soc. Transl.* **(2) 10**, 345-409.

Krasovskii, N. N. (1963). "Stability of Motion." Stanford Univ. Press.

Kuang, Y. (1993). "Delay Differential Equations with Applications to Population Dynamics." Academic Press, Boston.

Kuang, Y. (1993). Global stability in one or two species neutral delay population models." *Canadian Appl. Math. Quart.* **1**, 23–45.

Kuang, Y. (1991). On neutral delay logistic Gause-type predator-prey systems. *Dynamics Stability Systems* **6**, 173–189.

Lakshmikantham, V. and Leela, S. (1969). "Differential and Integral Inequalities," Vol. I. Academic Press, New York.

Lakshmikantham, V., Leela, S., and Vasundhara Devi, J. (2009). "Theory of Fractional Dynamic Systems" Cambridge Scientific Publishers, Cambridge, UK.

Lakshmikantham, V. and Rao, M. Rama Mohana (1995). *Theory of Integro-differential Equations*, Gordon and Breach Science Publishers, Lausanne, Switzerland.

LaSalle, J. P. (1968). Stability theory for ordinary differential equations. *J. Differential Equations* **4**, 57-65.

LaSalle, J. P. and Lefschetz, Solomon (1961). "Stability by Liapunov's Direct Method with Applications." Academic Press, New York.

Lefschetz, Solomon (1965). "Stability of Nonlinear Control Systems." Academic Press, New York.

Levin, J. J. (1963). The asymptotic behavior of a Volterra equation. *Proc. Amer. Math. Soc.* **14**, 434–451.

Levin, J. J. (1965). The qualitative behavior of a nonlinear Volterra equation. *Proc. Amer. Math. Soc.* **16**, 711-718.

Levin, J. J. (1972). On a nonlinear Volterra equation. *J. Math. Anal. Appl.* **39**, 458-476.

Levin, J. J. (1968). A nonlinear Volterra equation not of convolution type. *J. Differential Equations* **4**, 176–186.

Levin, J. J. and Nohel, J. A. (1960). On a system of integrodifferential equations occuring in reactor dynamics. *J. Math. Mechanics* **9**, 347–368.

Levin, J. J. and Nohel, J. A. (1963). Note on a nonlinear Volterra equation. *Proc. Amer. Math. Soc.* **14**, 924-929.

Levin, J. J. and Nohel, J. A. (1964). On a nonlinear delay equation. *J. Math. Anal. Appl.* **8**, 31-44.

Levinson, N. (1960). A nonlinear Volterra equation arising in the theory of superfluidity. *J. Math. Anal. Appl.*, **1**, 1-11.

Liu, Yicheng and Li, Zhiaxiang (2008). Krasnoselskii type fixed point theorems and applications. *Proc. Amer. Math. Soc.* to appear.

Liu, Yicheng and Li, Zhixiang (2006). Schaefer type theorem and periodic solutions of evolution equations. *J. Math. Anal. Appl* **316**, 237-255.

Londen, Stig-Olof (1972). On the solutions of a nonlinear Volterra equation. *J. Math. Anal. Appl.* **39**, 564-573.

Lurie, A. I. (1951). "On some nonlinear problems in the theory of automatic control." H. M. Stationery Office, London.

Lyapunov, A. M. (1992). The general problem of the stability of motion. *Int. J. Control.* **55**, 531-773. (This is a modern translation by A. T. Fuller of the 1892 monograph.)

Mainardi, Francesco (2010). "Fractional Calculus and Waves in Linear Viscoelasticity" Imperial College Press, London.

Mann, W. R. and Wolf, F. (1951). Heat transfer between solids and gases under nonlinear boundary conditions, *Quat. Appl. Math.* **9** 163-184.

Meir, A. and Keeler, E. (1969). A theorem on contractive mappings. *J. Math. Anal. Appl.* **28**, 326-329.

Miller, R. K. (1968). On the linearization of Volterra integral equations. *J. Math. Anal. Appl.* **23**, 198–208.

Miller, R. K. (1971a). "Nonlinear Volterra Integral Equations." Benjamin, New York.

Miller, R. K. (1971b). Asymptotic stability properties of linear Volterra integrodifferential equations. *J. Differential Equations* **10**, 485–506.

Natanson, I. P. (1960). "Theory of Functions of a Real Variable," Vol. II. Ungar, New York.

Nicholson, R. S. and Shain, I. (1964). Theory of stationary electrode polography, *Analytical Chemistry* **36**, 706-723.

Oldham, Keith B. and Spanier, Jerome. (2006) *The Fractional Calculus*, Dover, New York, 2006.

Olmsted, John M. H. (1959). "Real Variables." Appleton-Century-Crofts, Inc., New York.

Padmavally, Komarath (1958). On a non-linear integral equation. *J. Math. Mech.* **7**, 533-555.

Park, Sehie (2007). Generalizations of the Krasnoselskii fixed point theorem. *Nonlinear Anal.* **67**, 3401-3410.

Perron, O. (1930). Die Stabilitätsfrage bei Differentialgleichungen. *Math. Z.* **32**, 703–728.

Podlubny, I. (1999) "Fractional Differential Equations" Academic Press, San Diego.

Purnaras, I. K. (2006). A note on the existence of solutions to some nonlinear functional integral equations. *E. J. Qualitative Theory of Diff. Equ.* **No. 17**, 1-24. (http://www.math.u-szeged.hu/ejqtde/2006/200617.html)

Ritt, Joseph Fels (1966). "Differential Algebra." Dover, New York.

Roberts, J. H. and Mann, W. R. (1951). On a certain nonlinear integral equation of the Volterra type, *Pacific J. Math.* **1**, 431-445.

Rudin, W. (1966). "Real and Complex Analysis." McGraw-Hill, New York.

Sansone, G. and Conti, R. (1964). "Non-linear Differential Equations." Macmillan, New York.

Schaefer, H. (1955). Über die Methode der a priori Schranken. *Math. Ann.* **129**, 415–416.

Smart, D. R. (1980). "Fixed Point Theorems." Cambridge Univ. Press, Cambridge.

Somolinos, A. (1977). Stability of Lurie-type functional equations. *J. Differential Equations* **26**, 191–199.

Strauss, A. (1970). On a Perturbed Volterra Integral Equation. *J. Math. Anal. Appl.* **30**, 564-575.

Tavazoei, M. (2010) A note on fractional-order derivatives of periodic functions, *Automatica* **46** (5) 945-948.

Tavazoei, M. and Haeri, M. (2009) A proof for non existence of periodic solutioons in time invariant fractionaal order systems, *Automatica* **45** (8) 1886-1890.

Taylor, Angus E. and Mann, W. Robert (1983). "Advanced Calculus, Third ed." Wiley, New York.

Tricomi, F. G. (1985). "Integral Equations." Dover, New York.

H. F. Weinberger, H. F. (1965). *A First Course in Partial Differential Equations with Complex Variables and Transform Methods*, Blasidell, New York, 1965.

Windsor, Alistair (2010). A contraction mapping proof of the smooth dependence on parameters of solutions to Volterra integral equations. *Nonlinear Anal.* **72**, 3627 -3634.

Volterra, V. (1913). "Lecons sur les équations intégrales et les équations intégro-differentielles." Collection Borel, Paris.

Volterra, V. (1928). Sur la théorie mathématique des phénomès héréditaires. *J. Math. Pur. Appl.* **7**, 249–298.

Volterra, V. (1959). "Theory of Functionals and of Integral and Integro-differential Equations." Dover, New York.

Weinberger, H. F. (1965). "A First Course in Partial Differential Equations with Complex Variables and Transform Methods." Blasidell, New York.

Yoshizawa, T. (1966). "Stability Theory by Liapunov's Second Method." Math. Soc. Japan, Tokyo.

Zhang, B. (1995). Asymptotic stability in functional-differential equations by Liapunov functionals. *Trans. Amer. Math. Soc.* **347**, 1375–1382.

Zhang, B. (1997). Asymptotic stability criteria and integrability properties of the resolvent of Volterra and functional equations. *Funkcial. Ekvac.* **40**, 335–351.

Zhang, B. (2001). Formulas of Liapunov functions for systems of linear ordinary and delay differential equations. *Funkcial. Ekvac.* **44**, 253–278.

Zhang, B. (2009a). Boundedness and global attractivity of solutions for a system of nonlinear integral equations. *Cubo, A Mathematical Journal* **11**, No. 3, pp. 41-53.

Zhang, B. (2009b). Liapunov functionals and periodicity in a system of nonlinear integral equations. *Elect. J. Qualitative Theory of Differential Eq.* Spec. Ed. I, No. 1, 1-15.

Author Index

Subject Index

Made in the USA
Charleston, SC
10 September 2012